Springer Water

The book series Springer Water comprises a broad portfolio of multi- and interdisciplinary scientific books, aiming at researchers, students, and everyone interested in water-related science. The series includes peer-reviewed monographs, edited volumes, textbooks, and conference proceedings. Its volumes combine all kinds of water-related research areas, such as: the movement, distribution and quality of freshwater; water resources; the quality and pollution of water and its influence on health; the water industry including drinking water, wastewater, and desalination services and technologies; water history; as well as water management and the governmental, political, developmental, and ethical aspects of water.

Abdelkader Anouzla · Salah Souabi
Editors

Technical Landfills and Waste Management

Volume 1: Landfill Impacts, Characterization and Valorisation

Springer

Editors
Abdelkader Anouzla
Faculty of Science and Technology
University of Hassan II Casablanca
Casablanca, Morocco

Salah Souabi
Faculty of Science and Technology
University of Hassan II Casablanca
Casablanca, Morocco

ISSN 2364-6934 ISSN 2364-8198 (electronic)
Springer Water
ISBN 978-3-031-52632-9 ISBN 978-3-031-52633-6 (eBook)
https://doi.org/10.1007/978-3-031-52633-6

This Springer imprint is published by the registered company Springer Nature Switzerland AG
The registered company address is: Gewerbestrasse 11, 6330 Cham, Switzerland

Paper in this product is recyclable.

Preface

Waste has become one of the most critical environmental problems in countries worldwide. Over time, the amount of waste from households and industries illegally dumped in nature has increased. Consequently, this has led to a growing shortage of landfill areas and increased costs at landfills.

The world is evolving toward a circular economy, which emphasizes lowering waste and extending the useful life of commodities. The proliferation of so-called squatter settlements worldwide has led to poor living conditions, illustrated mainly by the absence of public hygiene standards in several countries. Robust demographic growth and massive rural exodus have led to an accelerated development process worldwide, characterized by intense urbanization, often with little or no control over the population.

Waste has become one of the most critical environmental problems in countries worldwide. Over time, the amount of waste from households and industries illegally dumped in nature has increased. Consequently, this has led to a growing shortage of landfill areas and increased costs at landfills.

This book presents a study that identifies, in several chapters, the problems of solid waste, as well as the characteristics and impacts of municipal solid waste. Indeed, several chapters examined how this trash might contaminate the land, waterways, groundwater, and other ecosystem elements. The evaluation of the environmental impact of landfills is an important topic that has gained increased attention in recent years due to growing concerns about the environment. This issue has been extensively discussed in the literature and has become a focal point for researchers and policy-makers alike. The significance of this subject is underscored by the need to address the potential consequences of improper waste management, particularly in developing countries where the management of solid waste is often unsustainable. This book provides a comprehensive overview of the significant impacts associated with waste mismanagement on a global scale, emphasizing the prevalence of these issues in developing nations. Furthermore, it delves into the potential analytes and composition present in MSW landfills, shedding light on the associated risk assessment. By conducting a thorough narrative literature review, this book aims to underscore the global significance of inadequate waste management practices and their implications for the environment and human health. Moreover, it serves as a valuable resource for scholars and stakeholders seeking to evaluate and estimate the potential risks posed by poor MSW management, thereby contributing to efforts to enhance sustainability on a global scale. In light of the complexities and multifaceted nature of the environmental impact of landfills, this book seeks to provide a comprehensive analysis that can inform and guide future research and policy development in this critical area.

Casablanca, Morocco Abdelkader Anouzla

Contents

Contributors

Issam A. Al-Khatib Institute of Environmental and Water Studies, Birzeit University, Birzeit, Palestine

Asitha T. Cooray Department of Chemistry, Faculty of Applied Sciences, University of Sri Jayewardenepura, Gangodawila, Nugegoda, Sri Lanka;
Faculty of Applied Sciences, Instrument Centre, University of Sri Jayewardenepura, Gangodawila, Nugegoda, Sri Lanka

Subhasish Das Department of Environmental Science, Pachhunga University College, Mizoram University (A Central University), Aizawl, Mizoram, India

Giorgio De Donno Dipartimento di Ingegneria Civile Edile e Ambientale, "Sapienza" Università di Roma, Rome, Italy

Pablo Emilio Escamilla-García Instituto Politécnico Nacional, Mexico City, Mexico

Milad Ghaffariraad M.Sc. in Environmental Engineering, School of Engineering, Civil Engineering Department, Urmia University, Urmia, Iran

Mehdi Ghanbarzadeh Lak School of Engineering, Civil Engineering Department, Urmia University, Urmia, Iran

Zakia Gueboudji Biotechnology, Water, Environment and Health Laboratory, Abbes Laghrour University of Khenchela, Khenchela, Algeria

HamidReza Jahangirzadeh Soureh M.Sc. in Environmental Engineering, School of Engineering, Civil Engineering Department, Urmia University, Urmia, Iran

Kenza Kadi Biotechnology, Water, Environment and Health Laboratory, Abbes Laghrour University of Khenchela, Khenchela, Algeria;
Faculty of Life and Nature Sciences, Abbes Laghrour University, Khenchela, Algeria

Pabasari A. Koliyabandara Department of Civil and Environmental Technology, Faculty of Technology, University of Sri Jayewardenepura, Pitipana, Homagama, Sri Lanka

Sudantha S. Liyanage Department of Civil and Environmental Technology, Faculty of Technology, University of Sri Jayewardenepura, Pitipana, Homagama, Sri Lanka

Maher Mahmoudi Faculty of Sciences of Gabes, University of Gabes, Gabes, Tunisia;
Laboratory of Plant, Soil and Environment Interactions (LIPSE), LR21LS01, University of Tunis El Manar, Tunis, Tunisia

Davide Melegari Dipartimento di Ingegneria Civile Edile e Ambientale, "Sapienza" Università di Roma, Rome, Italy

Kamel Nagaz Dryland and Oases Cropping Laboratory, Institute of Arid Regions, Mednine, Tunisia

Snigdha Nath Department of Environmental Science, Pachhunga University College, Mizoram University (A Central University), Aizawl, Mizoram, India

Valeria Paoletti Dipartimento di Scienze della Terra, dell'Ambiente e delle Risorse, Università degli Studi di Napoli Federico II, Naples, Italy

Ester Piegari Dipartimento di Scienze della Terra, dell'Ambiente e delle Risorse, Università degli Studi di Napoli Federico II, Naples, Italy

D. D. P. Preethika Department of Civil and Environmental Technology, Faculty of Technology, University of Sri Jayewardenepura, Pitipana, Homagama, Sri Lanka

Md. Mostafizur Rahman Department of Environmental Science, Bangladesh University of Professionals, Mirpur, Cantonment, Dhaka, Bangladesh;
Laboratory of Environmental Health and Ecotoxicology, Department of Environmental Sciences, Jahangirnagar University, Savar, Dhaka, Bangladesh

Sadia Sikder Department of Environmental Science, Bangladesh University of Professionals, Mirpur, Cantonment, Dhaka, Bangladesh

Konthoujam Khelchandra Singh Department of Environmental Science, Pachhunga University College, Mizoram University (A Central University), Aizawl, Mizoram, India;
Department of Forestry and Environmental Science, Manipur University (A Central University), Imphal, Manipur, India

Chamika Siriwardana Faculty of Applied Sciences, Instrument Centre, University of Sri Jayewardenepura, Gangodawila, Nugegoda, Sri Lanka

Sumpam Tangjang Department of Botany, Rajiv Gandhi University (A Central University), Doimukh, Arunachal Pradesh, India

Mohammad Toha Department of Environmental Science, Bangladesh University of Professionals, Mirpur, Cantonment, Dhaka, Bangladesh

Meththika Vithanage Ecosphere Resilience Research Center, Faculty of Applied Sciences, University of Sri Jayewardenepura, Gangodawila, Nugegoda, Sri Lanka

Mohammed Zari Chemical and Environmental Engineering Department, Faculty of Engineering, University of Nottingham, Coates Building, University Park, Nottingham NG7 2RD, UK;
Department of Environment, Faculty of Environmental Sciences, King Abdulaziz University, Jeddah 21589, Saudi Arabia

Chapter 1
Electrical and Electromagnetic Prospecting for the Characterization of Municipal Waste Landfills: A Review

Giorgio De Donno, Davide Melegari, Valeria Paoletti, and Ester Piegari

Abstract In this chapter, we review the main results of electrical and electromagnetic prospecting applied to the characterization and monitoring of municipal waste landfills in the last decade. Among all the geophysical surveys, these methods are the most used for subsurface investigations of landfills since they provide a cost-effective approach that allows for detailed and non-invasive imaging of the subsurface in terms of the electrical properties, down to depths which generally vary from a few tens of centimeters to several tens of meters. Nevertheless, the indirect geophysical mapping needs the direct even if punctual information from boreholes and wells for an accurate reconstruction of the contaminated zones. Electrical and electromagnetic methods are used for multiple purposes that include mapping landfill boundaries, measuring waste volume and composition, as well as identifying and tracking leachate plumes. In particular, electrical methods are widely used for leachate detection (both inside and outside the landfill) and for the geometrical reconstruction of the landfill using electrical conductivity and chargeability as the main proxies. Low-frequency electromagnetic methods are mostly used for a hydrogeological characterization and extensive screening of the high-conductive areas associated to the leachate accumulation. These methods have lower resolution compared to the electrical techniques but often allow greater depth of investigation. High-frequency electromagnetic surveys are instead mainly focused on the shallow part of the landfill for detection of defects on the covering liner and characterization of the covering layer. We discuss recent results related to the topic providing updated references in relation to the specific applications and emphasizing the importance of site-specific validation through direct information. At last, a special focus is given to novel trends, emerging techniques and data integration by machine learning-based approaches for mapping and monitoring of municipal solid waste landfills.

G. De Donno · D. Melegari
Dipartimento di Ingegneria Civile Edile e Ambientale, "Sapienza" Università di Roma, Rome, Italy

V. Paoletti · E. Piegari (✉)
Dipartimento di Scienze della Terra, dell'Ambiente e delle Risorse, Università degli Studi di Napoli Federico II, Naples, Italy
e-mail: ester.piegari@unina.it

A. Anouzla and S. Souabi (eds.), *Technical Landfills and Waste Management*, Springer Water, https://doi.org/10.1007/978-3-031-52633-6_1

Keywords Landfill monitoring · Geoelectrical prospecting · Electrical resistivity · Electromagnetic surveys · Data integration

1.1 Introduction

Electrical and electromagnetic (EM) methods have been increasingly used for environmental and geotechnical investigations, particularly for landfill characterization (Soupios and Ntarlgiannis 2017; Nguyen et al. 2018; Juarez et al. 2023). These methods have proven to be powerful tools for this purpose, as the strong electrical contrasts between waste, leachate, and surrounding formations make the electrical resistivity an appropriate proxy for landfill studies. Among the geophysical methods, electrical and electromagnetic prospections offer several advantages compared to other techniques and to direct investigations, such as:

- *non-invasive survey*: they do not require extensive excavation or disturbance allowing for the investigation of the landfill without additional waste generation
- *cost-effectiveness*: they are generally more cost-effective compared to traditional methods used for landfill characterization
- *rapid data collection*: they provide fast data collection which enables timely decision-making and facilitates the implementation of appropriate waste management strategies
- *high resolution capabilities*: mostly of them can provide detailed and high-resolution information about the electrical properties of the landfill
- *spatial coverage*: they can complement other investigation techniques, such as borehole sampling, well logging and laboratory analysis to improve the spatial coverage of the landfill.

Generally, EM methods are divided into low-frequency (LFEM, also known as FDEM and TDEM if signals are analyzed in the frequency- or in the time domain, respectively) and high-frequency (known as ground penetrating radar—GPR) methods, being the former capable to investigate deep portions of the subsurface but generally with low resolution, while the latter conversely allows for high-resolution surveys even if only restricted to the very shallow portion of the subsurface. Due to the above-mentioned limitations, the EM techniques have been often used during the recent years in conjunction with electrical methods. In fact, electrical methods permit the acquisition of tomographic high-resolution data also reaching significant depths and enabling the assessment of both the resistive behaviour (the so-called electrical resistivity tomography—ERT) and the capacitive effects (induced polarization—IP) of the waste mass and the surrounding formations. However, the objectives to be achieved, the site-specific features and the available budget can steer the choice towards a specific method or a combination of methods.

This chapter aims at presenting the current state of the art on the applications of electrical and EM prospections for the characterization of MSW landfills mainly focusing on the technological and methodological developments of these methods in recent years. For this reason, only papers and reviews published over the last decade and concerning electrical and EM methods are included in the present work. According to these eligibility criteria, we select 57 papers, which are listed in Table 1.1 with information about publication's year, authors' names, prospecting methods and aims of the geophysical surveys.

Overall, the selected papers covered one or more of the following aspects and issues concerning landfill characterization and management (see Soupios and Ntarlagiannis 2017 and references therein):

- *leachate detection*: leachate has different conductivity properties than the surrounding material. Electrical and EM methods can identify areas with higher conductivity due to leachate accumulation, helping potential contamination plumes to be located.
- *settlement and erosion detection*: electrical and EM surveys can identify layers and areas with anomalous electrical and electromagnetic responses, which may indicate voids, air pockets within the landfill, erosion, or areas of waste compaction. These phenomena could affect landfill stability or subsidence.
- *waste characterization*: different waste materials have varying electrical conductivity properties. In particular, EM methods can be used to differentiate between waste types based on their electric and magnetic responses, helping in waste characterization and sorting.
- *characterization of cover material*: Landfills require a final cover to control odors, reduce water infiltration, and prevent erosion. Electrical and EM methods can help to evaluate the thickness and homogeneity of the cover material.
- *liner and barrier assessment*: Electrical and EM methods can assess the integrity of liners and barriers designed to contain waste and prevent contamination. Changes in subsurface conductivity can indicate gaps or breaches in these containment systems.
- *infrastructure mapping*: Buried infrastructure such as pipes and utility lines within the landfill can be accurately mapped using EM methods. These data are important for preventing potential damage during excavation or construction activities.
- *long-term monitoring*: Electrical and EM surveys can track changes over time, allowing operators to monitor trends in waste distribution, moisture content, migration of gases, such as methane and other critical parameters, such as effectiveness of mitigation measures.
- *planning and design*: Electrical and EM methods can aid in the planning and design of new landfill sites. By understanding the subsurface conductivity variations, engineers can optimize waste placement and design appropriate containment measures.

Figure 1.1 shows the geographical distributions of the landfills investigated in the papers included in this review. We note that most of the landfills are located in Europe, followed by case histories from Asia, while there are no examples for Australia.

The histogram in Fig. 1.2 shows the number of papers concerning the application of a specific method. ERT is by far the most used method, but an increasing trend

Table 1.1 List of papers included in the present work which fulfil the eligibility criteria

Year	Authors	Location	Goal	Electrical	EM	Other methods	Direct information
2012a	Gazoty et al.	Denmark	Landfill characterization	2D ERT/TDIP full decay	Not applied	Not applied	Boreholes
2012b	Gazotv et al.	Denmark	Landfill characterization	2D TDIP full decay	Not applied	Not applied	Boreholes
2012	Fiandaca et al.	Denmark	Spectral content of waste polarization processes	2D TDIP full decay	Not applied	Not applied	Boreholes
2013	Fiandaca et al.	Denmark	Spectral content of waste polarization processes	2D TDIP full decay	Not applied	Not applied	Boreholes
2013	Yochim et al.	Canada	Estimation of water content in landfill	Not applied	Borehole GPR	Not applied	Not available
2013	De Carlo et al.	Italy	Detection liner damage	2D ERT	Not applied	MALM	Not available
2013	Ramalho et al.	Portugal	Imaging leachate accumulation	SP, VES	FDEM	Not applied	Waste sampling and analysis
2014a	Audebert et al.	France	Impact of geomembrane on ERT	2D ERT	Not applied	Not applied	Not available
2014b	Audebert et al.	France	Understanding leachate recirculation in bioreactors	2D ERT	Not applied	Not applied	Not available

(continued)

Table 1.1 (continued)

Year	Authors	Location	Goal	Electrical	EM	Other methods	Direct information
2014	Bellezoni et al.	Brazil	Contaminant migration	2D ERT	Not applied	Geotechnical and physical–chemical analysis	Monitoring wells
2014	Genelle et al.	France	Detection cover damage	2D ERT, SP	Not applied	Not applied	Boreholes
2014	Tsourlos et al.	Greece	Leak detection	2D ERT	Not applied	Not applied	Monitoring wells
2015	Yin et al.	Singapore	Landfill characterization	2D ERT	Not applied	Seismic	Boreholes
2016	Abdulrahman et al.	Malaysia	Imaging leachate accumulation	2D ERT/IP	Not applied	Not applied	Not available
2016a	Audebert et al.	France	Understanding leachate recirculation in bioreactors	2D ERT	Not applied	Not applied	Not available
2016b	Audebert et al.	France	Understanding leachate recirculation in bioreactors	2D ERT	Not applied	Not applied	Not available
2016	Cimar et al.	Turkey	Waste mass identification	2D ERT	Not applied	Gamma-ray spectrometric measurements	Not available
2016	Dumont et al.	Belgium	Gravimetric water content	2D ERT	FDEM	Not applied	Waste sampling and analysis
2017	Dumont et al.	Belgium	Landfill characterization	2D ERT	FDEM	HVNSR, MASW, Magnetics	Boreholes

(continued)

Table 1.1 (continued)

Year	Authors	Location	Goal	Electrical	EM	Other methods	Direct information
2017	Carpenter and Reddy	USA	Imaging leachate accumulation	2D ERT	FDEM, GPR	Seismic	Wells
2017a	De Donno and Cardarelli	Italy	Landfill characterization	2D ERT/IP	Not applied	Not applied	Not available
2017b	De Donno and Cardarelli	Italy	Landfill characterization	2D ERT/IP	Not applied	Not applied	Not available
2017	Feng et al.	China	Imaging leachate accumulation	2D ERT	Not applied	Not applied	Boreholes
2017	Maurya et al.	Denmark	Imaging leachate accumulation	2D/3D ERT	Not applied	Not applied	Water chemistry and ionic strength
2017	Yin et al.	Singapore	Landfill characterization	2D ERT	Not applied	MASW	Boreholes
2017	Wemegah et al.	Ghana	Landfill characterization	2D TDIP full decay	Not applied	Magnetics	Boreholes
2018	Balia	Italy	Landfill characterization	2D ERT	Not applied	Gravity, seismic	Not available
2018	Di Maio et al.	Italy	Landfill characterization	2D ERT/IP	FDEM, TDEM VLF	SR, ST	Waste sampling and analysis
2018a	Dumont et al.	Belgium	Understanding water infiltration in a landfill cover	2D ERT	Not applied	Not applied	Waste sampling and analysis

(continued)

Table 1.1 (continued)

Year	Authors	Location	Goal	Electrical	EM	Other methods	Direct information
2018b	Dumont et al.	Belgium	Assessing the efficiency of recirculation on bioreactor landfills	2D ERT	Not applied	Not applied	Temperature monitoring by boreholes
2018	Osinowo et al.	Nigeria	Imaging leachate accumulation	ERT	3D FDEM	Not applied	Not applied
2018	Baawain et al.	Oman	Imaging leachate accumulation	Not applied	TDEM	Not applied	Boreholes
2019	Høyer et al.	Denmark	Imaging leachate accumulation	2D TDIP full decay (show just ERT)	FDEM, TDEM	Not applied	Boreholes
2019	Hu et al.	China	Monitoring the dewatering of vertical and horizontal wells	2D ERT	Not applied	Not applied	Hydrogeological monitoring using pumping wells
2019	Ma et al.	China	Monitoring high gas pressures in landfill	2D ERT	Not applied	Not applied	Measurement of pore pressures by boreholes
2019	Mepaiyeda et al.	South Africa	Landfill characterization	2D ERT/IP	Not applied	Not applied	Not available
2020	Feng et al.	China	Ensuring the safety of a landfill for tunneling	2D ERT	TEM	Not applied	Waste sampling and analysis
2020	Flores-Orozco et al.	Austria	Mapping biogeochemically active zones	2D TDIP	Not applied	Not applied	TOC contents in leachate of solid waste samples

(continued)

Table 1.1 (continued)

Year	Authors	Location	Goal	Electrical	EM	Other methods	Direct information
2020	Helene et al.	Brazil	Imaging leachate accumulation	2D ERT	Not applied	Not applied	Not available
2020	Maryadi et al.	West Java	Groundwater contamination around landfill	Not applied	GPR	Not applied	Direct water quality assessment
2021	Ibraheem et al.	Germany	Landfill characterization	2D ERT	Not applied	Magnetics	Boreholes
2021	Jacome et al.	Canada	Imaging leachate accumulation	2D ERT/IP	Not applied	Not applied	Wells with measurements of methane concentrations
2021	Kondracka et al.	Poland	Landfill characterization	ERT	GPR	Seismic Refraction Tomography, MASW	Boreholes
2021	Narciso et al.	Belgium	Landfill characterization	Not applied	FDEM	Not applied	Not applied
2021	Van De Vijver et al.	Belgium	Landfill characterization	2D ERT/IP	FDEM	Sampling data	Not available
2022	Costanzo-Alvarez et al.	Canada	Imaging leachate accumulation	2D ERT	Not applied	Not applied	Boreholes with measurements of water quality
2022	Deidda et al.	Italy	Imaging leachate accumulation	Not applied	3D FDEM	Not applied	Not applied
2022	Steiner et al.	Austria	Quantitative water content estimation	ERT/FDIP	Not applied	SR	Waste sampling and analysis

(continued)

Table 1.1 (continued)

Year	Authors	Location	Goal	Electrical	EM	Other methods	Direct information
2022	Umar et al.	Malaysia	Imaging leachate migration	Not applied	GPR	Not applied	
2022	Zhang et al.	China	Landfill characterization	Not applied	GPR	Not applied	Not applied
2023	Adabanija	Nigeria	Imaging leachate accumulation	2D ERT	Not applied	Not applied	Hydrogeologic and geotechnical investigation
2023	Isunza Manrique et al.	Belgium	Landfill characterization	2D ERT/IP	Not applied	Not applied	Boreholes
2023	Martorana et al.	Italy	Imaging leachate accumulation	3D ERT/IP	Not applied	Not applied	Not available
2023	Moser et al.	Italy	Best electrode configuration	2D FDIP	Not applied	Not applied	Not applied
2023	Piegari et al.	Italy	Imaging leachate accumulation	2D ERT/IP	Not applied	Not applied	Boreholes
2023	Suknark et al.	Thailand	Understanding potential of landfill mining for RDF	2D ERT	Not applied	Not applied	Waste samples
2023	Sun et al.	China	Monitoring resistivity change with DL	2D ERT	Not applied	Not applied	Not available

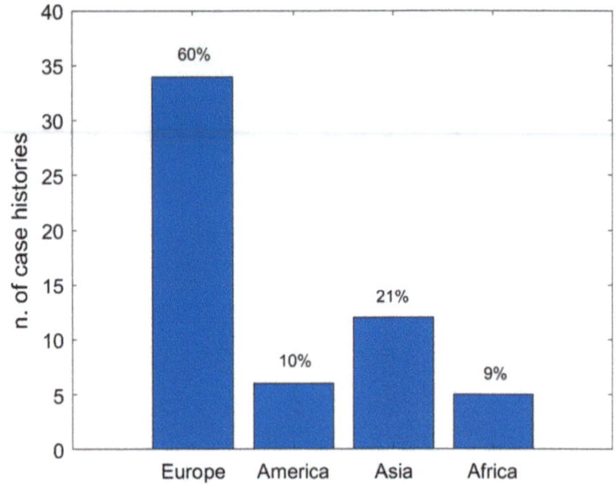

Fig. 1.1 Geographical distribution of the landfills considered for the review

in the use of all electrical and EM methods is found in the most recent years (from 2018).

In the following Sect. 1.2, we review the selected papers on the basis of the predominant method used for the characterization, that is electrical (Sect. 1.2.1) or EM (Sect. 1.2.2), while recent works presenting novel trends and applications of these methods for landfill characterization are included in Sect. 1.3.

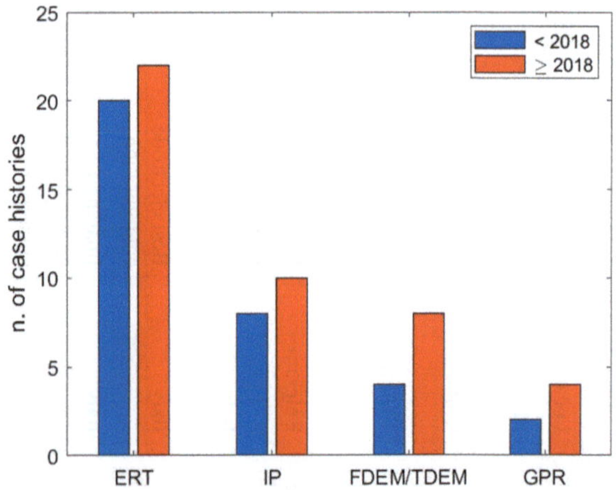

Fig. 1.2 Frequency of use of the electrical and EM prospecting methods for the considered case histories

1.2 Electrical and Electromagnetic Prospecting Applied to MSW Landfills

1.2.1 Electrical Methods

Electrical methods can sense the distribution of electrical properties of the subsurface, by injecting a direct-current (DC) and recording the resulting electric voltage differences (ERT and IP methods) or by passively recording electric voltage arising due to natural or anthropogenic causes (the so-called self-potential—SP).

For the electrical resistivity tomography (ERT) method, a DC (steady-state) electric current is generally injected on surface through a pair of electrodes and voltage differences are measured on other pairs at different distances from the current electrodes, whose magnitude is a function of the resistivity distribution of the investigated medium. Data acquisition involves multiple measurements of both current and potential at different electrode locations (multi-electrode methods) to investigate the electrical properties of the subsurface both laterally and in depth. Strating from the experimental observations of current and voltage, the ERT method allows for a 2D or 3D reconstruction of the electrical resistivity distribution by solving an inverse problem through which the final model is obtained iteratively by ensuring the best fit between observed and predicted data (Loke et al. 2013). Electrical resistivity tomography can provide high-resolution models of the subsurface widely used for numerous applications in MSW landfills, such as leachate imaging and monitoring, assessing liner integrity, definition of landfill geometry and geological settings (Soupius and Ntarlagiannis 2017; Nguyen et al. 2018; Martinho 2023). The main limitations are the ambiguity arising when investigating materials having similar resistivity values and the loss of resolution with depth, such that the detection of small bodies at significant depths is often unfeasible.

This problem can be overcome by combining ERT with other geophysical investigations or by integration with the induced polarization (IP) method. The IP phenomenon is the low-frequency polarization of rocks and soils occurring at the interface between mineral grains and the pore fluid, due to changes in the lithology and pore fluid chemistry, increase of clay content or both inorganic and organic contaminants (Binley and Slater 2020). The IP effect is often modelled by the chargeability, which is the ratio between the magnitude of the surface polarization effects and the volumetric (bulk) conduction (Binley and Slater 2020). The IP survey shares the same electrodes and geometry of the ERT acquisition, but also measuring the voltage drop that occurs after switching off the DC current. IP observed data can be inverted for an integral chargeability value or for complex multi-parameter models (i.e. Cole and Cole 1941), in order to achieve information on the capacitive behavior of the waste mass and the geological surrounding deposits (Binley and Slater 2020). Since the magnitude of the surface conduction is much lower compared to the bulk conduction contribution, IP measurements are challenging and low signal-to-noise ratio occurring in extreme conductive environments such as those encountered in MSW landfills often prevents the acquisition of good quality IP data.

Alternatively, or in combination with ERT and IP methods, the passive self-potential (SP) method can provide useful information related to the decomposition zones and to leachate transport throughout the site or externally. SP is a passive geoelectrical method which is based on the measurement (usually using many electrode pairs) of the potential difference arising due to electrochemical causes (redox potential) or due to electrokinetic phenomena mainly produced by groundwater and/or leachate flows internally or externally to the landfill site. Since non-polarizable electrodes (which requires extra-care when plugged-in) are needed to proceed with SP measurements and the magnitude of the recorded voltage is often quite low (often some mV or few tens of mV), this method has been only rarely applied for investigating MSW landfills in the last years.

In the following subsections, we summarize the main applications of ERT, IP and SP methods to landfill characterizations over the last decade.

1.2.1.1 ERT

In the last decade, the primary goal of implementing ERT techniques in MSW landfills has been leachate imaging and monitoring (see Table 1.1). Liquid waste is frequently associated with high ion concentrations in landfills, which corresponds to areas with low resistivities. Hence, the conductive distribution obtained from geoelectrical models makes possible to use ERT to identify leachate concentration zones. In general, it is easier to identify leachate because of the difference between its electrical characteristics and those of the surrounding media (liners, geological units, etc.). However, geophysical inverted models leave ambiguities for identifying contaminated zones, especially when clayey soils are present. These soils are frequently used as a low-permeability barrier at the bottom of landfills and, as a result, the lowest resistivity values could not match exclusively the zones with leachate plumes. In order to reduce the ambiguity and uncertainty of the interpretation, many studies have taken into account different data types, complementing geophysical techniques and other surveys (see Table 1.1).

As an example, Maurya et al. (2017) correlated resistivity with ionic strength, allowing to validate the level of leachate in a landfill. The aim of the study was to get a detailed site characterization and plume monitoring at a landfill site in Grindsted, southern Denmark. Geoelectrical data were acquired using a 2D ERT (gradient array and three separate profiles), while many water samples were collected from various boreholes and analyzed to determine the ionic strength. The outcomes are shown in Fig. 1.3, where a 2D inverted resistivity models built from all profiles is combined with an estimate of ionic strength obtained from different boreholes drilled at varying depths along the lines. The ionic strength and the low resistive anomaly exhibit great agreement, thus demonstrating a good correlation between geoelectrical data and ionic strength. Therefore, the thorough understanding of the leachate distribution, made possible by such integration, improved the accuracy and consistency of leachate plume mapping by ERT.

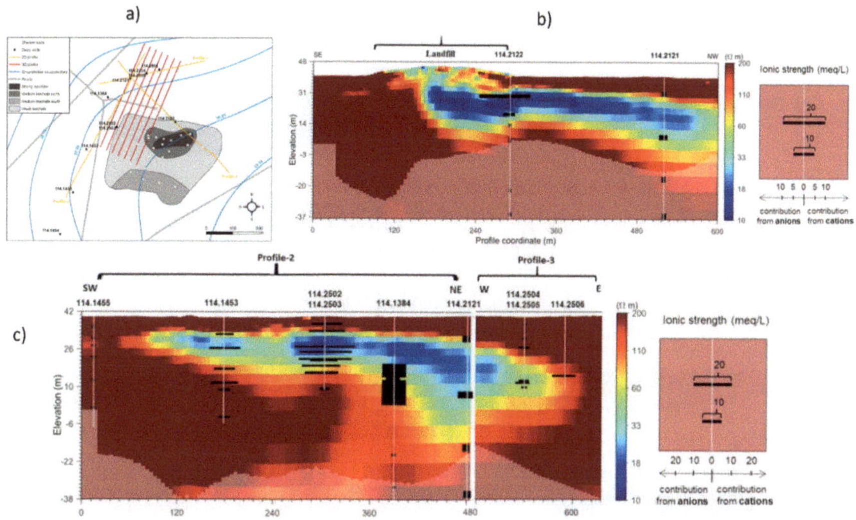

Fig. 1.3 **a** Map of Grindsted Landfill with locations of profiles for 2D (orange) and 3D (red) ERT surveys. **b** 2D inverted resistivity model along profile 1 with black bars showing the ionic strength as in the insert to the right. **c** Combined 2D inverted resistivity models of profile 2 and 3 with black bars showing the ionic strengths for the 7 boreholes close to the profiles (modified from Maurya et al. 2017)

Dumont et al. (2016) used the gravimetric water content as the main proxy to improve the effectiveness of leachate detection in the MSW landfill located at Mont-Saint-Guibert (Belgium). The presented approach is based on the evaluation of the empirical coefficients in the well-established empirical laws linking electrical and petrophysical properties (Archie 1942) by testing borehole samples in the laboratory and applying the necessary corrections. In this way, the volumetric water content is obtained from the inverted resistivity values and in turn, the gravimetric water content from the density of the samples. The resistivity and gravimetric water content sections (Fig. 1.4) show the effectiveness of this process, where high value of gravimetric water content corresponds to high-conductive zones and therefore the ERT results can be effectively validated.

Other papers involving ERT surveys aimed at characterizing the MSW landfill in terms of geometry and internal structure also integrating different geophysical data. Dumont et al. (2017) focused on the same landfill as per the previous study, but also executing electromagnetic, magnetic and seismic surveys. The purpose of this work was the geometrical characterization of the landfill, identifying its boundaries and the maximum depth, as well as leachate imaging. They demonstrated that a joint integration of different geophysical data can reduce the ambiguities arising when ERT is applied standalone.

Other uses of ERT encompassed the evaluation of the integrity of the bottom layer (mainly oriented to check liners integrity) to prevent leachate leakage outside the landfill site. In one of these works, De Carlo et al. (2013) investigated a dismissed

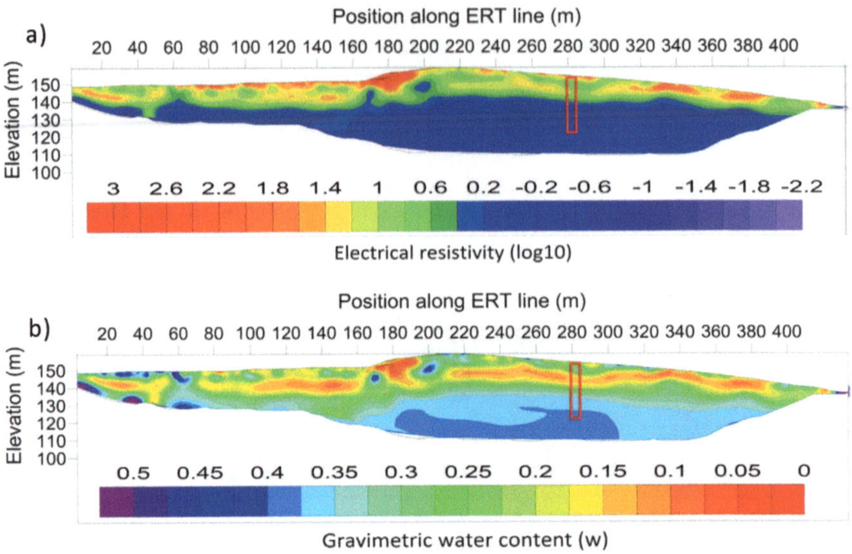

Fig. 1.4 Comparison between distributions of resistivity values in (**a**) and gravimetric water contents in (**b**) along the same profile (modified from Dumont et al. 2016)

landfill in Corigliano (Italy) using electrical methods, also reproducing two synthetic scenarios (damaged or undamaged bottom liner) in order to better understand the phenomena occurring at the case study. They endorsed the use of forward modeling both in the interpretation phase and in the design phase of any survey. Audebert et al. (2014a) focused on the influence of geomembrane on geoelectrical data acquisition made at the Champs-Jouault landfill (western France). The authors showed through a numerical modeling method that the geomembrane and the ERT line must be separated by a minimum amount to provide the good conditions of use of the traditional inversion methods.

1.2.1.2 IP

IP surveys are the primary technique to support ERT as they just make possible to acquire additional diagnostic parameters only increasing the acquisition time but using the same electrodes and equipment. Similarly to ERT, leachate monitoring is the main goal for IP surveys conducted at MSW landfill sites, as leachate (chargeable) can be distinguished from surrounding unsaturated layers or liners (not chargeable). Many applications in the last decade employed the integral chargeability or the normalized chargeability (chargeability divided by the resistivity) as the main proxy for assessing the IP contribution (see Table 1.1). As an example, De Donno and Cardarelli (2017a) showed the benefits of ERT and IP 3D data inversion on a MSW landfill site located in Central Italy, where the integrated interpretation of the

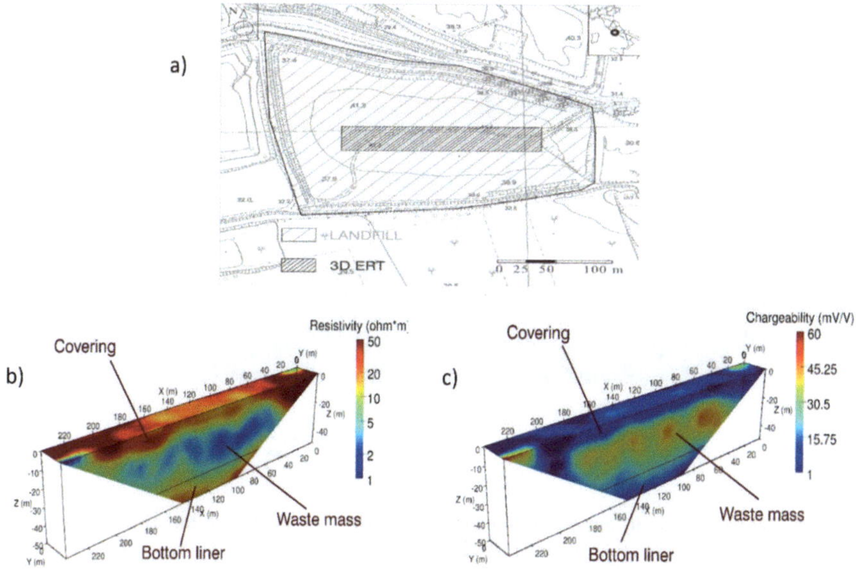

Fig. 1.5 Map of the geophysical survey area inside the landfill. **a** Volumetric view of the inverted 3D ERT and IP data (**b**, **c**) respectively (modified from De Donno and Cardarelli 2017a)

resistivity and chargeability models (Fig. 1.5) enabled an accurate identification of the areas of leachate accumulation.

Recently, the development of full-decay IP inversion methods (e.g. Fiandaca et al. 2012; 2013), which take into account the whole information contained in the IP acquisition, has paved the way for an improvement of the geophysical reconstruction in MSW landfills. Therefore, the works by Gazoty et al. (2012a, b) are included in our review although being outside the considered time range, as they were the precursors of such new application. In fact, by using an IP investigation with full decay inversion, the authors provided a four-parameter Cole–Cole model, which can also be correlated with petrographic parameters. Specifically, in Gazoty et al. (2012a) ERT and IP surveys were carried out to identify the geometric bounds and leachate levels of a landfill in Denmark. The final reconstruction (only resistivity and intrinsic chargeability sections are shown in Fig. 1.6 for the sake of simplicity) demonstrated the usefulness of this approach as highlighted by the straightforward correlation with the available borehole data which is particularly effective for shallow layers.

Strating from this pioneering approach, this research topic has been further developed recently (e.g. Høyer et al. 2019; Flores-Orozco et al. 2020; Moser et al. 2023). In particular, Flores-Orozco et al. (2020) found an excellent correspondence between the polarization response observed at the Heferlbach landfill (Vienna, Austria) and the typical biodegradability of leachate by integrating spectral information derived from a full decay inversion and analysis on well samples.

In the period of interest there have been only few examples of integration between IP and other geophysical techniques (see Table 1.1). In an exhaustive example Di

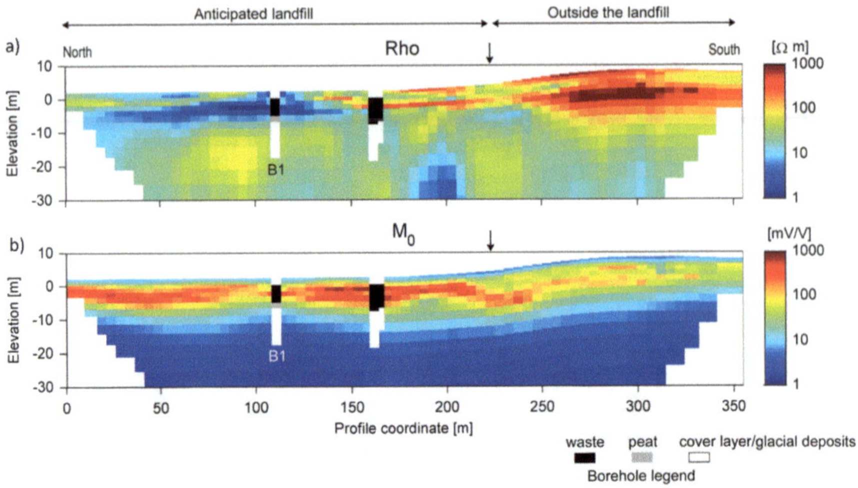

Fig. 1.6 2D resistivity (**a**) and intrinsic chargeability (**b**) models, where the available borehole information is superposed (Gazoty et al. 2012a)

Maio et al. (2018) integrated ERT and IP techniques along with electromagnetic (TDEM, FDEM, VLF), seismic (SR, ST) and waste sample analyses, yielding a complete and effective characterization of a landfill in Sardinia (Italy).

1.2.1.3 SP

As an alternative to or in combination with ERT and IP, the SP technique was particularly used in the first decade of the 2000s for MSW landfill characterization, even though only few papers have been published in the last decade (see Table 1.1), mainly due to the complexity of such anthropogenic environments, which can compromise the SP acquisition. In Ramalho et al. (2013), an SP survey is carried out for identification of groundwater flow in the landfill. Another example of effective application of SP can be found in Genelle et al. (2014), where the authors assessed the integrity of the covering layer of a landfill in France using SP data in conjunction with ERT.

1.2.2 Electromagnetic Methods (EM)

EM methods are active geophysical methods that are classified on the basis of the frequency range of the controlled sources. The low-frequency EM induction methods (LFEM), also known as frequency-domain EM (FDEM) or time-domain EM (TDEM) depending on the domain where signals are analyzed, use EM fields generated within a low-frequency range (~ 1–100 kHz), where the diffusive regime is the dominant mechanism of energy transport. The Ground-Penetrating Radar (GPR) uses frequencies > 1 MHz, for which wave propagation obeys diffraction, reflection and refraction laws. The basic principles of such methods are briefly summarized.

The FDEM method, also known as electromagnetic induction technique (EMI) uses two electrical coils to send a primary EM signal in the underground and to detect a secondary EM signal that is modified according by the electrical and magnetic properties of the subsurface, with respect to the primary field. The output is made by two components that are in-phase and out-of-phase (quadrature) with the primary field. The quadrature component is used to calculate the bulk ground electrical resistivity, while the in-phase component is connected to magnetic susceptibility. FDEM in landfill environments is essentially used for detecting landfill boundaries and high conductivity zones likely correlated to contamination (Di Maio et al. 2018 and references therein). A simultaneous presence of high values of quadrature and in-phase components suggest that the waste is compound by ferrous material (magnetizable) with leakage effects that provide an increase of the electrical conductivity (Godio and Naldi 2009).

The TDEM method involves sending a transient electromagnetic pulse into the ground and measuring the resulting electromagnetic response connected to the variations in soil electric conductivity. The method induces electric current flow within the subsurface and, after the transmitted signal is shut off, measures a voltage signal that is returned from the underground materials. The returning signal is measured as a function of time, and data are inverted to recover a layered earth model of resistivity (e.g., Di Maio et al. 2018). In landfill studies, this technique is used for discriminating between layers having low resistivity and consequently for defining leachate intrusion and diffusion. Its application to landfills environment is less widespread than the FDEM method.

Ground Penetrating Radar (GPR) is a high-frequency electromagnetic method that sends EM pulses into the ground sensing diffractions and reflections mainly due to variations of soil relative electric permittivity (the so-called dielectric constant). A normally sufficient dielectric constant gradient exists between the landfill and the hosting materials, which allows characterizing the distribution of wastes in a buried landfill by this technique (Zhang et al. 2022). For landfill purposes, the method is especially used for mapping and monitoring landfill sites and delineating contamination plumes (Kondracka et al. 2021), usually in combination with the electrical method.

1.2.2.1 FDEM and TDEM

The FDEM and TDEM methods are effective tools for studying the impact caused on natural environments by landfills as they produce spatially dense data over large areas with reduced time and costs. When the landfill is capped with impermeable liner, those techniques represent one of the few non-invasive solutions for the investigation of the body landfill and for the development of a site conceptual model (Deidda et al. 2022 and references therein).

Ramalho et al. (2013) presented a study on an old domestic and industrial waste sealed landfill site in Gaeiras, Portugal, where leachate overproduction became an environmental urgent problem. The authors conducted an integrated geophysical investigation including FDEM and Radio Frequency-EM measures (RF-EM). Their EM data allowed the identification of zones with accumulated leachates with very high apparent electrical conductivity, showing preferential pathways for leachates. This case study showed that only a multidisciplinary approach (in this case carried out by spontaneous potential, vertical electrical soundings and magnetic prospecting as well) to this kind of environmental problems can lead to satisfactory assessments. The integrated application of the geophysical methods clearly shows that this landfill was sited in a very unsuitable location. Supported by these results, the company in charge of the landfill has taken remediation measures with geotechnical solutions.

Dumont et al. (2016) studied a municipal solid waste landfill in a former sand quarry in Mont-Saint-Guibert landfill, Belgium, using ERT and borehole electromagnetics. Their study showed that bulk electrical resistivity values measured on waste sample and in the EM borehole present an excellent correlation and that the two methods can be used to estimate the moisture content over large areas. Dumont et al. (2017) conducted a waste site delimitation at the same landfill in Belgium by electromagnetic (in-phase and out-of-phase) and magnetic (vertical gradient and total field) methods that clearly showed the transition between the waste deposit and the host formation. Inside the landfill perimeter, the EM mapping also allowed the differentiation between the new and the old deposit area and the identification of zones characterized by specific electrical resistivity and magnetic susceptibility signatures.

Di Maio et al. (2018) presented a physical and structural characterization of an old municipal landfill and of encasing rocks at Sassari, Sardinia, Italy, obtained by an integrated analysis of data from a multi-methodological geophysical exploration, among which FDEM and TDEM data. All the FDEM sections highlighted a very clear decreasing of resistivity (5–10 Ωm) in the central portion of the survey area, with a gradual increasing towards its borders, related to the maximum thickness of the conductive waste in the central sector of the survey area and the presence of the resistive limestone bedrock at the edges of the area (Fig. 1.7). TDEM data fit well with the FDEM results showing a highly conductive layer (1–10 Ω m) located at a depth of about 30–35 m from the ground level, associated with particularly conductive facies of the substratum.

Osinowo et al. (2018) studied the level of contamination caused by the decomposition of wastes by defining the lateral distribution and the vertical limit of leachate at the Awotan landfill, Nigeria. The leachate caused zones of anomalous conductivity distribution within the subsurface and the authors studied them through the analyses of EM and ERT data. The good agreement between EM and ERT data over the Awotan landfill site delineated the region of high contamination in the form of low subsurface resistivity distribution, suggesting the importance of integrating the electromagnetic and electrical resistivity investigation techniques for environmental studies.

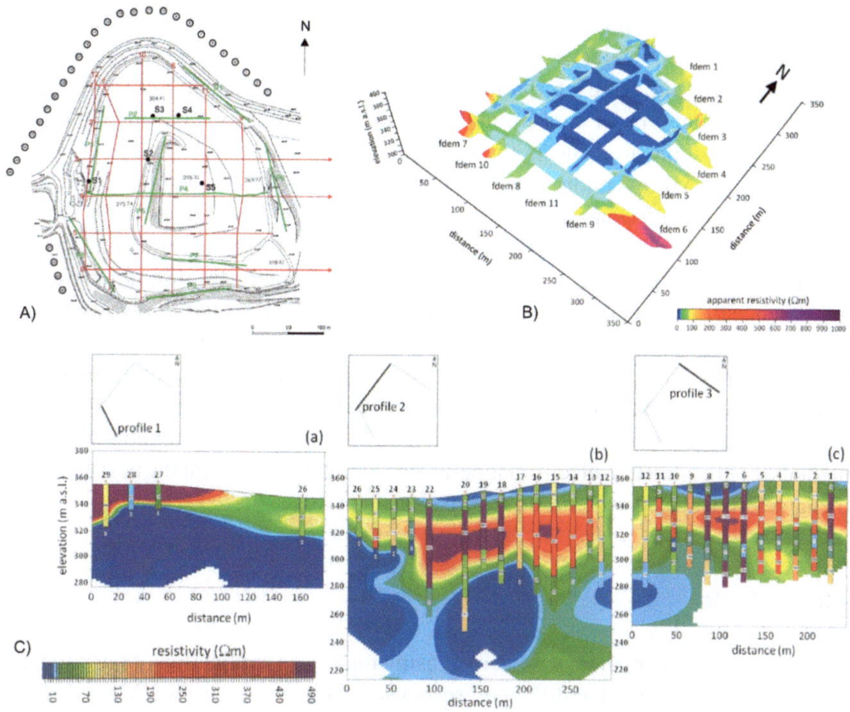

Fig. 1.7 **A** The landfill area with location of the 2D cross sections and waste drilling sites. Red continuous lines: electric and electromagnetic (FDEM) profiles; green continuous lines: seismic profiles; numbered circles: TDEM electromagnetic soundings; S2 and S5: boreholes for seismic tomography; S1–S5: boreholes for waste, leachate and biogas sampling. **B** Apparent resistivity pseudosections provided by the FDEM survey along six profiles oriented in W-E direction and five profiles oriented in N-S direction; **C** Resistivity sections reconstructed by correlating the 1D inversion results (represented by columns in the three sections) of the TDEM soundings performed along the south-western (**a**), north-western (**b**) and north-eastern boundaries (**c**) of the landfill (modified from Di Maio et al. 2018)

Baawain et al. (2018) performed a TDEM survey as well as drilling investigations to identify possible contamination of a dumping site in Barka, Oman. The method was applied to evaluate conductivity of the contaminated plumes in hot and arid/semiarid region. The TDEM survey results suggested that the low-resistivity zones (range 40–80 Ωm) may be contaminated areas. The combined results of drilling wells, piezometers, and TDEM apparent resistivity maps showed a coincidence of the migrated leachate plume and water table. It also showed that the plume migration, due to pores clogging resulted from solids or particulate contaminants, follows a preferential flow path, driven by previous disposal of domestic wastewater and waste oil from auto-garages through tankers.

The migration of landfill leachate through a lacustrine sandy aquifer and wetlands has been studied by Høyer et al. (2019) in a small area surrounding the former landfill of Pillemark at the island of Samsø, Denmark. The use of FDEM and TDEM surveys provided information about the depth to the good conductor. Due to data uncertainties and equivalences of the geophysical inversions, different methods did not always lead to the same interpretation. This emphasized the importance of considering the different type of data strengths and weaknesses, as well as the conceptual understanding of the area during the interpretation.

A rigorous EMI inversion approach has been developed by Deidda et al. (2022) to identify the heterogeneities in the waste body of a closed landfill in Ugento, Southern Italy. Unlike other geophysical techniques, EMI measurements can be conducted on the top of a capped landfill, by overcoming the limitations due to the presence of an impermeable liner on the top surface. The landfill investigated by the authors is characterized by very high-conductivity targets with high concentration of leachate and/or metallic objects (Fig. 1.8). As high conductivity makes the electromagnetic model nonlinear, the authors used non-linear inversion of Maxwell's equations. The inverted model provided detailed information unattainable with other methods, confirming that electromagnetic measurements represent the best technique to characterize closed systems such as capped landfills.

1.2.2.2 GPR

In the last twenty years there was an increasing interest in application of GPR to landfills. As there are several landfill types, response characteristics to the EM wave differ, and this complicates data interpretation. The GPR effectiveness can be influenced by factors such as the type of waste, soil conditions, and the depth of investigation required. The high soil conductivity caused by leachate migration in contaminated areas below the water table can indeed weaken the GPR signal and make it difficult to detect leachate plumes. Yochim et al. (2013) used GPR (with antennas of 25, 50, and 100 MHz) as a tool for estimating in-situ water content within two landfills in Ontario, Canada. The large degree of subsurface heterogeneity and the electrically conductive clay cap covering landfills, both of which affect the transmission of the electromagnetic pulses, made the use of GPR to quantify the in-situ water content in the landfill a challenging matter. The authors showed that the borehole GPR technology has higher effectivity with respect to the surface one to estimate the water content within wastes. Maryadi et al. (2020) investigated the Cipayung Landfill, West Java, to detect the groundwater contamination level around the landfill using GPR and direct water quality assessment. There, poor waste management has resulted in leachate accumulation that continues to this day infiltrating to the groundwater. Ground penetrating radar, with a frequency of 50 MHz and a maximum penetration of 40 m, was useful to detect the change of permittivity and conductivity of layers below the surface and allowed to mark off the contaminated groundwater, with a clear electromagnetic signal attenuation, to the fresh one. The study suggested that contamination occurred in radius up to hundreds of meters from the landfill site and this result was confirmed by direct water quality assessment. Umar et al. (2022) investigated the impact of the leachate on the subsurface geological structure at Rimba Mas landfill, Malaysia by GPR using an antenna of 250 MHz. Leachate migration

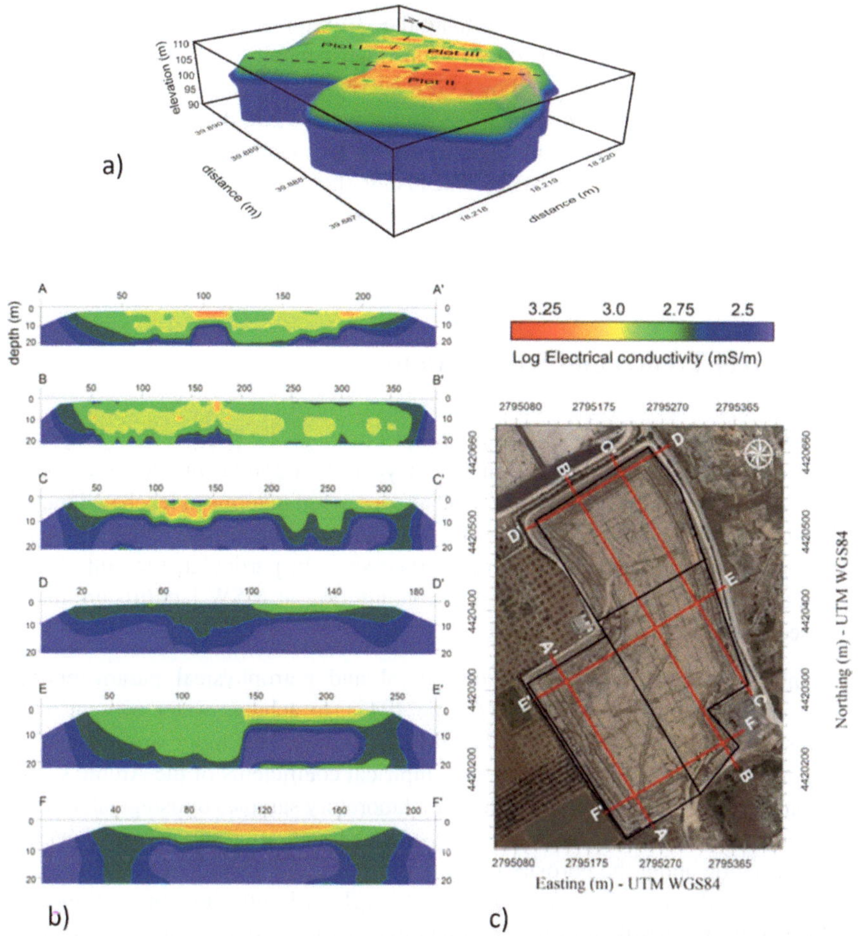

Fig. 1.8 3D visualization of the electromagnetic model produced from the inversion procedure of FDEM data. **b** Cross sections extracted from the volume in (**a**); **c** The landfill area with location of the 2D cross sections within the inversion model (modified after Deidda et al. 2022)

was detected near the groundwater monitoring station by a clear distinction between the water table and the leachate permittivity. Kondracka et al. (2021) compared the advantages and disadvantages of four geophysical methods (ERT, Seismic Refraction Tomography, Multichannel Analysis of Surface Waves and GPR) for waste characterisation at the landfill of Dabrowa Górnicza, Poland, and found that the GPR signal was greatly attenuated by the surface cover (loamy silt and sandy clay) and the actual maximum penetration was only 2.5 m, regardless of the used antenna frequency (250 and 500 MHz). Thus, apart from the considerable attenuation of the signal within the waste itself, a main issue when applying GPR to waste dump studies seem to be the type and thickness of the surface cover and the field conditions. Dawrea et al. (2021)

carried out a theoretical study to estimate the most appropriate petrophysical relationship between dielectric permittivity and water content to estimate in-situ water saturation in landfills investigated by a GPR survey. Zhang et al. (2022) investigated the applicability of the GPR technology to estimate waste depth and volume of a construction waste landfill, which previously hosted ponds and cultivated land (by antennas of 100 MHz and 250 MHz). The results showed that GPR effectively delineated boundaries between underground waste and the surrounding strata allowing authors to produce a pseudo three-dimensional model of the whole buried landfill.

1.3 Novel Trends and Applications

Apart from standard use of electrical and electromagnetic prospecting techniques, new trends have been emerging in recent years to improve the accuracy of the geophysical landfill reconstruction (typically described in terms of leachate imaging, geometry of the landfill, liners integrity). The new approaches are essentially based on data integration and advanced data processing. In particular, the cutting-edge methods oriented to characterization and monitoring of MSW landfills are mainly focused on the following three topics:

- exploring the link between geoelectrical and petrophysical parameters, i.e. through the Archie's equation (Archie 1942), to achieve water content of the waste mass after inversion of ERT, EM and IP data. The major weakness of this approach is the need to estimate the empirical coefficients of the Archie's equation (a, m and n) from measurements on laboratory samples or using values taken from literature; in addition, the assessment of saturation is unfeasible without information about the porosity of the waste mass (e.g. Dumont et al. 2018a).
- applying advanced inversion procedures, such as jointly inverting geophysical (generally electrical and seismic) data directly for petrophysical parameters (i.e. waste porosity, saturation, cation exchange capacity) or in terms of waste matrix (water or leachate and air fractions), or performing geostatistical joint inversion. These methods allow a quantitatively assessment of the composition of the waste mass, thus improving the effectiveness of the geophysical reconstruction, even though they share with the previous works the same limitation about the estimation of empirical coefficients (e.g. Steiner et al. 2022).
- developing machine learning (ML)-based methods to obtain a quantitative integration among different geophysical parameters, and primarily to improve the accuracy in imaging leachate accumulation zones. These emerging techniques can allow data fusion regardless of empirical relationships with petrophysical parameters, even though a preliminary calibration (training) is required for supervised algorithms or, when unsupervised methods are employed, the results have to be validated or constrained by direct information (well data, boreholes, etc.) (e.g. Piegari et al. 2023).

In the following sections, we briefly review the most significant papers covering the above-mentioned three topics published in the last decade.

1.3.1 Petrophysical Parameters

One of the most important petrophysical parameters for MSW landfill characterization is the moisture content, whose distribution is a straightforward proxy for highlighting leachate accumulation zones. For this goal, Dumont et al. (2016) aimed at characterizing the relationship between gravimetric water content and electrical resistivity, and assessing geoelectrical methods as tools to characterize the gravimetric water distribution in a landfill located at Mont-Saint-Guibert landfill (Belgium). Audebert et al. (2016a, b) recorded two ERT time-lapse data sets on different waste deposit cells at the MSW landfill site of Champs-Jouault (France), in order to compare the hydrodynamic behaviour of leachate flow between the two cells. They showed that leachate hydrodynamic behaviour is comparable from one waste deposit cell to another with: (i) a high leachate infiltration speed at the beginning of the infiltration, which decreases with time, (ii) a horizontal anisotropy of the leachate infiltration shape and (iii) a very small fraction of the pore volume used by the leachate flow. Dumont et al. (2018b) investigated the changes in water content in a landfill cover layer with long time-lapse electrical resistivity tomography (ERT) profiles during a rainfall event, applying site-specific Archie's laws with empirical parameters. Hu et al. (2019) performed pumping tests in vertical and horizontal wells at the Chang'an and Tianziling landfills (China), measuring the leachate level drawdown around the wells and the electrical resistivity of the surrounding waste. They found site-specific empirical relationships between resistivity of the waste samples and volumetric moisture content which are well-fitted with Archie's law. More recently, in four sanitary landfills in Thailand Suknark e al. (2023) have analyzed the potential of landfill mining for refuse-derived fuel (RDF) production based on waste electrical resistivity, including the influence of waste age and soil cover. Linear and multivariate regression analyses were used to constrain the data correlation based on the waste's physical characteristics, such as moisture content. The main finding is that resistive materials, waste compaction, and pore pressure were all positively correlated with the RDF fraction, whereas conductive materials and moisture content were negatively correlated with the RDF fraction.

1.3.2 Advanced Inversion Strategies

In the last years, there have been few attempts to upgrade the standard standalone inversion procedures to improve the effectiveness of the geophysical reconstruction of the MSW landfills. Audebert et al. (2014b) proposed a new methodology called MICS (Multiple Inversions and Clustering Strategy), which allows ERT users to

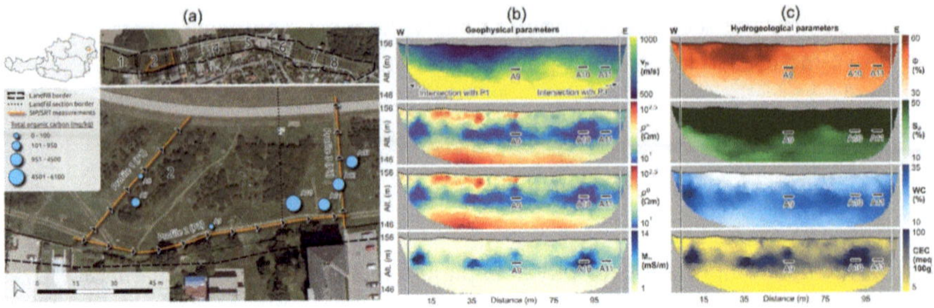

Fig. 1.9 **a** Location of the Heferlbach landfill (Austria); **b** joint inversion imaging results for data collected along P2 expressed in terms of seismic velocity (v_P), electrical resistivity (ρ^0, ρ^∞) and normalized chargeability (M_n); **c** joint inversion imaging results for data collected along P2 expressed in terms of porosity (Φ), air saturation (S_a), water content (WC) and cation exchange capacity (CEC). The black dots along the surface of each model represent the sensor/shot positions. Horizontal lines at the position of sampling points A9, A10 and A11 indicate the bottom of the landfill as observed during excavations for the collection of waste samples (modified from Steiner et al. 2022)

improve the delimitation of the infiltration area in leachate injection monitoring. The procedure consists in a multiple inversion that allows to take a wide range of resistivity models into account and a clustering strategy to improve the delineation of the infiltration front. A promising work by Narciso et al. (2021) introduced an iterative geostatistical inversion of FDEM data applied to a landfill in Belgium, for the characterization of waste deposits in order to evaluate the associated environmental risks. Steiner et al. (2022) jointly inverted electrical (both ERT and IP) and seismic tomography data in terms of three phases (solid, fluid, gas) and quantitatively assessed saturation, porosity, water content and the cation exchange capacity of the waste mass (also considering the surface conductivity contribution) at the Heferlbach landfill (Austria). Their results (Fig. 1.9) reveal a high water content within the MSW unit for areas characterized by a strong polarization response, which can be related to an increased biogeochemical activity as evidenced by the detected methane production.

1.3.3 Machine Learning-Based Approaches

Recently, some authors explored the potential of ML-based algorithms for an accurate monitoring of leachate accumulation within MSW landfills, which is a key factor for solid waste management and groundwater protection. Isunza Manrique et al. (2023) presented both a probabilistic inversion approach and a ML algorithm of multi-layer perceptron (MLP) for classification problem including uncertainty estimation from multiple geophysical datasets acquired at the Onoz landfill (Belgium). They found that the anthropogenic-geologic scenario from the probabilistic approach is more realistic as it agrees with the trial pit logs, in contrast to the MLP algorithm whose performance is largely related to the number of training data. Piegari et al.

(2023) proposed a clustering approach (K-means) for mapping leachate contamination through an effective integration of resistivity, chargeability and normalized chargeability data extracted from tomographic inverted model. The results of the application of the proposed procedure (Fig. 1.10) to two MSW landfills in Central and Southern Italy provided an easy and less ambiguous identification of the leachate accumulation zones, which were also validated through well-logged piezometric levels.

Sun et al. (2023) designed a novel deep network (named LDI-MVFNet) for multi-view fusion to invert the real resistivity distribution of the medium caused by leachate so as to infer the distribution of leachate. Processing of field data acquired at the MSW landfill in Jiangsu Province (China) showed that LDI-MVFNet can improve data noise suppression and inversion accuracy, thus reflecting more accurately the real distribution of resistivity than that from classical inversion.

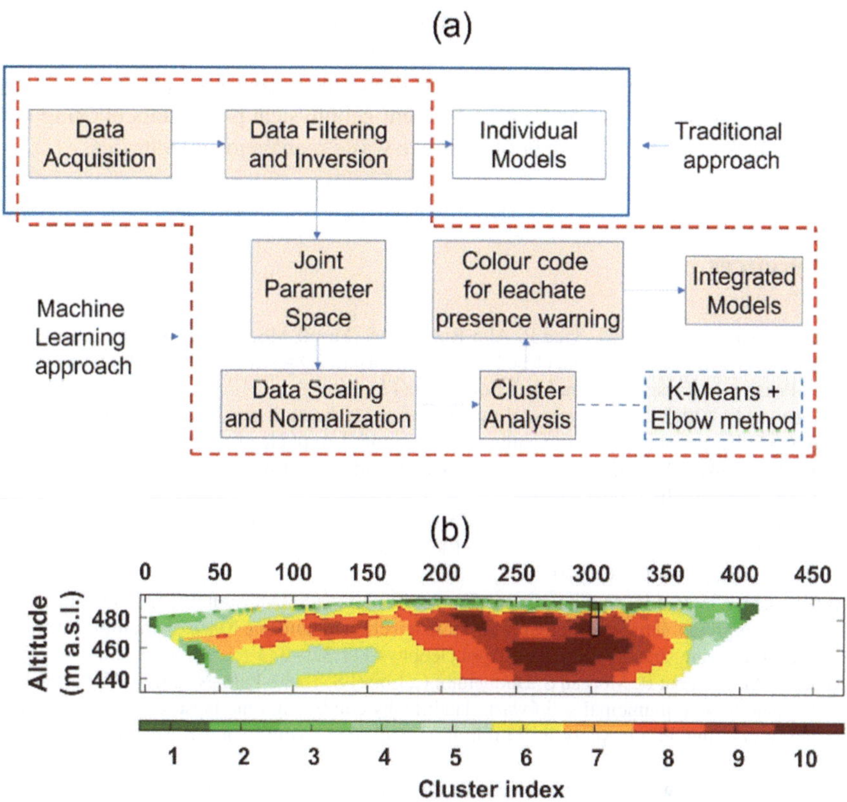

Fig. 1.10 **a** Sketch of the traditional and ML-based approaches. The red dotted line encloses the procedure proposed by Piegari et al. (2023). **b** Integrated section after cluster analysis for an example ERT/IP line. Each cluster index has been associated to a colour. The darkest red areas identify the most hazardous areas in terms of leachate accumulation. The piezometric levels are superposed to the models (modified from Piegari et al. 2023)

1.4 Conclusions

The overview of the most significant case histories in the last ten years highlights the potential of electrical and electromagnetic methods for characterization of MSW landfills. Over the last decade, there is an overall increasing trend in the use of such prospecting techniques both in terms of number of applications and different goals of the surveys. In particular, ERT is the most used method as it permits a wide range of applications concerning landfill characterization and management, such as site analysis, definition of landfill geometry and imaging and monitoring of leachate contamination. Nevertheless, a multi-methodological approach has to be preferred to reduce ambiguities and uncertainties related to a standalone application of geophysical methods as well as validation of the geophysical results through direct information (boreholes, wells) is mandatory to ensure the effectiveness of the investigation at landfill sites.

Finally, new trends have emerged in recent years trying to explore the link between geoelectrical and petrophysical parameters, apply advanced inversion procedures and develop machine learning-based methods. These emerging techniques have allowed for a quantitative integration among different geophysical parameters opening new perspectives in the development of effective waste management strategies.

References

Abdulrahman A, Nawawi M, Saad R, Abu-Rizaiza AS, Yusoff MS, Khalil AE, Ishola KS (2016) Characterization of active and closed landfill sites using 2D resistivity/IP imaging: case studies in Penang, Malaysia. Environ Earth Sci 75(4):347

Adabanija MA (2023) Spatio-temporal monitoring of leachates dispersion beneath a solid wastes dump in basement complex of southwestern Nigeria. J Appl Geophys 210:104953

Archie GE (1942) The electrical resistivity log as an aid in determining some reservoir characteristics. Trans AIME 146(01):54–62

Audebert M, Clément R, Grossin-Debattista J, Günther T, Touze-Foltz N, Moreau S (2014a) Influence of the geomembrane on time-lapse ERT measurements for leachate injection monitoring. Waste Manage 34(4):780–790

Audebert M, Clément R, Touze-Foltz N, Günther T, Moreau S, Duquennoi C (2014b) Time-lapse ERT interpretation methodology for leachate injection monitoring based on multiple inversions and a clustering strategy (MICS). J Appl Geophys 111:320–333

Audebert M, Clément R, Moreau S, Duquennoi C, Loisel S, Touze-Foltz N (2016a) Understanding leachate flow in municipal solid waste landfills by combining time-lapse ERT and subsurface flow modelling–Part I: Analysis of infiltration shape on two different waste deposit cells. Waste Manage 55:165–175

Audebert M, Oxarango L, Duquennoi C, Touze-Foltz N, Forquet N, Clément R (2016b) Understanding leachate flow in municipal solid waste landfills by combining time-lapse ERT and subsurface flow modelling–Part II: constraint methodology of hydrodynamic models. Waste Manage 55:176–190

Baawain MS, Al-Futaisi AM, Ebrahimi A, Omidvarborna H (2018) Characterizing leachate contamination in a landfill site using Time Domain Electromagnetic (TDEM) imaging. J Appl Geophys 151:73–81

Balia R (2018) Old municipal and industrial waste landfills: some examples of possible contribution of geophysical survey techniques for their assessment before reclamation. Detritus 1:110–115

Bellezoni RA, Iwai CK, Elis VR, da Silva Paganini W, Hamada J (2014) Small-scale landfills: impacts on groundwater and soil. Environ Earth Sci 71:2429–2439

Binley A, Slater L (2020) Resistivity and induced polarization: theory and applications to the near-surface earth. Cambridge University Press

Carpenter PJ, Reddy KR (2017). Geophysical Imaging of Landfill Interiors: Examples from Northern Illinois, USA. Springer, Singapore, 1–11.

Cınar H, Altundaş S, Ersoy E, Bak K, Bayrak N (2016) Application of two geophysical methods to characterize a former waste disposal site of the Trabzon-Moloz district in Turkey. Environ Earth Sci 75:1–16

Cole KS, Cole RH (1941) Dispersion and adsorption in dielectrics I. Alternating current characteristics. J Chem Phys 9:341–351

Costanzo-Alvarez V, Aldana M, Trigo-Ferre R, Jacome M, Galatro D, Izarra C, Amon CH (2022) Combining a geoelectrical survey with integrated groundwater quality data to map the spatial distribution and temporal variations of a leachate plume in a closed landfill (Southern Ontario, Canada). Environ Earth Sci 81(20):498

Dawrea A, Zytner RG, Donald J (2021) Enhanced GPR data interpretation to estimate in situ water saturation in a landfill. Waste Manage 120:175–182

De Carlo L, Perri MT, Caputo MC, Deiana R, Vurro M, Cassiani G (2013) Characterization of a dismissed landfill via electrical resistivity tomography and mise-à-la-masse method. J Appl Geophys 98:1–10

De Donno G, Cardarelli E (2017a) VEMI: a flexible interface for 3D tomographic inversion of time-and frequency-domain electrical data in EIDORS. Near Surface Geophys 15(1):43–58

De Donno G, Cardarelli E (2017b) Tomographic inversion of time-domain resistivity and chargeability data for the investigation of landfills using a priori information. Waste Manage 59:302–315

Deidda GP, De Carlo L, Caputo MC, Cassiani G (2022) Frequency domain electromagnetic induction imaging: an effective method to see inside a capped landfill. Waste Manage 144:29–40

Di Maio R, Fais S, Ligas P, Piegari E, Raga R, Cossu R (2018) 3D geophysical imaging for site-specific characterization plan of an old landfill. Waste Manage 76:629–642

Dumont G, Pilawski T, Dzaomuho-lenieregue P, Hiligsmann S, Delvigne F, Thonart P, Robert T, Nguyen F, Hermans T (2016) Gravimetric water distribution assessment from geoelectrical methods (ERT and EMI) in municipal solid waste landfill. Waste Manage 55:129–140

Dumont G, Robert T, Mark N, Nguyen F (2017) Assessment of multiple geophysical techniques for the characterization of municipal waste deposit sites. J Appl Geophys 145:74–83

Dumont G, Pilawski T, Hermans T, Nguyen F, Garré S (2018a) The effect of initial water distribution and spatial resolution on the interpretation of ERT monitoring of water infiltration in a landfill cover. Hydrol Earth Sys Sci Discussions:1–26

Dumont G, Robert T, Nguyen F (2018b) Electrical resistivity tomography and distributed temperature sensing monitoring to assess the efficiency of horizontal recirculation drains on retrofit bioreactor landfills. Geophysics 83(2):B13–B23

Feng SJ, Gao KW, Chen YX, Li Y, Zhang LM, Chen HX (2017) Geotechnical properties of municipal solid waste at Laogang Landfill, China. Waste Manage 63:354–365

Feng SJ, Zhao Y, Zhang XL, Bai ZB (2020) Leachate leakage investigation, assessment and engineering countermeasures for tunneling underneath a MSW landfill. Eng Geol 265:105447

Fiandaca G, Auken E, Christiansen AV, Gazoty A (2012) Time-domain-induced polarization: full-decay forward modeling and 1D laterally constrained inversion of Cole-Cole parameters. Geophysics 77(3):E213–E225

Fiandaca G, Ramm J, Binley A, Gazoty A, Christiansen AV, Auken E (2013) Resolving spectral information from time domain induced polarization data through 2-D inversion. Geophys J Int 192(2):631–646

Flores-Orozco A, Gallistl J, Steiner M, Brandstätter C, Fellner J (2020) Mapping biogeochemically active zones in landfills with induced polarization imaging: the Heferlbach landfill. Waste Manage 107:121–132

Gazoty A, Fiandaca G, Pedersen J, Auken E, Christiansen AV (2012a) Mapping of landfills using time-domain spectral induced polarization data: the Eskelund case study. Near Surface Geophys 10(6):575–586

Gazoty A, Fiandaca G, Pedersen J, Auken E, Christiansen AV, Pedersen JK (2012b) Application of time domain induced polarization to the mapping of lithotypes in a landfill site. Hydrol Earth Syst Sci 16(6):1793–1804

Genelle F, Sirieix C, Riss J, Naudet V, Dabas M, Bégassat P (2014) Detection of landfill cover damage using geophysical methods. Near Surface Geophysics 12(5):599–612

Godio A, Naldi M (2009) Integration of electrical and electromagnetic investigation for contaminated site. Am J Environ Sci 5:562–569

Helene LPI, Moreira CA, Bovi RC (2020) Identification of leachate infiltration and its flow pathway in landfill by means of electrical resistivity tomography (ERT). Environ Monit Assess 192:1–10

Høyer AS, Klint KES, Fiandaca G, Maurya PK, Christiansen AV, Balbarini N, Bjerg PL, Hansen TB, Møller I (2019) Development of a high-resolution 3D geological model for landfill leachate risk assessment. Eng Geol 249:45–59

Hu J, Wu XW, Ke H, Xu XB, Lan JW, Zhan LT (2019) Application of electrical resistivity tomography to monitor the dewatering of vertical and horizontal wells in municipal solid waste landfills. Eng Geol 254:1–12

Ibraheem IM, Tezkan B, Bergers R (2021) Integrated interpretation of magnetic and ERT data to characterize a landfill in the north-west of Cologne, Germany. Pure Appl Geophys 178(6):2127–2148

Isunza-Manrique I, Caterina D, Nguyen F, Hermans T (2023) Quantitative interpretation of geoelectric inverted data with a robust probabilistic approach. Geophysics 88(3):B73–B88

Jacome M, Costanzo-Alvarez V, Aldana M, Patraskovic P, Drielsma C, Galatro D, Amon C (2021) A methodology to characterize a sanitary landfill combining, through a numerical approach, a geoelectrical survey with methane point-source concentrations. Environ Technol Innov 21:101225

Juarez MB, Mondelli G, Giacheti HL (2023) An overview of in situ testing and geophysical methods to investigate municipal solid waste landfills. Environ Sci Pollut Res 30(9):24779–24789

Kondracka M, Stan-Kłeczek I, Sitek S, Ignatiuk D (2021) Evaluation of geophysical methods for characterizing industrial and municipal waste dumps. Waste Manage 125:27–39

Loke MH, Chambers JE, Rucker DF, Kuras O, Wilkinson PB (2013) Recent developments in the direct-current geoelectrical imaging method. J Appl Geophys 95:135–156

Ma P, Ke H, Lan J, Chen Y, He H (2019) Field measurement of pore pressures and liquid-gas distribution using drilling and ERT in a high food waste content MSW landfill in Guangzhou, China. Eng Geol 250:21–33

Martinho E (2023) Electrical resistivity and induced polarization methods for environmental investigations: an overview. Water Air Soil Pollut 234(4):215

Martorana R, Capizzi P, Pirrera C (2023) Unconventional arrays for 3D electrical resistivity and induced polarization tomography to detect leachate concentration in a waste landfill. Appl Sci 13(12):7203

Maryadi M, Champai FMS, Triananda IN, Fitriana DN (2020) Analysis of groundwater contamination level in residential areas around Cipayung landfill using ground penetrating radar. IOP Conf Ser: Mater Sci Eng 874:012013

Maurya PK, Rønde VK, Fiandaca G, Balbarini N, Auken E, Bjerg PL, Christiansen AV (2017) Detailed landfill leachate plume mapping using 2D and 3D electrical resistivity tomography-with correlation to ionic strength measured in screens. J Appl Geophys 138:1–8

Mepaiyeda S, Madi K, Gwavava O, Baiyegunhi C, Sigabi L (2019) Contaminant delineation of a landfill site using electrical resistivity and induced polarization methods in Alice, Eastern Cape, South Africa. Int J Geophys 2019:1–13

Moser C, Binley A, Flores-Orozco A (2023) 3D electrode configurations for spectral induced polarization surveys of landfills. Waste Manage 169:208–222

Narciso J, Azevedo L, Van De Vijver E, Van Meirvenne M (2021) Geostatistical electromagnetic inversion for landfill characterization. In: Symposium on the application of geophysics to engineering and environmental problems (SAGEEP) proceedings, pp 35–38

Nguyen F, Ghose R, Manrique II, Robert T, Dumont G (2018) Managing past landfills for future site development: a review of the contribution of geophysical methods. In: 4th International symposium on enhanced landfill mining, Mechelen, 05–06 Feb 2018

Osinowo OO, Falufosi MO, Omiyale EO (2018) Integrated electromagnetic (EM) and electrical resistivity tomography (ERT) geophysical studies of environmental impact of Awotan dumpsite in Ibadan, southwestern Nigeria. J Afr Earth Sc 140:42–51

Piegari E, De Donno G, Melegari D, Paoletti V (2023) A machine learning-based approach for mapping leachate contamination using geoelectrical methods. Waste Manage 157:121–129

Ramalho EC, Dill AC, Rocha R (2013) Assessment of the leachate movement in a sealed landfill using geophysical methods. Environ Earth Sci 68:343–354

Soupios P, Ntarlagiannis D (2017) Characterization and monitoring of solid waste disposal sites using geophysical methods: current applications and novel trends. In: Modelling trends in solid and hazardous waste management, pp 75–103

Steiner M, Katona T, Fellner J, Flores-Orozco A (2022) Quantitative water content estimation in landfills through joint inversion of seismic refraction and electrical resistivity data considering surface conduction. Waste Manage 149:21–32

Suknark P, Buddhawong S, Wangyao K (2023) Investigating the effect of waste age and soil covering on waste characteristics prior to landfill mining using an electrical resistivity tomography technique. J Environ Manage 339:117898

Sun X, Qian X, Nai C, Xu Y, Liu Y, Yao G, Dong L (2023) LDI-MVFNet: a multi-view fusion deep network for leachate distribution imaging. Waste Manage 157:180–189

Tsourlos P, Vargemezis GN, Fikos I, Tsokas GN (2014) DC geoelectrical methods applied to landfill investigation: case studies from Greece. First Break 32(8):81–89

Umar MS, Ghazali MD, Samad AM, Zainon O, Zainuddin K, Marzuki F (2022) Implementation of Ground penetrating radar in assessing leachate in sanitary landfill. In: 2022 IEEE 20th student conference on research and development (SCOReD), 8–9 Nov 2022, Bangi, Malaysia

Van De Vijver E, Manrique II, Bobe C, Caterina D, Hermans T, Wille E, Nguyen F (2021) Geophysics in support of dynamic landfill management. In: SEG international exposition and annual meeting, p D011S135R001https://doi.org/10.1190/segam2021-3594135.1

Wemegah DD, Fiandaca G, Auken E, Menyeh A, Danuor SK (2017) Spectral time-domain induced polarisation and magnetic surveying–an efficient tool for characterisation of solid waste deposits in developing countries. Near Surface Geophys 15(1):75–84

Yin K, Tong HH, Noh O, Wang J-Y, Giannis A (2015) Mapping refuse profile in Singapore old dumping ground through electrical resistivity, s-wave velocity and geotechnical monitoring. Bull Environ Contam Toxicol 94(3):275–281. https://doi.org/10.1007/s00128-014-1427-y

Yin K, Tong H, Giannis A, Wang JY, Chang VWC (2017) Multiple geophysical surveys for old landfill monitoring in Singapore. Environ Monit Assess 189:1–13. https://doi.org/10.1007/s10661-016-5722-4

Yochim A, Zytner RG, McBean EA, Endres AL (2013) Estimating water content in an active landfill with the aid of GPR. Waste Manage 33(10):2015–2028

Zhang T, Zhang D, Zheng D, Guo X, Zhao W (2022) Construction waste landfill volume estimation using ground penetrating radar. Waste Manag. Res. 40(8):1167–1175

Chapter 2
Characteristics and Impacts of Municipal Solid Waste (MSW)

Mehdi Ghanbarzadeh Lak, Milad Ghaffariraad, and HamidReza Jahangirzadeh Soureh

Abstract The term "Solid Waste" based on the Resource Conservation and Recovery Act (RCRA), means any garbage or refuse, sludge from water and wastewater treatment plants, or air pollution control facilities, and other discarded material, resulting from industrial, commercial, mining, and agricultural operations, and from community activities. It is important to note that the definition of solid waste is not limited to the physical solid state. Many solid wastes are liquid, semi-solid, or contain gaseous material. Municipal Solid Waste (MSW) more commonly known as trash or garbage, originally included any discards in urban areas, may contain product packaging, grass clippings, furniture, clothing, bottles, food scraps, newspapers, appliances, and batteries. The comprehensive waste management system generally encompasses four stages: (1) waste generation, (2) collection and transport, (3) physical/biological/thermal treatment, and (4) landfilling. Considering all environmental, economic, social, and political aspects, one could evaluate and select the best option to handle any of the four above-mentioned stages in a comprehensive waste management system. Furthermore, identifying the physical, chemical, and biological characteristics of the waste stream is essential in the selection process. The first step in determining the characteristics of MSW is to identify the main sources of MSW in the study area. Resources can be classified based on generation tonnage, location, and existing laws and regulations. Subsequently, the tonnage of waste in resources should be estimated. MSW generation per capita depends on various factors such as the level of public welfare, type of economic activities, population distribution, climate, culture, and customs. The main physical/chemical/biological properties of MSW include density, material composition, particle size

M. Ghanbarzadeh Lak (✉)
School of Engineering, Civil Engineering Department, Urmia University, Urmia, Iran
e-mail: m.ghanbarzadehlak@urmia.ac.ir

M. Ghaffariraad · H. Jahangirzadeh Soureh
M.Sc. in Environmental Engineering, School of Engineering, Civil Engineering Department, Urmia University, Urmia, Iran
e-mail: st_m.ghaffariraad@urmia.ac.ir

H. Jahangirzadeh Soureh
e-mail: st_h.jahangirzade@urmia.ac.ir

A. Anouzla and S. Souabi (eds.), *Technical Landfills and Waste Management*, Springer Water, https://doi.org/10.1007/978-3-031-52633-6_2

distribution, moisture content, and heating value. Identifying waste characteristics would be especially useful to design and to operate recycling or energy recovery facilities. Preparation of samples with appropriate size (to estimate optimal accuracy), separation of components, and necessary analysis are essential to determine waste characteristics.

Keywords MSW · Density · Material composition · Particle size distribution · Moisture content · Heating value

2.1 Municipal Solid Waste Management Definitions, Importance, and Emerging Challenges

In a general sense, the term "waste" is used for all materials that are considered of little or no value from the perspective of producers or consumers. More precisely, solid waste (SW) refers to all solid discards resulting from human and animal activities, which are typically solid (meaning they contain a relatively small amount of liquids that cannot easily flow) and are not useful. Virtually all human activities lead to the generation of waste in various forms, whether solid, liquid, gas, or combinations thereof. Solid waste is generated during raw material mining, refining processes, manufacturing, and when products are used by consumers. The bulk of solid waste generated in communities is the result of agricultural and mining activities. Some types of waste, such as residues from sewage treatment and electric power generation, are also considered significant solid waste types requiring special attention. Municipal solid waste (MSW) includes both durable (such as discarded furniture and appliances) and nondurable goods, containers and packaging materials, food wastes, yard trimmings, and other organic waste materials originating from residential, commercial, institutional, and industrial sources. Other types of waste that generally do not fall under the definition of MSW include industrial waste from manufacturing and processing activities, construction and demolition debris, agricultural wastes, hazardous wastes containing oil and gas residues, and mineral wastes resulting from the extraction and processing of mineral materials. Regardless of the type of waste generated, individuals, private organizations, and government agencies are responsible for finding solutions to minimize waste generation, control harmful emissions resulting from waste management, and recover materials and energy from the waste stream. In other words, waste disposal must be planned in a way that does not endanger human health and minimizes the environmental impacts caused by waste management.

In contemporary society, waste management issues have risen to the forefront of environmental concerns, particularly in developed communities. The production of substantial quantities of solid waste, characterized by more complex components than past waste outputs (for instance, Tetra Pak packaging containers are now prevalent compared to traditional glass bottles), represents a hallmark of improved living standards within an industrialized society. The generation of over two billion tons of solid waste globally in 2016, with at least one-third of it being environmentally unmanaged, and the projected production of over 3.4 billion tons of waste by 2050 (Kaza et al. 2018), has sounded the alarm for crises across various environmental dimensions. Furthermore, the production and consumption of goods in contemporary modern life have led to the emergence of handling special waste categories that, in some instances, exhibit resistance to biological decomposition, are toxic to ecosystems, combustible, corrosive, or potentially explosive (in other words, hazardous). Urgent action regarding solid wastes that have historically been mismanaged, such as unsanitary landfill sites and untreated leachate ponds naturally forming in proximity to such sites based on local topography (Ghaffariraad and Ghanbarzadeh 2021; Ghanbarzadeh et al. 2018, 2012; Sabour et al. 2010), is recognized as a significant challenge in mitigating the environmental damages incurred over decades of economic growth without considering environmental preservation principles.

In recent years, the occurrence of certain environmental incidents and disasters has raised public awareness concerning issues and challenges related to solid waste disposal sites. This has underscored the necessity for the enforcement of stringent regulations, allocation of adequate financial resources, and assignment of knowledgeable workforce dedicated to waste management sector. One of the most renowned of these events is the tragic Love Canal incident in the Niagara region of New York, where the storage of hazardous waste by the Hooker Chemical Company in a 28 ha disposal site during the 1940s resulted in a significant environmental catastrophe. In the 1890s, William T. Love, a former railroad attorney, devised a plan to construct a preplanned urban community consisting of parks and residences on the shores of Lake Ontario, which he named it “the Model City, New York”. In the late 1890s, work commenced on Love’s plan, which included the creation of a navigable canal to bypass Niagara Falls. However, the withdrawal of investor support due to an economic crisis at the time and the efforts of environmental groups advocating for legislation in Congress to preserve Niagara Falls and restrict water diversion from the Niagara River led to only partial development of the Love Canal (approximately 1.6 km of the canal was excavated, measuring about 15 m wide and 3–12 m deep). With the abandonment of this project, the canal gradually filled with water. During the summer, children would swim in it, and in the winter, it served as a skating rink. In the 1920s, the city of Niagara Falls used the area around this canal as a disposal site for municipal waste, a practice that continued until around 1948. Finally, in the 1940s, the Hooker Chemical Company purchased this canal and its banks (covering an area of approximately 16 ha). After drainage and lining the canal bottom with a thick clay liner, it became a disposal site for byproducts of chemical manufacturing processes, including dyes, fragrances, rubber solvents, and synthetic resins (stored in 210 L barrels).

In early 1952, when the future development plan for the site was finalized, Hooker ceased using the Love Canal as a waste disposal site. Over the ten-year operation of this disposal center, approximately 19,800 tons of chemical materials, primarily consisting of caustics, alkaline substances, fatty acids, and chlorinated hydrocarbons, were deposited at this site. Barrels containing these chemical substances were placed at depths of 6–7.5 m below the ground, and to prevent leakage, the upper portion was sealed with an impermeable clay cover. Over time, vegetation grew on the surface of the disposal center.

The need for additional land for new schools in the 1950s prompted the Niagara Falls City School District to purchase the aforementioned site from Hooker Chemical Company for a nominal fee of just one dollar. Subsequently, schools were constructed, and later, the land was sold to the private sector, leading to the development of several residential communities in its vicinity. Company officials believed that through this action, they would reduce their future liabilities regarding the buried chemical materials. Moreover, as part of the property transfer agreement, the possibility of any claims against Hooker due to potential harm to individuals resulting from the presence of industrial waste was effectively nullified (in fact, this condition was included as part of the payment made to the company in exchange for ownership transfer).

This transfer essentially relieved Hooker of its responsibility for the secure containment and maintenance of hazardous waste, shifting all responsibilities to the new owners, who often lacked the necessary expertise in waste management. Consequently, the new property owners, in their construction activities, disregarded waste containment engineering regulations in various ways. Weather conditions during the rain-heavy years of the early 1960s allowed previously stabilized chemical substances to leach outside the site. By the late 1970s, when hazardous wastes had infiltrated beneath the homes in the area (as reported by residents who observed pools of oily or colorful fluids in their yards or basements), over 900 families were compelled to relocate. Many of them suffered from chronic health issues and exhibited increased white blood cell counts and leukemias. This led to the imposition of hundreds of millions of dollars in damages on federal officials for cleanup costs and compensation for damages (Rhyner et al. 1995; State of New York, Department of Health 1978; Worthley and Torkelson 1981).

The bitter experience of Love Canal catalyzed the enactment of the Comprehensive Environmental Response, Compensation, and Liability Act (CERCLA), commonly known among experts as the Superfund Law. According to CERCLA, taxes were imposed on chemical and petroleum industries, and extensive powers were granted to federal authorities to respond directly to releases or threats of releases of hazardous substances that might endanger public health or the environment. Additionally, a list of Superfund priority sites was compiled, with Love Canal being the first site on that list (The three sites: Love Canal, New York; Times Beach, Missouri; and Valley of the Drums, Kentucky, hold special significance in the environmental history of the United States and its influence on the passage of CERCLA). One of the key provisions of the law was the "retroactive liability," whereby the original producers of hazardous waste would not be able to evade responsibility for their past waste production.

Despite what has been mentioned, most discussions these days do not focus on irreparable local incidents (such as Love canal environmental catastrophe) but mainly on the increasing amounts of solid wastes in developing countries. The images circulated at various times in social media about the dumping of waste in the main streets of Beirut (Lebanon), Naples (Italy), and Paris (France) due to protests by residents near the current disposal centers or labor strikes in the waste management industry, as well as the reports about the efforts of municipal authorities to find a new suitable location for the disposal of municipal waste, confirm this assertion. In today's world, the challenges faced by cities and towns in finding alternative solutions to landfills that have either reached their maximum capacity or will soon do so force citizens to deal with waste management issues daily. There was a time when citizens could simply toss all their household waste into a trash bin on the sidewalk for collection. However, nowadays, to preserve landfill space, residents often have to separate their waste into two or more bins for recycling. Furthermore, the disposal of some materials, such as garden waste, is the responsibility of the producer by law and is not collected with household waste in many parts of the world.

In addition, waste disposal costs have significantly increased in recent years. According to the World Bank (Kaza et al. 2018), in 2016, in low-income countries, the average cost of a basic solid waste management system, including collection, transportation, and sanitary disposal, was at least $35/ton, and in communities with high-income levels that utilize advanced waste management and recycling approaches, costs can go up to an average of $50 to $100/ton. In other words, the budget allocated to municipal solid waste management (MSWM) is a significant cost item in cities and usually accounts for nearly 20% of municipal budgets in low-income countries, over 10% in middle-income countries, and about 4% in high-income countries (Kaza et al. 2018). The imposition of volume-based waste management costs, although successful in countries such as Austria, South Korea, and the Netherlands, is still unusual because it requires strong planning and execution coordination. In low-income countries, citizens and commercial establishments usually pay a fixed fee for managing their waste. Most communities have incurred substantial financial commitments to provide the financial resources for new solid waste processing and disposal facilities. Collection costs have also increased due to the need for complex systems heavily dependent on human labor to collect separated waste fractions. Despite the difficulties and high costs, citizens often enthusiastically support recycling activities to conserve resources and improve environmental quality. This, in turn, leads to other issues, such as finding and creating markets for products made from recycled materials.

Currently, the construction of municipal wastewater treatment plants is facing relatively little opposition from neighboring residents. This is because the benefits of discharging less polluted water into rivers and other water sources are well recognized by the public, and wastewater treatment plants are considered important urban infrastructure. In contrast, it is almost impossible to find cases where the construction of solid urban and industrial waste processing and disposal facilities did not face public opposition. These oppositions are often related to people's concerns about the potential dangers of proximity to these facilities, undesirable environmental effects resulting from the construction of these facilities, the devaluation of properties, and the feeling that communities with such facilities become second-rate or "dumping grounds". Regarding health and environmental hazards, it is possible to point to issues such as (1) emissions of air pollutants and ashes in the incineration process;

(2) leachate leakage and bio-gas emissions in landfills; (3) sludge generated from the wastewater treatment process; (4) emissions of air pollutants, wastewater, and residual waste pollutants in the composting process; and (5) residual waste in the recycling process. In this context, some dangerous organic compounds and heavy metals (such as lead, mercury, and cadmium), which enter the waste stream through solvents, paints, plastics, vehicle fuels, inks, cleaners, fluorescent lamps, and pesticides, are of the greatest concern. Disposal of these materials in large quantities is controlled by laws that have been enacted for hazardous materials. However, alternative legal methods for disposing of such hazardous materials, which enter the waste stream from small household, agricultural, and commercial producers, have not been predicted due to the limited amounts produced.

2.2 An Overview of Integrated Solid Waste Management

2.2.1 Introduction

The rapid increase in population and urbanization, coupled with continuous economic development and the growing demand for goods and services, has led to a significant rise in the generation of MSW, especially in developing countries. This poses significant challenges to environmental protection and sustainable development. One effective solution to address such issues is the adoption of an efficient MSWM system (Yean Yng Ling and Song Anh Nguyen 2013).

An Integrated solid waste management system consists of a hierarchy of choices after the waste generation process, including: (1) source reduction (both quantity and toxicity) at the generation points, (2) reuse, (3) recycling (will be more efficient when source separation done), (4) composting, (5) incineration with energy recovery, (6) land disposal of residuals (landfilling), and (7) incineration without energy recovery.

Implementing this approach linearly implies that, first of all, efforts should be concentrated on reducing the inflow of materials into the waste stream. Once materials have entered the waste stream, the top priority should be reusing or recycling them to create new products. Waste disposal without material and energy recovery should be the last option to consider (Rhyner et al. 1995). Subsequently, all elements within the integrated waste management system will be addressed. Besides, the complementary actions/definitions including waste generation, temporary storage, collection, transfer, and transportation of waste, will be discussed in more detail in the coming subchapters.

2.2.2 Waste Generation

MSW includes substances such as durable and non-durable goods, containers, packaging materials, food waste, tree branches and leaves, and other organic and inorganic materials discarded in residential, commercial, institutional, and industrial areas. Examples of MSW types originating from these sources are listed in Table 2.1. Residential waste is generated in residential units and by the individuals residing in them. Commercial waste sources encompass retail and wholesale outlets and service activity centers in the community. Institutional waste is generated by schools, hospitals, and government facilities. Industrial waste in this context results from activities conducted in administrative centers and industrial operations, excluding residuals from manufacturing and production processes. The classification of materials typically used in identifying MSW components is detailed in Table 2.2 (Rhyner et al. 1995).

2.2.3 Source Reduction of Waste (Quantity and Toxicity)

According to the definition provided by the United States Environmental Protection Agency (1989), source reduction is defined as: "A set of actions taken during the design, production, and use of products to reduce the quantity and toxicity of waste generated at the end of their useful life". Many manufacturers are sensitive to public concerns about product toxicity. To address these concerns, many have made changes to the design or formulation of their products, resulting in their products containing very minimal amounts of materials that might pose a hazard when they enter the waste stream or eliminate such materials. A prominent example in this field is the change in the structure of household batteries, which were the main sources of mercury and cadmium in urban solid waste (Hurd 1992; Rhyner et al. 1995). The removal of mercury from zinc-carbon cell batteries (Hurd 1992; Johnson and Hirth 1990) and the use of lithium as a replacement for cadmium in rechargeable batteries (Rhyner et al. 1995) are among the actions of product manufacturers in the field of reducing waste from the source (reducing toxicity). Consumers can also take effective steps

Table 2.1 Sources of MSW

Source	Types of solid waste
Residential	Appliances, newspapers, clothing, disposable tableware, food packaging, cans, bottles, food scraps, and yard trimmings
Commercial	Corrugated boxes, food waste, office papers, disposable tableware, and yard trimmings
Institutional	Office papers, cafeteria and restroom wastes, classroom wastes, and yard trimmings
Industrial	Corrugated boxes, lunchroom wastes, office papers, and wood pallets

to reduce the toxicity of waste by separating and keeping toxic wastes separate from other wastes at the source of production. Collection of household hazardous waste on certain days or delivery of lead-acid batteries to designated centers are examples of such actions.

Source reduction is a concept applied to both producers and consumers. For producers, source reduction is a planned approach to minimizing the use of raw materials in manufacturing processes, and product delivering, and reducing the generation of waste at the end of a product's useful life. To quantify the results of source reduction efforts, data on waste generation and product manufacturing are required. First, the amount of raw materials needed for producing a unit of the product and the quantity of waste generated should be determined. Then, the type and quantity of packaging materials used for product sales should be examined to reduce the outcomes of source reduction actions.

Indeed, the discussion of consumer or household waste reduction measures can be somewhat complex. People engage in different activities, have different habits, and have different income levels, lifestyles, and attitudes toward possessions. Information in the literature on this topic and in educational programs can provide ideas about the types of choices and activities that reduce waste. However, it can be difficult to quantify or express the quantitative results of these programs. On the other

Table 2.2 Explanations related to the contents of MSW

Category		Description
Paper and paperboard	High-grade paper	Office paper and computer paper
	Mixed paper	Mixed colored papers, magazines, glossy paper, and other papers not fitting the categories of high-grade paper, newsprint, and corrugated
	Newsprint	Newspaper
	Corrugated	Corrugated boxes, corrugated and brown (kraft) paper
Yard waste		Branches, twigs, leaves, grass, and other plant material
Food waste		All food waste excluding bones
Glass		Clear and colored glass
Plastics		All types of plastics, codes 1 -7, PETE or PET, HDPE, PVC, LDPE, PP, PS, other
Ferrous metals		Iron, steel, tin cans, and bi-metal cans
Nonferrous metals		Primarily aluminum, aluminum cans, copper, brass, and lead
Wood		Lumber, wood products, pallets, and furniture
Rubber		Tires, footwear, wire cords, gaskets
Textiles		Clothing, furniture, footwear
Leather		Clothing, furniture, footwear
Miscellaneous		Other organic and inorganic materials, including rock, sand, dirt, ceramics, plaster, bones, ashes, etc.

hand, manufacturers have several options available to reduce waste, such as minimizing process waste, developing products with longer life or repairability, developing products that use less material, product packaging using fewer materials, and using recyclable materials (Rhyner et al. 1995).

2.2.4 Reuse of Discards

In some cases, people continue to discard unwanted materials mainly because they do not aware of anything better to do. Draperies that are too short, or a closet door that was scratched by a favorite pet, are examples of these wastes. We are conditioned to think of things that are old, empty, worn, broken, ugly, or marred as useless, so we throw them away without much thought about the consequences. The process of reusing starts with the assumption that the used materials that flow through our lives can be a resource rather than refuse.

The following are some examples of reuse.

- Containers can be reused at home or for school projects.
- Reuse wrapping paper, plastic bags, boxes, and lumber.
- Give outgrown clothing to friends or charity.
- Buy beverages in returnable containers.
- Donate broken appliances to charity or a local vocational school, which can use them for art classes or for students to practice repairing.
- Offer furniture and household items that are no longer needed to people in need, friends, or charity.
- Sheets of paper that have been used on only one side can be used for note-taking or rough drafts.
- Old, outdated furniture can be reupholstered or slipcovered. Have padding added to the furniture to give it a new look.
- Old towels and sheets can be cut in small pieces and used for dust cloths.
- Books and magazines can be donated to schools, public libraries, or nursing homes.
- Packing materials, such as polystyrene, plastic quilting, and similar materials, can be saved and reused again for packing.
- Carry a reusable tote bag or take bags to the store when you go shopping. There are attractive nylon mesh bags available that can be stored easily in the glove compartment of your car. Durable canvas bags, which take very little space to tuck away when not in use, can also be used.
- If you buy prepared microwaveable dinners, save the plates for outdoor parties or for children.
- Old tires can be used in the garden and in the play yard.

2.2.5 *Source Separation of Discarded Materials*

Today, in most developed countries, environmental preservation is part of the common culture. In these countries, waste is considered as a public resource, and this perspective has led to a reduction in per capita waste generation. The importance of waste reduction lies in reducing the pressure on natural ecosystems and saving costs related to raw material production, recycling, waste management, and energy conservation (Yean Yng Ling and Song Anh Nguyen 2013).

Another solution to combat the waste crisis is recycling, which is the most important concept in waste management and is seen as a green step towards sustainable development. The best and most ideal form of recycling is proportional to source separated waste recycling (Matsumoto 2011). Given that the sources of recyclable materials are mainly households and industries, implementing source separation schemes by households is of great importance. Source separation is one of the challenging issues in developed countries that has gained attention in recent decades (Nguyen et al. 2015). The production of dry waste has outpaced urbanization rate in the world. While many countries are looking for solutions to reduce waste, the amount of waste generated on Earth is increasing annually. Each person contributes a significant financial burden to society by generating 1 g of waste. The disposal of municipal waste imposes substantial financial burdens upon urban management. Beyond the immediate fiscal costs, there exist concealed expenditures that impact not only present-day inhabitants but also future generations. In the absence of a well-devised operational solution, it not only engenders mounting challenges for municipal governance but may also ultimately culminate in environmental devastation in the future (Nguyen et al. 2015).

2.2.5.1 Source Separation—A Fundamental Step in Waste Recycling

To address a wide range of issues, serious measures must be taken to optimize various stages of waste recycling. Establishing a systematic arrangement for waste collection (e.g., development a routine collection schedule for separately collection of wet discards and commingled recyclables) is a fundamental requirement. The task of separating wet and dry wastes at homes could enhance the quality of recycling products, and protect recycling department employees health, as well (Kim 1998). Raising public awareness about waste separation, along with increasing interaction between the public and municipal authorities, prevents the improper use of national financial resources and increases environmental hygiene (Matsumoto 2011).

2.2.5.2 Keys to the Success of Waste Separation Programs

In the literature, the influential variables affecting source separation behavior of public refer to factors similar to the following cases. Several factors have been identified as significant contributors to recycling behavior, encompassing gender, age (with older individuals often placing a higher value on waste), socioeconomic status (including income, education level, and occupation), awareness of the importance of recycling and the consequences of improper waste disposal, the influence of media, political leanings and ideologies, satisfaction with municipal services, social indices, a sense of social belonging, perceived usefulness and belief in the benefits of recycling, intrinsic motivations, the implementation of incentive schemes, and individual participation levels.

The reduction and segregation of waste are not attainable on a large scale without raising public awareness, widespread public participation, and government support. The success of such initiatives depends on public belief in their efficacy, trust in executive authorities, and recognition of the positive outcomes of waste segregation programs. Depending on the level of public awareness, individuals' engagement and participation in waste segregation efforts can vary (Kwatra et al. 2014).

Education, cultural promotion, and public participation are deemed the most crucial steps toward organizing waste management and bridging the existing gaps. One key facet of public involvement is active and self-initiated participation, wherein people take the initiative, and relevant authorities provide guidance. Participation in the separation of household waste falls under this category of involvement. However, cultural promotion will have a profound impact here. A prerequisite for cultural promotion is the utilization of region-specific approaches, as we are dealing with individuals, each possessing unique ethical characteristics. Consequently, a combination of various methods and programs should be employed (Matsumoto 2011).

2.2.5.3 Implementation Methods in Waste Separation

Throughout the world, many methods and programs have been introduced to encourage people to separate waste from its source. The results of these studies indicate that planning, proper design of waste storage sites, and awareness of novel waste management technologies can lead to greater acceptance of waste separation programs. Table 2.3 lists the strategies proposed in different countries for source separation of waste.

2.2.6 Temporary Storage

This step in MSWM covers tasks from on-site storage at the point of generation to the time of delivery to waste collection systems, making it inclusive of a wide spectrum of container and temporary storage vessel types. At first glance, the issues

Table 2.3 Executed waste source—separation programs

No.	Design type	Description
1	Door-to-door collection	Garbage collection by special machines that are done from the door of houses
2	Providing special containers (in different colors)	Containers are usually provided in 3 colors to separate wet, dry, and paper waste
3	Providing bags with different colors	Bags with different colors, each used to collect a specific type of waste
4	Providing branded bags with labels	The labels on these bags specify the type of waste that should be placed in them in addition to a brief description of waste segregation
5	Placing special containers in high-traffic areas, public bins	Containers and buckets in specific sizes and different colors are used in high-traffic areas
6	Placement of unique multi-purpose stations in places accessible to citizens	In these stations, citizens separate their waste or put the waste they have already separated in their place
7	Placing special single-purpose stations (mostly for paper) at specific points	This plan is more useful for separating paper from other waste
8	Placement of waste exchange stations	These stations are usually located near residential areas and citizens can deliver their separate waste and use incentive schemes for it
9	Placing special containers and bags in offices and schools	In this plan, in addition to paper, other wastes are also separated
10	Construction of chambers for waste separation in residential complexes	In this plan, some residential complexes separate their waste in the chambers that are closest to these complexes, and usually, the management of these chambers is with the citizens themselves
11	Creating special separation sections in each residential unit	In this plan, the task of segregation is the responsibility of the residents
12	Providing multi-purpose bins	Each bin consists of several parts, usually 4 parts, each part is for a type of waste
13	Implementation of incentive plans	Such as buying separate waste, and providing book or goods voucher
14	Implementation of the semi-underground system using large bins	Most of which are located underground and are mostly used in high-traffic areas of the city

related to the storage of waste and the required systems may seem straightforward. However, the direct technical, economic, social, health, and environmental effects of this activity on the implementation of other solid waste management stages have led to this stage occupying a special place among various solid waste management activities. Factors that must be considered at this stage include the type and capacity of containers, their location, and the collection method.

2.2.6.1 Temporary Waste Storage Methods

The type and capacity of waste storage containers play a crucial role in the selection of the final collection system as part of comprehensive waste management. Quantitative and qualitative characteristics of generated solid waste, such as production volume, waste type, density, and moisture content, are among the influential factors in selecting the type and capacity of temporary storage containers. The temporary waste storage containers can be categorized based on various approaches (Fig. 2.1). For instance, from a recyclable waste management perspective, they are divided into two categories: (1) containers for wet waste and (2) containers for dry waste (containers for dry waste can be further subdivided into specific containers for glass, paper, plastic, etc.). If the collection method is considered, the categorization will include containers for manual and mechanized collection. As for waste transportation, storage containers can also be classified as stationary and mobile containers.

The materials used in the construction of temporary storage containers and their dimensions also lead to other classifications. In terms of material, there is a wide range of containers, from paper and plastic bags to compressed plastic and metal containers, each serving a distinguishable purpose. Furthermore, containers can be classified into small, medium, and large categories based on their size. On the other hand, the use of compacting equipment in the construction of some temporary storage containers introduces another factor for their classification. Containers with stationary compacting devices are mainly used for storing very bulky industrial waste in commercial centers.

The design team for the temporary storage system considers various parameters, including climate conditions, the geographical location of the area, urban fabric, accessibility, waste generation per capita, and its physical properties. After conducting technical, economic, social, and environmental assessments, they determine the type of temporary storage containers and their installation point.

2.2.6.2 Placement of Temporary Waste Storage Containers

In urban areas customers are usually required to place the discarded materials by the curbside for pickup on collection day. Alternatives to curbside pickup are alley and backyard pickup. Alley pickup is usually possible only in older residential or commercial areas. Sometimes, when two houses are adjacent, a concrete four-wall structure is usually built between them or in another suitable location for storing materials. This location can be covered or open, and usually enclosed by a fence. If continuous maintenance is not performed, it can become an unsightly and unsanitary place.

The placement of waste storage in industrial and commercial establishments depends on the available space and accessibility for the collection service. In new designs, a specific space is allocated for this purpose. Since these containers belong to the owners of the facilities, their placement is determined through consultation with waste collection institutions.

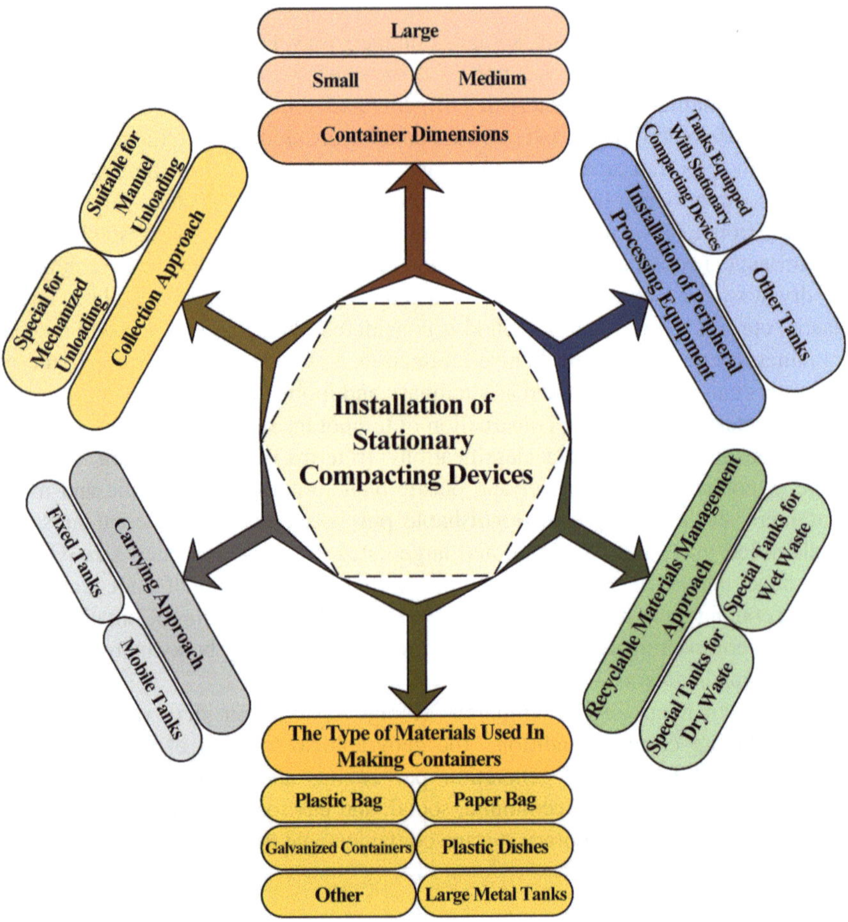

Fig. 2.1 Categorization of temporary waste storage containers based on different approaches

2.2.6.3 Factors Affecting the Selection of Solid Waste Temporary Storage System

The performance of the solid waste storage system, like any other system, is influenced by various internal and external factors, including.

- The type of waste collection system.
- The frequency of waste collection.
- Climatic and weather conditions.
- The quantity and quality of waste.
- The characteristics of urban fabric.
- The presence or absence of source separation policies.

The frequency of waste collection is mainly related to the capacity of the storage containers. However, the frequency, in turn, is highly variable depending on the geographic and climatic conditions prevailing in the collection area (Considering the limitations caused by the decay of organic materials in hot weather). Only materials resistant enough against climatic conditions are used in storage container fabrication. For instance, paper or plastic bags are not suitable for humid and pluvious regions. In the design of manual collection systems, an effort should be made to design the size of the containers so that they can be manually removed from the ground and emptied into specialized collection machinery. In densely populated residential, administrative, and commercial areas lacking specific storage environments for solid waste, the frequency of waste collection should be higher compared to other urban areas, considering traffic loads. Therefore, the capacity of solid waste storage containers must be increased proportionally to the increased collection frequency. The quality of generated waste also plays a significant role in selecting the size and capacity of solid waste storage containers. In source separation systems, the size of the containers is determined by the physical nature of the waste stored in them.

The size of common containers used in apartments and houses varies widely. The determining factor for the capacity of containers is the collection frequency and the amount of waste generated and stored in them. In general, higher collection frequencies can be considered for waste collection from apartments, grocery stores, restaurants, and hotels. Based on this, the collection and disposal of excess waste at these locations can vary widely, ranging from a maximum of once per day to a minimum of three times per week.

In the design and construction of special solid waste storage tanks, several key considerations should be taken into account.

- The tanks should be constructed with a strong and straightforward design.
- The tank's interior surfaces should be smooth and devoid of sharp angles and corners.
- To prevent waste residues from lingering inside the tank, it should be free of protrusions and grooves that can trap materials.
- The tank material should be resistant to corrosion and temperature variations.
- The tank should have an aesthetically pleasing design and be cost-effective.

Today, waste management authorities emphasize the standardization of temporary storage containers for technical and hygiene reasons. Non-standard containers, often used by households for storing excess waste in materials such as cardboard, cartons, plastic bags, metal, or wood, can lead to waste dispersal by animals and wind before collection, resulting in unsightly surroundings. Moreover, the use of non-standard containers can disrupt waste collection systems, and distribute the excess waste during collection operations by open-top vehicles. This, in turn, can increase the costs of collecting additional waste. Therefore, the use of non-standard containers is not approved by waste collection authorities. Suitable containers for storing solid waste should typically be equipped with lockable lids or a mechanism that can be attached to the body to prevent waste from remaining in the container. Ideally, to prevent lid retention and the scattering of excess waste by animals or wind, these containers should be equipped with a locking mechanism.

2.2.7 Waste Collection

The waste collection and transfer process involve several stages, including waste collection by workers, truck movement in the vicinity, and the transfer of waste to intermediate transfer stations or final disposal centers. In general, waste collection is a challenging, complex, and costly task. Approximately 60–80% of the budget for solid waste management in a community is allocated to temporary storage, collection, and transport operations. Therefore, any improvement in the waste collection system can significantly reduce the costs of solid waste management.

2.2.7.1 Solid Waste Collection

The most common methods of waste collection in residential areas are Curb Side, Alley, Setout, Setout-Setback, and Backyard Carry. In the curbside and allay methods, citizens bring the filled bins or plastic bags to the tables or designated collection points in the alley and street and return the empty bins after emptying. In the setout method, a worker brings the tanks to the collection point. In the next step, the workers empty the contents of the tanks into a special truck and the citizens bring the empty tanks back to the buildings and houses. In the setout-setback method, the third worker is responsible for returning the empty tanks. In the backyard carry pickup method, the worker first pours the solid waste into the cart or wheeled tank and then pours the contents into a special truck. This method leaves the existing tank in place in the backyard.

The most prevalent waste collection vehicles are trucks capable of compacting waste and manual loading from the rear and sides. These trucks typically require one, two, or three workers, including the driver. A standard waste collection truck has a capacity of 14–18 m^3 (15–20 cubic yards) and can carry 4–5 tons of waste to disposal sites or intermediate transfer stations. Larger waste collection vehicles with automatic loading and compacting capabilities can empty and replace stationary containers used in residential apartments and shopping centers. Some waste collection vehicles employ a system of exchanging containers. Large—empty containers are transported to a designated waste collection point, filled, and then replaced with empty ones. The filled container is then taken to a transfer station or disposal site for unloading. After emptying the container, it is relocated to a new location.

In many communities and countries, the frequency of waste collection is limited to once or twice a week. The daily routes of waste collection trucks are carefully planned and optimized for efficiency. Various methods have been employed to optimize the routes taken by these vehicles.

Waste collection rates vary depending on the country's income level and the region. In high-income and North American countries, waste collection rates are close to 100% (Figs. 2.2 and 2.3). While, in upper-middle-income countries, this rate is around 82%, whereas in lower-middle-income countries, it's about 51%. Furthermore, in low-income countries, it's approximately 39%. In low-income countries, uncollected waste is often managed independently by households and may be disposed of, burned, or composted. Improving waste collection services is vital for reducing pollution and, consequently, improving human and environmental health. Waste collection rates are significantly higher in urban areas than in rural areas because waste management is typically an urban service. In countries with average to low incomes, the rate of waste collection in urban areas is typically more than twice as high as in rural areas (Fig. 2.4).

2.2.8 *Waste Transfer and Transportation*

In cases where waste disposal centers are distant from the collection points the idle time of waste collection machinery workers will significantly increase, which is not

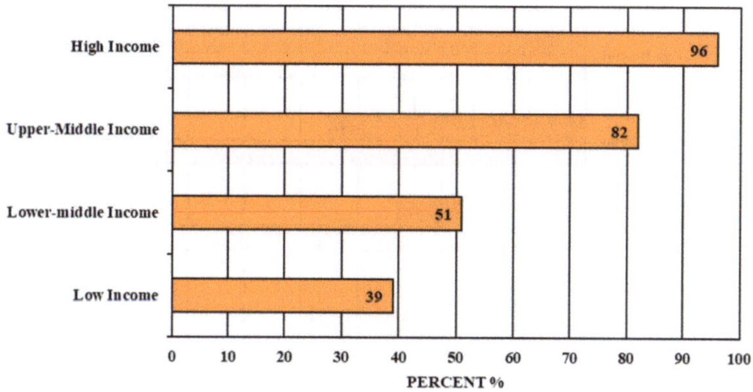

Fig. 2.2 Waste collection to income level rate

affordable. Therefore, to address this issue, transfer stations are established in suitable locations, and large trailers or trucks are used for waste transportation. According to reports, the use of these large trucks becomes cost-effective when the average round trip distance is more than 50 km (30 miles). Among the key considerations for designing a transfer station, the location, the type of station, ease of access, and its environmental impact are paramount (Henry and Heinke 1989).

The proximity of transfer stations to waste generation centers is of utmost importance. Having well-connected roads to the transfer station and secondary transportation routes is crucial. On the other hand, the location of a transfer station should also meet environmental criteria. If multiple transfer stations and disposal centers exist, it's essential to allocate and determine the optimal amount of waste to be transported from each transfer station to each disposal center.

Transfer stations as an alternative to direct waste transportation are generally recommended when the cost of transporting waste from the generation point to

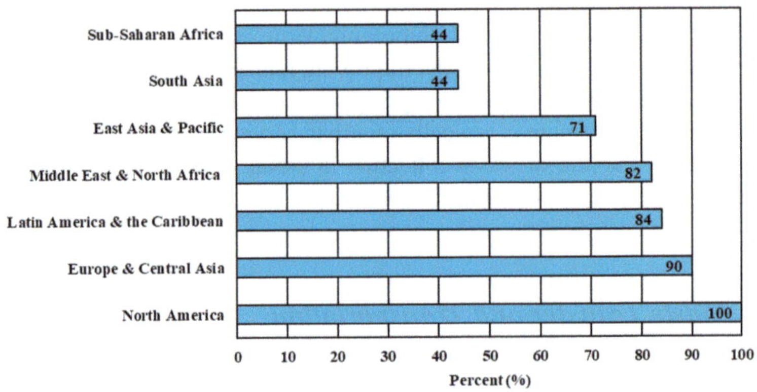

Fig. 2.3 Waste collection to region rate

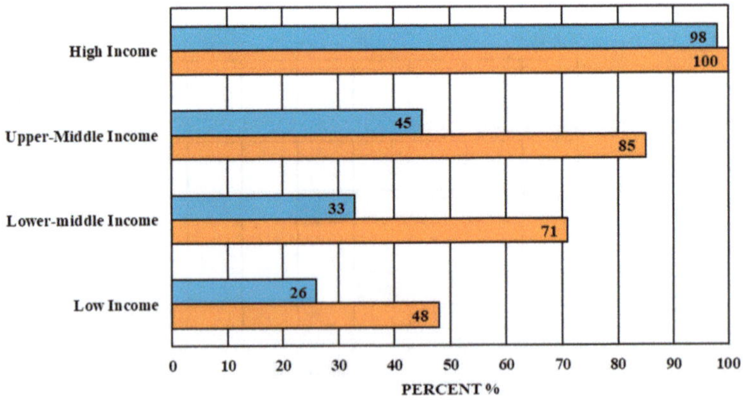

Fig. 2.4 Waste collection in rural and urban areas to income level rate

the disposal center is higher than the cost of transporting it from the generation point to the transfer station and then to the disposal site. For instance, considering certain assumptions regarding costs, it can be concluded that in distances exceeding 28 km (17.5 miles) between the waste generation source and the disposal center, the construction and operation of transfer stations are economically viable. Furthermore, the benefits of using transfer stations include reduced fuel consumption, minimized negative impacts on city traffic congestion, reduced air pollution, lower depreciation of machinery, and the flexibility to allocate more time by service providers for waste collection instead of transportation. Furthermore, by using these stations, a portion of the waste processing operations can be performed on-site.

2.2.8.1 Types of Transfer Stations

Transfer stations come in various types, the following.

Direct Dumping Without Floor Storage

These stations typically have two levels. At the upper level, waste collection vehicles, often garbage trucks, directly dump their contents into the lower level, where open-top trailers are parked for loading. This method of operation is highly efficient due to its minimal equipment requirements and the workforce needed for loading trailers. Additionally, the speed of operations is notable, as the loading of trailers can begin immediately upon receiving waste, making it readily accessible. These factors contribute significantly to the efficiency necessary for the transportation of solid waste.

Direct Dumping with Floor Storage

These stations have three levels. Waste collection vehicles unload their cargo at the upper level, which can include both separated and unseparated waste. This waste is then transferred to the middle level, typically using conveyor belts or heavy machinery like wheel loaders or bulldozers. Finally, waste is loaded onto open-top trailers located at the third level (lower level).

Compactor-Equipped Transfer Stations

Nowadays, one of the most common types of transfer stations is equipped with compactors. These stations can operate with or without floor storage. In these stations, incoming waste is either directly fed into compactors and then transferred to specialized trailers or, after floor storage, moved by loaders or other heavy equipment into compactors before being loaded onto trailers.

Deep Pit Transfer Stations

These transfer stations offer high storage capacity and the ability to transfer solid waste 24 h a day. They reduce the need for specialized waste-hauling vehicles and provide efficient on-site waste processing within the station's infrastructure. To prevent damage caused by direct dumping into the pit, some waste must remain at the bottom, causing increased anaerobic decomposition and the production of unpleasant odors.

Combined Transfer Stations

Modern transfer stations are often designed to serve multiple purposes. A station can combine elements of direct dumping and pit-style operation. In this manner, during peak traffic times and when storage demands are high, the transfer station operates in a pit-like configuration. Solid waste, following separation and the completion of other necessary processing steps, is directly loaded onto specialized transportation machinery. This overall approach contributes to enhancing the efficiency of the systematic solid waste transfer systems, both in technical and economic dimensions. Some examples of transfer station types are presented in Fig. 2.5.

2.2.9 *Recycling*

The recycling process, the third component of the integrated waste management approach, garnered significant attention from environmental groups and the general public in the early 1970s. However, during that time, the common definition of recycling was limited to a singular activity the separation and collection of materials such as paper, metal, glass, and plastic, rather than a comprehensive dynamic system as illustrated in Fig. 2.6. In many cases, volunteers and organizations would engage in separate material collection efforts, even though, due to the low economic value of these segregated materials, they were eventually sent to disposal facilities. They were confronted with the complexity of the markets for secondary materials and the reluctance of manufacturers to use recycled materials as raw materials in the production of new products. Perhaps the most crucial lesson learned from the experiences of the early 1970s was the necessity to expand markets for segregated materials to achieve success in the recycling process.

Today, recycling has become a fundamental requirement in the waste management practices of both developed and developing countries. This is because waste separation and recycling play a crucial role in the economic growth and environmental improvement of nations. Recycling and reusing waste materials not only alleviate urban management challenges but also have the potential to contribute to national income. In developed countries, waste is often referred to as "dirty gold". A significant portion of dry waste holds economic value, and recycling it can not only boost the country's economy and create income-generating opportunities in urban areas but also prevent the depletion of raw resources and national reserves. Among the other advantages of focusing on waste separation, one can highlight the following: resource conservation for sustainable development, energy savings, cost savings from recycling efforts, contribution to the national economy through reusing materials in the production cycle, reduction in transportation costs, lowering waste disposal volume, more efficient landfill management, improved waste purity, production of high-quality compost (Rousta et al. 2015).

Understanding the challenges and opportunities presented by the recycling process requires basic knowledge of the types of consumer goods, production processes, industrial dynamics, and the role of secondary materials industries in meeting primary producers' needs. In the recycling context, distinguishing between pre-consumer and post-consumer recyclable materials is of great importance. This classification is

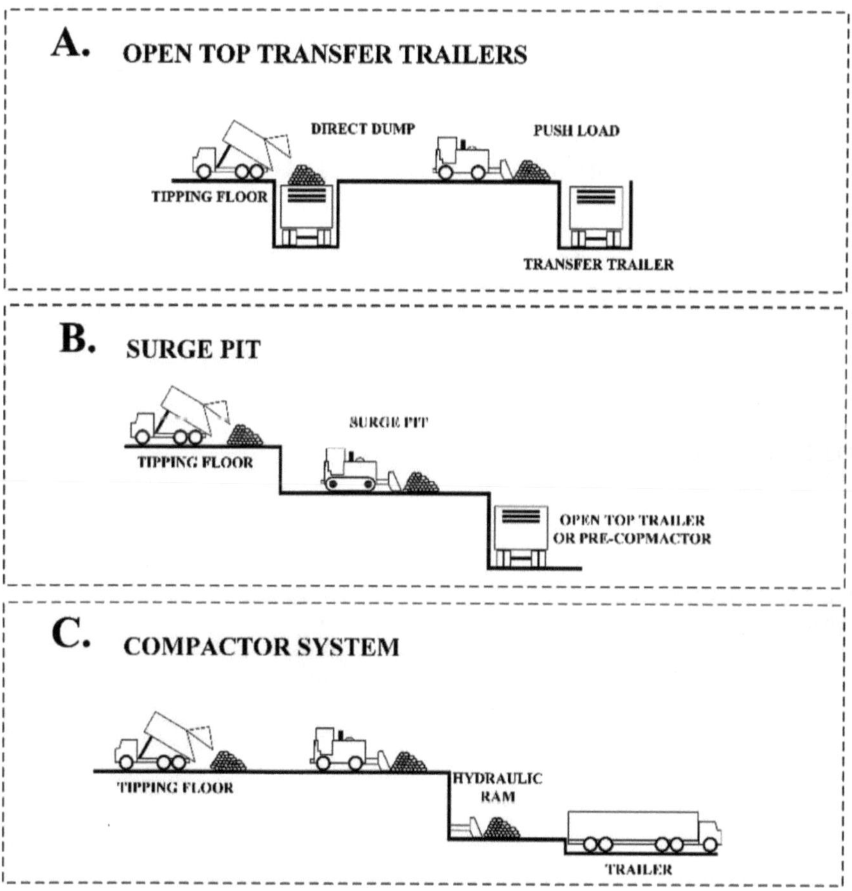

Fig. 2.5 Types of transfer stations

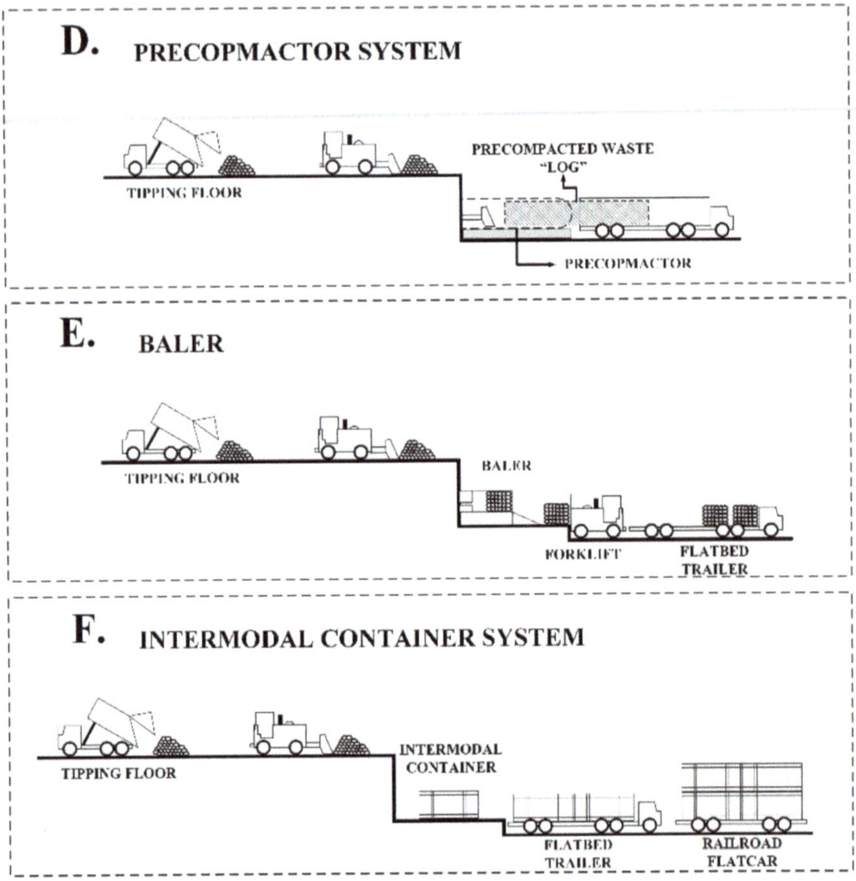

Fig. 2.5 (continued)

Fig. 2.6 Recycling process symbol

based on the source of the material. Post-consumer recyclable materials include those products that are disposed of by an industry or consumer at the end of their expected end use. Pre-consumer materials are materials or by-products that are produced in major factories and are generally reused. In some cases, recoverable pre-consumer materials are referred to as internal or primary waste materials. The importance of distinguishing between pre-consumer and post-consumer recyclables becomes apparent when there is a need to purchase specific products containing post-consumer recyclable materials according to necessity.

2.2.9.1 Recyclable Materials Market

A market is an institutionalized entity where pricing factors governing activities are established, serving as an intermediary between buyers and sellers of a specific commodity. In the context of recycling, market infrastructure comprises two levels: intermediary markets and end-use markets.

Intermediate markets, which are part of secondary materials industries, serve as significant intermediaries between suppliers of collected materials and the demands of manufacturers, which constitute end-use markets. Individuals working in the intermediate markets typically include collectors, processors, brokers, and converters. The precise definitions of these terms vary depending on the type of recyclable material and their specific functions. Their functions include securing recyclable materials through purchase or procurement, sorting and classifying materials to meet raw material standards, processing materials, converting them into specific physical forms (e.g., baling, shredding, pressing), and selling the resulting products. While brokers often do not engage directly in material processing or management, they provide essential services to recyclable material producers and consumers by overseeing supply and demand situations and utilizing this awareness to create buying and selling opportunities.

Like many commodities, the supply of recyclable materials and the demand for them often follow cyclical and seasonal patterns over time. Sometimes demand leads supply, and at other times, supply leads demand. These cycles can significantly influence pricing.

Manufacturers have both economic and environmental incentives to use recyclable materials instead of virgin raw materials. Environmental benefits include reduced energy and water consumption, especially when aluminum, steel, paper, and glass are produced using recycled materials. On the other hand, the economic benefits of recycling, such as energy cost savings controlling pollution, and saving raw materials' costs are apparent to manufacturers. One of the main reasons some manufacturers do not use recyclable materials is that many of them have made significant investments in machinery and equipment that use virgin raw materials, and they often own the resources of these raw materials or have long-term contracts to ensure adequate supply. However, recycling technologies and approaches vary significantly depending on the type of recyclable material.

2.2.10 Composting

The general model of composting is depicted in Fig. 2.7. Organic materials, primarily food scraps, enter the composting facility after being separated in the processing and recycling section. Along with garden waste (after shredding on-site), they are stored in the storage-receiving area. Storage in standardized piles prevents the activity of anaerobic microorganisms and, consequently, the production of unpleasant odors. Afterward, the materials are moved into the processing section and subjected to aerobic decomposition. The processing time varies depending on the selected method and can range from a few weeks to several months. The final product is matured in storage piles for about 90 days in the maturation section. Afterward, it undergoes screening. Since the breakdown of woody materials in the incoming stream occurs slowly, these materials are separated in the screening unit to maintain the flow's porosity, and they are returned to the receiving area. The final stage involves storing the finished compost and sending it to sales centers or for use in predetermined applications.

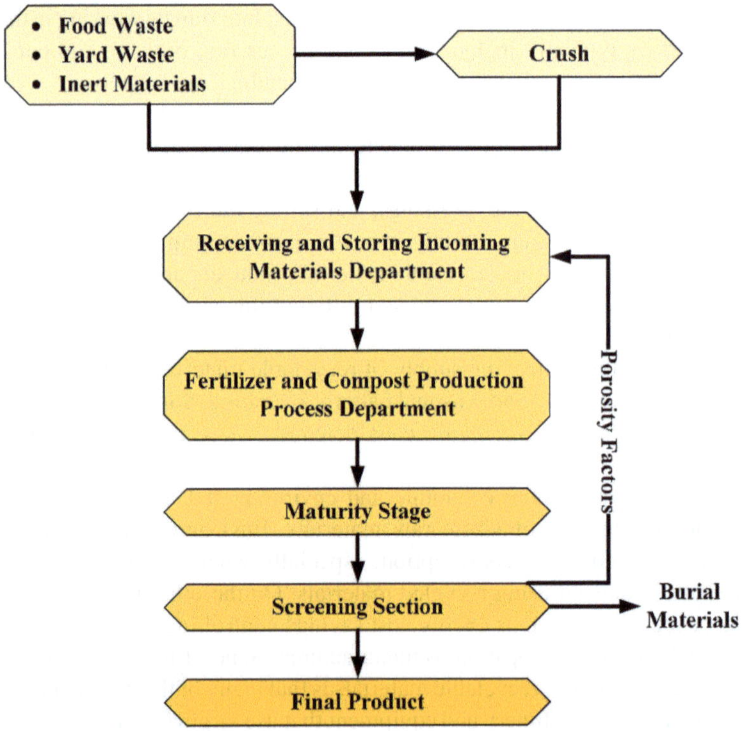

Fig. 2.7 Flow diagram of compost model

2.2.11 Waste Incineration for Energy Recovery

The use of waste incineration for energy recovery is motivated by two main factors: the lack of suitable land for waste disposal (landfilling) and the high energy cost associated with waste disposal. Incineration reduces the space required for disposal but does not eliminate the need for disposal facilities. Typically, only about 50% of the MSW stream is combustible. Through the incineration process, the combustible portion can be reduced to approximately 90% of its volume and 70% of its weight. Therefore, the residual waste must still be managed, often through disposal in landfills.

Incineration involves controlled burning of waste at high temperatures in specially designed facilities for efficient and complete combustion. Complete combustion includes converting all carbon to carbon dioxide (CO_2), hydrogen to water (H_2O), and sulfur to sulfur dioxide (SO_2). By-products and unintended emissions produced during the incineration process include ashes, gases, and thermal energy. Incineration is used to burn waste either to reduce its volume, eliminate specific chemical compounds, change its chemical characteristics, remove hazardous materials, or recover energy. Some specific types of waste incinerators are used to incinerate industrial sludges and wastewater treatment sludges. Currently, waste incineration is considered the environmentally acceptable method for disposing of certain hazardous wastes (Fig. 2.8).

There are two common types of facilities used for incinerating MSW: mass-burn incinerators built on-site and modular, pre-fabricated incinerators. These units differ in terms of design, construction, processing capacity, air pollution control requirements, service life, and costs. In the United States, Europe, and Japan, there is a wealth of experience in operating mass incineration units with or without energy recovery equipment, including recycling facilities. Historically, large cities typically use mass-burn incinerators with capacities ranging from 90 to 450 tons per day, while

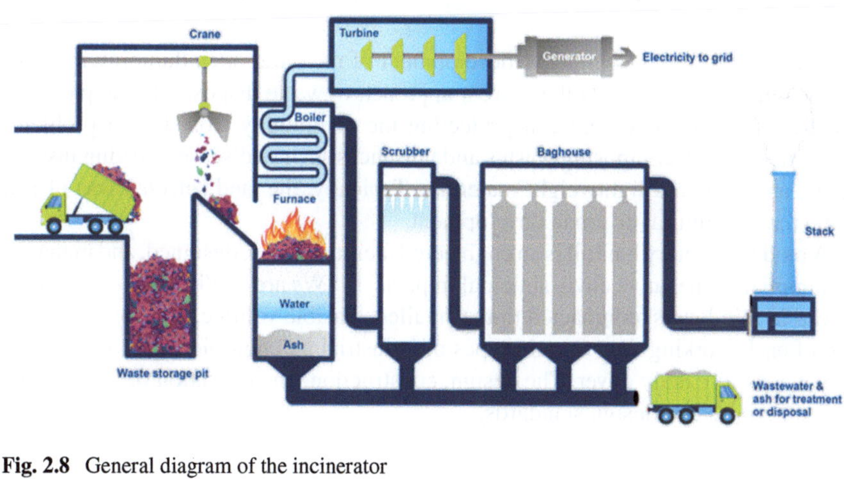

Fig. 2.8 General diagram of the incinerator

smaller communities employ modular units with daily capacities ranging from 22 to 90 tons. There are other types of waste incineration units, including fluidized bed incinerators and fluidized bed boilers. However, both of these require pre-processing of MSW to remove non-combustible materials before incineration. Municipal waste incinerators are designed to burn waste with diverse compositions, while industrial waste incinerators are designed to incinerate more homogeneous fuels.

Most modern incinerators are equipped to recover energy. In Europe, for example, steam and hot water generated in incineration plants are often used for central heating in houses and commercial centers. In the United States, approximately 75% of incineration facilities have energy recovery capabilities, and the energy produced is used for district heating or electricity generation. The revenue generated from the sale of steam or electricity can offset some of the high costs associated with constructing and operating these units.

Another method of obtaining energy from waste is through the processing of MSW and the production of refuse-derived fuel (RDF). RDF is a homogenized and shredded mixture of the combustible organic portion of MSW, it can be used as a supplemental fuel in industrial boilers or waste-to-energy facilities, and in some specially designed waste incinerators. Currently, mass-burn incinerators and RDF systems collectively make up about 90% of the current waste-to-energy processing capacity in the United States.

Air pollution and ash disposal are major concerns raised by groups opposing waste incineration. Waste incinerators must be equipped with electrostatic precipitators, scrubbers, or other equipment to reduce emissions to acceptable levels set by regulatory bodies. Ash generated from waste incineration or RDF facilities also needs to be tested to determine whether it is hazardous or not. In cases where ash is found to be hazardous, hazardous waste disposal regulations must be applied.

2.2.12 *Land Disposal of Residuals (Landfilling)*

Sanitary landfilling, as a modern development of the older open dumping methods on the outskirts of cities, is the current approach to waste disposal. In the past, this practice was primarily aimed at protecting the community from health problems associated with decomposing wastes and nuisances such as disease-carrying insects, odors, and scattered lightweight materials. Typically, the land selected was of low value for agricultural or urban development.

A modern sanitary landfill is an engineered facility chosen, designed, and managed to minimize its negative environmental impacts. MSWs are confined to a small area, spread in thin layers, compacted to the smallest feasible volume, and covered at the end of each working day. Certain types of industrial and non-biodegradable wastes may not require daily cover. The design, construction, and operation of landfills are subject to laws and design standards.

Landfills are an essential component of any MSWM system. Efforts to reduce waste generation, recycle, incinerate, and compost can decrease the amount of material sent to landfills, but any experienced waste management professional will attest that some residual materials will always remain that require disposal.

Many of the challenges that municipalities face in identifying and developing new landfill sites are not necessarily due to a shortage of suitable land in areas with appropriate soil and hydrogeological conditions but are often due to public opposition. Communities oppose the construction of new landfill sites for reasons such as negative consequences (such as dust, noise, traffic, and odors), aesthetic concerns, and environmental considerations (such as groundwater pollution from landfill leachate, emissions of landfill gas to adjacent properties, and the use of agricultural lands). While public opposition and government regulations have posed significant challenges in solid waste management, they have also had a positive influence by exerting pressure on communities and industries to explore waste reduction, recycling, and alternative disposal methods more thoroughly.

2.2.12.1 Site Selection for Landfills

Before searching for a suitable location for a landfill, the dimensions of the site for the intended service area, typically planned over a 10–20-year period, should be determined. Investment in the development and operation of smaller disposal facilities is rarely economically justifiable. The dimensions of the site are determined based on the amount of waste that will be disposed of over the useful life of the disposal facility. The amount of waste that will be disposed of depends on factors such as population, industrial, commercial, and administrative characteristics of the service area, the waste generation rate, preliminary design, and future development potential.

Finding a suitable location for a landfill typically begins by excluding unsuitable areas from an environmental perspective. In general, six types of land are considered environmentally unsuitable for a landfill: floodplains, wetlands, land near airports, areas with geological faults, earthquake-prone areas, and any otherwise unstable site. Local legislative authorities may impose additional restrictions on suitable landfill locations, such as requiring a minimum distance from highways, parks, ponds, lakes, or other bodies of water, or prohibiting the selection of areas that might be considered a threat to underground water resources or habitat of endangered plant and animal species. When a site does not meet all criteria, it is possible to apply for permits from local legislative authorities to amend or waive certain restrictions.

2.2.12.2 Landfill Design

Landfills must be designed in a manner that preserves the quality of groundwater resources. To date, a significant number of landfills have been constructed as either traditional or non-engineered facilities for the disposal of solid urban waste.

A traditional landfill lacks a liner, allowing leachate to escape into the surrounding environment as it passes through the landfill bed. The design of these landfills incorporates a mechanism where leachate is partially treated as it traverses through unsaturated soil beneath the site. An ideal geophysical structure for a traditional landfill is an area with thick, impermeable soils (such as clayey soils), an underlying impermeable rock layer without an aquifer, and a broad unsaturated zone above the stable surface.

The second type of landfill is a sanitary or engineered landfill, which includes liner systems and a leachate collection system. Initially, engineered landfills were only used when site conditions did not permit the use of traditional landfills (e.g., insufficient thickness of underlying soils or the presence of fractures in the rock stratum) or when hazardous waste was disposed of. The primary challenge with such landfills lies in the necessity of leachate collection and treatment, both during the active phase and post-closure maintenance.

2.2.13 Waste Incineration Without Energy Recovery

The process of this method is similar to energy recovery waste incineration, as described in Sect. 2.2.11. The difference lies in the absence of energy recovery in this process, where the primary goal is the disposal of waste without considering the potential for producing by-products through incineration.

2.3 Impacts of MSW Management

2.3.1 Introduction

Today, urbanization, population growth, migration, and changes in consumption patterns have led to an increase in the rate and diversity of MSW production. This has transformed the proper management of MSW into a significant social and governance challenge for urban administrators. As the role of efficient MSW management in achieving sustainable development is undeniable, mismanagement in this area can have irreversible social, health, environmental, and economic consequences (Abubakar et al. 2022). Therefore, the following text discusses some of the disasters resulting from improper MSW management.

2.3.2 *Some Disasters Caused by Improper Management of MSW in the World*

2.3.2.1 Closure of the Naameh Landfill in Beirut, Lebanon

The Naameh landfill in Lebanon served as the primary disposal site for the country's waste since 1997, accumulating over 15 million tons of waste by 2014. Due to the environmental and social repercussions and the objections of residents of the Naameh region in southern Beirut to the dumping and storage of waste in their area, the landfill was closed on July 7, 2015, with the responsible waste collection and disposal company in Beirut suspending its operations for a week. In response to this situation, some individuals set fire to piles of garbage in the streets, while others scattered toxic substances and disinfectant powders on heaps of waste. This crisis severely disrupted the tourism industry in Lebanon, in addition to its health and environmental implications (Ghadban et al. 2017).

2.3.2.2 Waste Management Crisis in Naples, Italy

Since the 1980s, the main landfills in the city of Naples (Pianura) faced a crisis due to excessive disposal of hazardous industrial waste. As this situation continued, municipal workers, on December 21, 2007, refrained from collecting waste due to the landfills' capacity being exhausted. This management crisis led to the accumulation of massive piles of waste in the city streets, triggering public protests. Ineffective waste management in Naples imposed irreparable social, environmental, and health effects on the residents of this region.

2.3.2.3 The Leuwigajah Landfill Disaster in Bandung, Indonesia

At 2:00 AM on February 21, 2005, a massive explosion occurred at the Leuwigajah landfill in Bandung, Indonesia. This explosion was the result of heavy rainfall and the sudden release of biogas between the layers of the landfill. During this incident, waste debris was scattered up to a radius of 1000 m around the disposal site. Unfortunately, this explosion trapped 143 people and 71 houses under the avalanche of garbage (Lavigne et al. 2014).

2.3.2.4 The Times Beach Tragedy in the United States

In the early 1970s, the Bliss Company took responsibility for the disposal of toxic waste. Subsequently, the company used a mixture of these hazardous wastes in addition to used engine oils for dust control on dirt roads and horse-riding trails. In March 1971, 62 horses died in one of the oil-sprayed areas. Investigations revealed that the oil used by the Bliss Company contained high levels of dioxin, despite the company's claims of ignorance. On December 4, 1982, Times Beach experienced its worst flood, leading to the widespread dispersion of dioxin throughout the city's communication networks, reaching a concentration of 0.3 ppm (Environmental Protection Agency (EPA) 2022).

2.3.3 Undesirable Effects of Improper MSW Management

MSW management encompasses a series of interconnected operations, including waste generation, collection, transportation, processing, and disposal, all aimed at reducing the undesirable effects of waste on human health, and the environment, improving the quality of life, and promoting economic development (Hirpe and Yeom 2021). Disruptions in any of these stages can pose challenges to the waste management chain.

2.3.3.1 Effects of Inadequate Waste Collection and Transportation Systems

Many problems associated with waste collection systems are related to a lack of facilities and equipment. In many developing countries like India, Pakistan, Brazil, and South Africa, inefficient waste collection systems have led to the release of garbage into streets, open spaces, and coastlines. Uncollected waste burdens cities with social, economic, and environmental costs. Foul odors, the spread of diseases such as malaria and dengue, blockages, and urban sewage overflow are some of the outcomes of improper waste collection systems (Adnan et al. 2020; Rousta et al. 2015).

2.3.3.2 Effects of Improper Recycling Processes

One essential requirement for a successful recycling program is accurate data on waste generation rates and waste quality. In many areas, there is insufficient information about the characteristics of locally produced waste, leading policymakers to design recycling programs and install related equipment based solely on scientific studies. Furthermore, in some regions, the government's failure to provide financial incentives to producers has resulted in reduced public participation in recycling schemes. Mismanagement in recycling process design can impose financial losses on municipal management systems (Moh and Abd Manaf 2014).

2.3.3.3 Effects of Improper Composting Processes

Composting is a biologically-based waste disposal method that involves breaking down organic waste into simpler forms using natural microorganisms such as bacteria and fungi. Despite the benefits of this process in agriculture, composting can release higher amounts of CO_2 gas compared to other similar processes, leading to potential environmental consequences. Additionally, some studies have linked compost with adverse health effects such as nasal congestion, sore throats, dry coughs, bronchial asthma, allergic rhinitis, and lung inflammation (Abubakar et al. 2022; Akmal and Jamil 2021; Saffron et al. 2003).

2.3.3.4 Effects of Improper Waste Incineration

Waste burning can occur in controlled settings (incineration) or open spaces (open burning). Both incineration methods (whether with or without energy recovery facilities), and open burning processes are recognized as sources of GHG emissions and have environmental impacts (Eggleston et al. 2006). On the other hand, open burning has been reported as a major cause of respiratory diseases, including infections, chest inflammation, allergies, asthma, anemia, and weakened immunity. Improper burning of plastic waste can produce chlorine gas and dioxins, which are harmful to human health and may lead to hemoglobin deficiency and cancer (Abubakar et al. 2022).

2.3.3.5 Effects of Improper Waste Disposal

Despite various methods of waste reuse like recycling and composting, landfills continue to be a significant part of waste management hierarchy. Waste decomposition in landfills can lead to the release of greenhouse gases and leachate production (Ghaffariraad and Ghanbarzadeh 2021). Methane and carbon dioxide are the most important gases released from landfills, however some low-concentration toxic gases can also escape. Leachate produced may leak into the surrounding soil and groundwater, causing environmental problems. Additionally, improper landfill siting can adversely affect the lives of nearby residents due to odor, smoke, dust, noise, and the degradation of the landscape. Furthermore, studies indicate that landfill construction can devalue nearby land and have negative economic effects (Danthurebandara et al. 2012).

2.4 Characteristics of Waste

2.4.1 Waste Quantity

The quantity of solid waste generated in an urban community, often expressed as per capita production (the weight of solid waste produced per person or active unit), depends on various factors. These factors include the community's income level, consumption habits, climate (temperature and precipitation), culture, social norms, the availability and prices of raw materials, economic activities diversity, and population distribution. In other words, it is not possible to provide a universal relationship to estimate it. Estimating waste quantity is the first step in waste management studies. In regions or areas where the input quantities to processing or disposal facilities, such as transfer stations, incinerators, and landfills, have been measured and recorded in the past, current planning can be done more confidently based on the quantity of waste produced. Conversely, in areas where accurate waste quantity information is not available, determining and estimating expected quantities becomes a significant challenge.

The best method for determining the quantity of waste produced is by installing weighing equipment at waste processing/disposal facilities and measuring the weight of machinery when entering and exiting these facilities. Therefore, the net weight of the vehicle load is calculated by the difference in weight when entering and exiting the facility. In regions with high waste production, it is usually not feasible to weigh all waste collection vehicles. In such cases, the maximum possible weighing approach is practical, and the average weight of waste generated can be calculated using Eqs. 2.1 and 2.2. For this purpose, it is initially necessary to classify the collection/transport machinery based on the trailer attachment status and the gross weight of the vehicle. Heavy vehicles are categorized into two types: rigid and articulated. An articulated vehicle comprises a tractor to which a semi-trailer is attached. A rigid truck may also carry a trailer, but in this context, it is not considered articulated. Furthermore, the term "gross weight of the vehicle" refers to the maximum allowable weight of the entire vehicle during loading, including fuel, passengers, cargo, and trailer weight.

$$\text{Weight} = \sum_{i=1}^{n} NT_i \times \overline{W_i} \tag{2.1}$$

$$\overline{W_j} = \sum_{j=1}^{m} \frac{W_j}{nt_j} \tag{2.2}$$

where "weight" is the total weight of waste received at the processing/disposal center during the specified period (tons), NT_i is the total number of trucks from class i that entered the center during the same period, $\overline{W_i}$ is the average weight of waste transported by trucks class i, n is the total number of trucks in any classes, W_j is the weight of trucks from class j weighed during the period (tons), nt_j is the number of trucks from class j weighed during the period, m is the total number of trucks from class j that were weighed during the period. When using Eq. 2.2, it should be noted that in the presence of outliers (observations that are significantly larger or smaller than the mean), using the median instead of the mean will lead to more robust results. Calculating the skewness of the data can be used to decide between using the median and the mean. When the skewness index falls between −1.5 and + 1.5, using the mean is appropriate.

In centers without weighing equipment, the quantity of waste is often recorded based on entry volume. Since estimating the volume of waste inside closed and covered vehicles is not possible, the reported volume corresponds to the capacity of the waste trucks when full. In this case, the quantity of waste can be estimated using the average specific weight of waste inside the vehicles (Eq. 2.3).

$$\text{Weight} = \sum_i \forall_{\text{in truck}\, i} \times D_{\text{in truck}} \tag{2.3}$$

where "weight" represents the weight of waste received at the facility (tons), $\forall_{\text{in truck}\, i}$ represents the volume (full capacity) of the ith vehicle entering the facility (m^3), and $D_{\text{in truck}}$ represents the average specific weight of waste in the truck (ton/m^3). (Methods for determining the specific weight of waste will be presented in subsequent sections.)

In centers without weighing equipment, a compromise solution can be implemented by contracting with local facilities equipped with heavy machinery weighing equipment. Periodic weighing operations can be performed during certain periods of the year and the results extrapolated to the entire year. It is recommended to conduct weighing operations during at least two complete working weeks in waste disposal centers (one week during minimum waste production periods and one week during maximum waste production periods). For even more accurate results, weighing operations can be repeated in each season of the year (i.e., a complete working week during each season). This way, after conducting statistical operations and obtaining the average waste production in one day, an estimate of the annual production can be achieved. Choosing a week for testing during each season will yield acceptable results because the variations in waste quantity during different days of the week are greater than monthly fluctuations.

In accordance with the World Bank's report in 2016 (Kaza et al. 2018), the average per capita solid waste production worldwide (excluding waste generated in industrial, hospital, and special categories such as electronic and construction debris) was approximately 0.74 kg/cap./day (with a range of variations from 0.11 to 4.54 kg/cap./day). The results of this report indicate a strong correlation between the quantity of waste production, the income level of the community, and the urban population. In regions with low-income levels (based on the Gross National Income per capita index provided by the World Bank in 2015, which is $1025 or less), approximately 93 million tons of waste are generated annually. While in high-income regions (USD 12,476 or higher), 683 million tons of waste were produced. In other words, high-income countries account for around 34% of the global waste production, even though their population constitutes only 16% of the world's total population (in comparison to low-income countries with 9% of the population and 5% of waste production). However, it is predicted that the rate of waste production growth in low-income countries will be significantly higher than that in high-income countries. In Sub-Saharan Africa, for instance, waste production is expected to triple by the year 2050. The lower waste production growth in high-income countries could be attributed to reduced demand for raw materials (due to prior development) and, as a result, decreased waste generation.

Example 2.1 According to the statistical information from the United States, the average household size (household dimension) decreased from 2.75 individuals in 1980 to 2.63 individuals in 1990. During this decade, annual urban waste production increased by 2.61% (from 137.3 to 177.6 million tons per year—note that the term "ton" here refers to US short ton equals to 2000 lbs.). The population in 1980 was 227 million people, and in 1990, it was 249 million people. Calculate the contribution of each of these factors, population growth, per capita waste production increase, and changes in household dimension, to the overall increase in solid waste generation in the United States.

Solution: Waste production increase can be attributed to two factors: population growth and changes in consumption and waste generation patterns (or changes in per capita waste production). These two factors, however, have a mutual influence on each other, and for problem-solving purposes, this assumption has been accepted. Regarding population change, the increase in waste production is equivalent to the product of the surplus population and the per capita production rate in 1980 (meaning that if the consumption habits of people remain constant during these years, we would expect this amount to be added to the waste stream). As for per capita change, the increase in waste production will be the product of the population in 1990 and the increase in per capita from 1980 to 1990.

The share of population growth in the annual urban waste production increase:

$$100 \times \frac{\overbrace{(249 - 227) \times 10^6}^{\text{surplus population}} \times \overbrace{\frac{137.3 \times 10^6}{227 \times 10^6}}^{\text{per capita WG in 1980}}}{\underbrace{(177.6 - 137.3) \times 10^6}_{\text{the amount of increase in waste production over ten years}}} = 0.33\%$$

The share of the per capita waste generation (WG) growth in the annual urban waste production increase:

$$100 \times \frac{\overbrace{249 \times 10^6}^{\text{population in 1990}} \times \overbrace{\left(\frac{177.6 \times 10^6}{249 \times 10^6} - \frac{137.3 \times 10^6}{227 \times 10^6}\right)}^{\text{the per capita increase over ten years}}}{\underbrace{(177.6 - 137.3) \times 10^6}_{\text{the amount of increase in waste production over ten years}}} = 0.67\%$$

Therefore, the change in people's consumption and waste generation patterns is responsible for two-thirds of the increase in the amount of waste, and the other third is due to the increase in population. This finding specifies the necessity of culture building and increasing public awareness in order to achieve the high goals of waste management, i.e. reducing waste production. Similar calculations can be repeated for the number of households. According to the information of the issue, the number of households and household waste production per capita in 1980 was equivalent to 82,545,455 households and 3326.65 pounds per year per household and in 1990 it was 94,676,806 and 3751.71, respectively. By following the previous trend, the contribution of the increase in the number of households and per capita household waste production, in the increase in the amount of total waste generated, is almost equal (50%).

2.4.2 *Physical Composition of Waste*

MSW is a heterogeneous mixture (in terms of the percentage of various waste materials present) and the composition of these materials varies from one region to another and also over time (as an example, changes in the amount of different waste materials in the United States municipal waste stream from 1960 to 2018 are shown in Fig. 2.9 (US Environmental Protection Agency 2018)). Industrial wastes, although generally possessing the same diversity as municipal wastes, will be homogeneous when generated from a specific source. Information obtained from physical analysis is generally used in calculating waste management service costs. However, the management sector can extract other parameters such as the primary source of specific waste types and changes in waste quantities over time from this information. In energy recovery projects planning, categorization and separation usually include combustible and non-combustible materials, while for recycling operations planning, detailed information about the recyclable components in the waste stream is required. It is worth noting that increasing the number of separation categories will result in a significant increase in the required number of samples for statistically reliable results. Consequently, the time and costs involved will also increase.

So far, various methods have been developed to determine the composition of solid waste, including methodologies such as material flow analysis (mass balance) and direct sampling. In the material flow method, the concept of the law of mass conservation is used to track the amount of material passing through a specific region or system over a defined period. In this method, the boundaries of the system are defined, and input and output data of the system are used to establish a mass balance and shaped as a table. For example, if we consider the boundaries of a country as the system's boundaries, the data related to annual paper waste generation would be the output of the waste flow system, and the input of the system would be the input flow of paper products into the consumption cycle after corrections related to exports, imports, and the useful life of products (within one year). This method, due to the impossibility of modeling the behavior of waste producers in discarding potentially unused items or keeping items for potential future use, is subject to considerable uncertainty.

In the direct sampling method, after selecting a suitable sample (representative in terms of statistical characteristics) from the waste stream, the separation of materials and their weighing is carried out. Samples can be obtained by randomly selecting several households and commercial centers in the city and collecting their daily waste generation, or they can be selected from the incoming load at transfer stations or waste processing/disposal facilities. There are three main approaches to sampling waste at the source (households and commercial establishments).

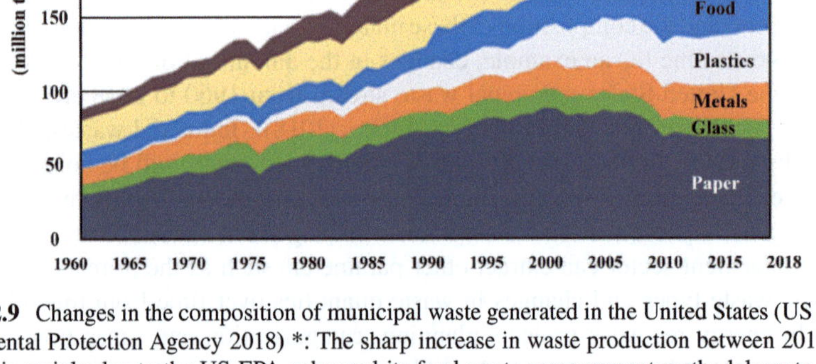

Fig. 2.9 Changes in the composition of municipal waste generated in the United States (US Environmental Protection Agency 2018) *: The sharp increase in waste production between 2017 and 2018 is mainly due to the US EPA enhanced its food waste measurement methodology to more fully account for all the ways wasted food is managed throughout the food system. **: The "All other" category mainly includes wood, rubber, leather, and textiles

(1) Simple Random Sampling: In this approach, the probability of sampling is the same for all units, and it is used in cases where the waste is homogeneous, such as waste generated in a multi-dwelling residential complex.
(2) Stratified Random Sampling: This approach is suitable for heterogeneous wastes. In this method, waste sources are classified into several homogeneous categories, such as multi- dwelling residential complexes and single-household buildings. Random sampling is then performed from each category.
(3) Systematic Random Sampling: This type of sampling is less common. In this method, the first sample is chosen randomly, but subsequent samples are selected according to a predefined pattern, such as selecting waste from units with even numbers.

In situations where waste is sampled from processing/disposal sites or transfer stations, the sampling method is typically random, utilizing mechanisms like the quartering procedure. It shoud be noted that due to moisture transfer between items in the waste body, the results of sampling will have more uncertainty compared to sampling from the source. This means that when waste is mixed in collection vehicles, wastes with high moisture content, such as food waste, lose their moisture, while absorbent materials, especially paper, will absorb this moisture. ASTM (American Society for Testing and Materials) has developed a standardized test method for determining the composition of municipal waste based on samples taken from a disposal facility using manual separation. This method is known as ASTM Method D5231-92 (1992).

2.4.2.1 Size of the Samples Required in the Physical Analysis of Waste

In physical waste analysis, the size of required samples for waste sampling is determined by operational constraints and the heterogeneity of the waste. The more heterogeneous the waste, the larger the sample size required, as it becomes necessary to capture components that may be present in small percentages within the waste stream. The sample size is also proportional to the number of samples. The fewer the number of samples, the larger they need to be. Mixing the waste stream can facilitate the collection of larger and more homogeneous samples. For heterogeneous waste, such as household waste, obtaining the required sample size may necessitate collecting several tons of waste. Figure 2.10 illustrates the waste sampling and homogenization process in various locations within an urban area. After homogenization, samples are further divided into smaller portions for analysis. In cases where reducing the size of the waste is not inconsistent with the main sampling objective, the waste can be further subdivided, cut, shredded, or crushed to achieve greater homogenization. Since the storage of waste for an extended period can affect the quality of the sample, physical analysis should be conducted promptly after sampling.

The personnel responsible for waste analysis should receive prior training and be well-informed. They will randomly select a specific number of fully loaded waste collection vehicles each day based on the required sample size (the calculation of which is provided later) and direct them to the separation area. Each sample should be approximately 100 kg. After recording general information such as the date and time, weather conditions for that day (e.g., temperature and relative humidity), details of the waste collection contractor, the serviced area, and specifications of the waste collection vehicle (including vehicle class, total mileage, container volume, fuel type, and consumption), the vehicle's load will be emptied at the predetermined sampling location. It is essential to discharge the waste onto a suitable surface (such as a concrete or asphalt-paved area). The designated area should be covered to prevent the sun's radiation from causing moisture evaporation during the day. To obtain a

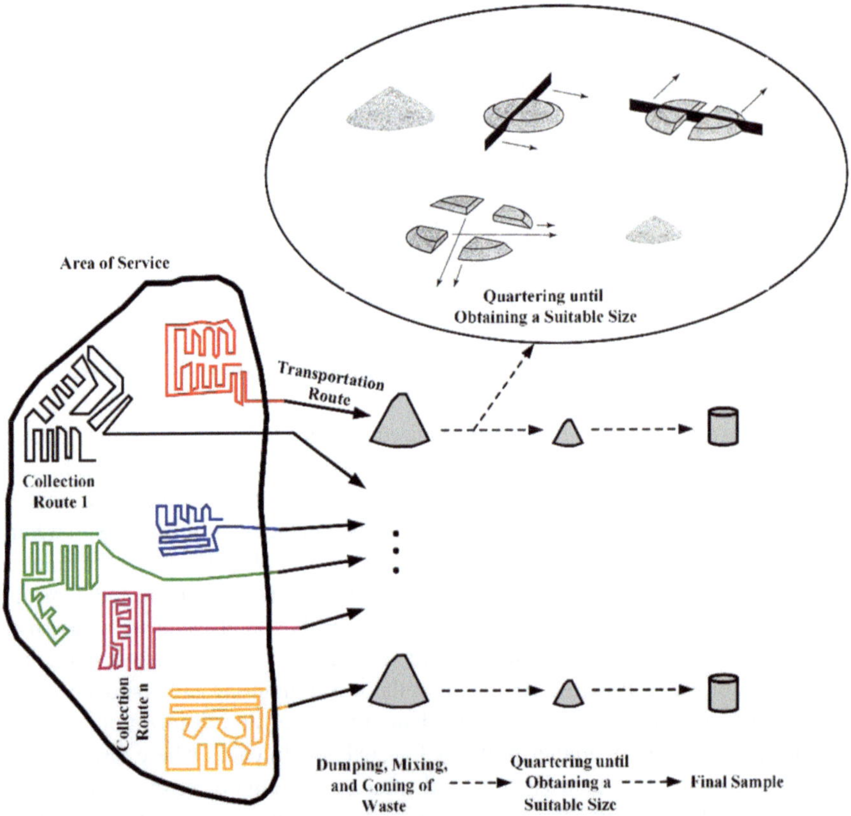

Fig. 2.10 Municipal waste sampling involves the process of coning waste received from various sources, creating a composite, dividing it into smaller samples through quartering, and preparing a final sample weighing approximately 100 kg

sample with a minimum weight of 91–136 kg (200–300 pounds) following ASTM D 5231 standard, the discharged waste is initially piled in heaps (at least four times the required sample weight), and then one-quarter of the heap volume, as shown in Fig. 2.10, is separated. The separated volume should be piled again, and one-quarter of it should be separated once more. This process continues until the weight of the separated sample reaches at least 100 kg.

2.4.2.2 Calculating the Required Sample Size for Physical Waste Analysis

Statistical analysis of the results of previous waste physical analysis done in a given area, can provide valuable information about the average percentage of waste components and their standard deviations. Under such circumstances, it becomes possible to calculate the number of required samples to achieve the desired level of precision. However, in areas where previous data related to physical waste analysis is not available, an initial estimate of the number of required samples can be obtained based on the information in Table 2.4 (derived from weekly sampling in various regions of the United States according to ASTM D 5231).

Once preliminary sampling is conducted or data from Table 2.4 is utilized to determine the average percentages ($\overline{x}$) and standard deviations (s) of each waste component (expressed as decimal numbers), the required number of samples (N) to reach the desired level of precision (e, expressed as a decimal number) can be calculated using Eq. 2.4. Finally, the maximum value of N related to the governing

Table 2.4 Average percentage (weight) and standard deviation of waste components in waste samples from different regions of America based on ASTM D 5231

Components	Waste composition mean percentage (% by weight)	Standard deviation
Mixed papers	0.22	0.05
Newspaper	0.10	0.07
Cardboard	0.14	0.06
Plastics	0.09	0.03
Yard wastes	0.04	0.14
Food wastes	0.10	0.03
Wood	0.06	0.06
Other organic wastes	0.05	0.06
Ferrous metals	0.05	0.03
Aluminum	0.01	0.004
Glass	0.08	0.05
Other inorganic compounds	0.06	0.03
Total	1.00	

component is reported as the required number of samples, and the total sample weight is obtained by multiplying it by 100 kg (The term "governing component" refers to the dominant or controlling component, for which the quantity N is maximized).

$$N = \left[\frac{t^* \times s}{e \times \overline{x}}\right]^2 \tag{2.4}$$

where, t* represents the corresponding Student's t-distribution for the expected confidence level (Table 2.5). At the beginning of the process, the value of t* considered to be equal to 1.960 for a 95% confidence level or 1.645 for a 90% confidence level (corresponding to $N = \infty$). After substituting the values into Eq. 2.4, a new N is calculated, and when the difference between the new and previous N is less than 10%, the operation will be terminated, and the larger value will be chosen.

The method described in this section assumes a normal distribution for the percentages of waste components and is recommended for use in cases with an acceptable number of samples.

Table 2.5 The t-statistic values as a function of sample size and confidence levels

Number of samples	Confidence interval		Number of samples	Confidence interval	
	90%	95%		90%	95%
2	6.314	12.706	23	1.717	2.074
3	2.92	4.303	24	1.714	2.069
4	2.353	3.182	25	1.711	2.064
5	2.132	2.776	26	1.708	2.060
6	2.015	2.571	27	1.706	2.056
7	1.943	2.447	28	1.703	2.052
8	1.895	2.365	29	1.38	2.048
9	1.86	2.306	30	1.699	2.045
10	1.833	2.262	31	1.697	2.042
11	1.812	2.228	36	1.69	2.030
12	1.796	2.201	41	1.684	2.021
13	1.782	2.179	46	1.679	2.014
14	1.771	2.16	51	1.676	2.009
15	1.761	2.145	61	1.671	2.000
16	1.753	2.131	71	1.667	1.994
17	1.746	2.12	81	1.664	1.99
18	1.74	2.11	91	1.662	1.987
19	1.734	2.101	101	1.66	1.984
20	1.729	2.093	121	1.658	1.98
21	1.725	2.086	141	1.656	1.977
22	1.721	2.08	∞	1.645	1.96

Example 2.2 As an illustrative example for explaining the calculation procedure of the required sample size in conducting a physical analysis, let's assume we would like to determine the amount of glass present in MSW with a precision of 10% and a confidence level of 90%. If we use samples with an approximate weight of 130 kg, what amount of waste should be separated?

Solution: Using the data from Table 2.4, we can obtain the average percentage of glass and its standard deviation. By substituting these values into Eq. 2.4, the required sample size will be 106 samples.

$$N_2 = \left[\frac{1.645 \times 0.05}{0.10 \times 0.08}\right]^2 = 106$$

$$N_3 = \left[\frac{1.659 \times 0.05}{0.10 \times 0.08}\right]^2 = 108$$

Considering that the newly obtained quantity ($N_3 = 108$) deviates by less than 10% from the previous quantity ($N_2 = 106$), the calculations are terminated, and the number of 108 was reported as the required sample size. Consequently, for 130 kg samples, approximately 14 tons of waste must undergo classification and separation.

2.4.2.3 Selection of Constituent Components and Conducting a Physical Waste Analysis

In the initial step of the sampling process and conducting physical waste analysis, determining which waste components need to be separated is of significant importance. In other words, it is necessary to identify the constituents that make up the waste composition in detail and inform this data to the separation personnel. Sometimes, the number of these classifications can be extensive, making the separation process time-consuming and costly. Therefore, depending on the ultimate goal (such as designing waste processing and recycling or disposal facilities), a realistic and feasible list can be prepared (similar to Table 2.6). In (Leonard 2003), the US EPA has considered 23 categories for separating waste components. In this reference, the primary objective is to estimate the greenhouse gas emissions (GHG) in various waste management processes. Here, the selection of a higher quantity for separation arises from the differences in emissions from each component during the production and release of GHGs. It is worth noting that waste separation operations are intensive processes with high operational volumes and are carried out manually by human labor. Therefore, increasing the number of separation categories and consequently, the number of required samples (please refer to the example in Sect. 2.4.2.2) will require more resources and workforce to separate various types of waste from the waste stream. As an estimation, nine workers can separate approximately 1600 kg of waste per day (in twelve groups) (Rhyner et al. 1995).

After determining the number and size of the required samples from the waste stream to identify the present components, the samples are manually separated at a pre-designated site near the entry gate of the transfer station or processing/disposal center. The necessary conditions for this separation site include.

- Having an area of approximately 1000 ft^2 (92 m^2) with a minimum width of 16 ft (5 m).
- Having suitable flooring.
- No restrictions on machine access to the separation site.
- Adequate protection against heavy rain and strong winds.
- Equipped with heating in cold weather.
- Located away from high-traffic paths and areas with heavy machinery operations.

In the separation site, first, large bulky items (such as cardboard and large wooden pieces), as well as bags containing a single type of waste (e.g., garden waste), are separated from the waste pile and placed in designated containers for weighing. The separation team places various types of waste constituents from the sample into containers located at the sides of the separation area. Once filled, these containers are moved to the scale location for weighing, and finally, their contents are discharged to be sent to the disposal center at the end of the operation.

According to ASTM D 5231 standard, the weighing equipment must have a minimum capacity of 200 lb. (91 kg) with an accuracy of 0.1 lb. (0.045 kg). If 30-gallon (115 L) containers are used for separation, the net weight of their contents is usually less than 100 lb. However, the separation team must ensure that they transfer the containers to the weighing location before the weight reaches approximately 50 kg because otherwise, moving the container becomes more challenging and may pose risks to the workers.

One noteworthy point in the separation process is how composite items are categorized. Composite items refer to a portion of waste contents made up of different and dissimilar materials, such as disposable diapers (mostly consisting of fibers

Table 2.6 An example of constituent components of MSW in the separation and physical analysis process

Number	Waste components	Number	Waste components
1	Food/putrescible	9	Sanitary wastes
2	Paper and cardboard	10	Special/hazardous wastes
3	Plastics	11	Leather and rubber
4	Textiles	12	Construction/Demolition
5	Ferrous metals	13	Electronic wastes
6	Non-ferrous metals	14	Composite packaging wastes
7	Wood and yard trimmings	15	Bulky wastes

and plastic), beverage containers made of two different metals, and electrical items like cables where a metal wire is encased in plastic insulation. When dealing with composite items, if possible, different parts of them should be separated, and each part should be placed in the corresponding storage container. If this is not practical, composite items will be categorized by the separation team manager following these instructions (in order).

1. If there are many similar composite items in the waste stream (such as aluminum electrical items with plastic coatings), after weighing the total weight of these materials and estimating the weight percentage of each of their constituent materials (in the previous example, metal and plastic), calculate the weight corresponding to each category (e.g., the weight of plastic) and place the equivalent weight of composite items in the corresponding container.
2. If only a small number of similar composite items exist, place them in a container that corresponds to the material that makes up the majority of the weight of the composite item (for example, place beverage cans made mostly of iron in the container for ferrous metals).
3. If composite items contribute a significant weight percentage to the sample, create a separate category for them, for example, composite roof shingles.
4. If none of the above methods are suitable, place the items (or a fraction of them) in a storage container labeled as "Other—Non-Combustible" or "Other—Combustible."

Manual separation operations will continue until the maximum remaining particle size is approximately 0.5 in (12.7 mm). At this stage, by visually estimating the percentage presence of each of the waste components, the remaining materials are proportionally placed in the appropriate storage containers.

In Figs. 2.11 and 2.12, the composition of municipal solid waste in Urmia in 2017 is presented in terms of weight composition and percentage presence in the waste stream. Out of a total of 583.3 tons of waste generated per day, food and putrescible waste accounted for over 68% of the weight of the city's waste stream. According to the World Bank report in 2016 (Kaza et al. 2018), the global average for food and green waste in the waste stream is reported to be 44%. However, it has been observed that the presence of such waste in the waste stream of countries is influenced by their income levels. In countries with low income, on average, 56% of the waste stream consists of food and green waste, while in high-income countries, it averages 32%. Additionally, from a regional perspective, an average of 58% of the waste stream in the Middle East and North African countries falls into this category of waste.

Today, the issue of food loss and the alarming statistics of food waste on a global scale (referred to as Food Loss and Waste or FLW) has challenged global food security, access to healthy food, the economy, and environmental sustainability. An accurate estimate of the amount of FLW is not available, but studies indicate that FLW accounts for approximately one-third of the total food globally (estimated at around 1.3 billion tons per year) (Kaza et al. 2018). The loss of resources, including land, water, labor, and energy used in food production, the adverse effects on climate change (as food production and distribution activities contribute significantly to greenhouse gas emissions and methane is released during the decomposition of wasted food), the reduction in the income of food producers, increased costs for consumers, and reduced access to healthy food are examples of the consequences of the growing trend in FLW worldwide. The causes of increased FLW vary in different regions of the world and are generally related to consumer behavior and government policies and regulations. Typically, FLW occurs in low-income countries during production, transportation after harvesting, storage, and processing, primarily due to management and technical constraints. In countries with medium and high incomes, FLW occurs more during the distribution and consumption stages, although it may also occur in the early stages, such as when agricultural subsidies lead to overproduction.

Improving coordination and creating enhanced alignment among different actors in the food supply chain is a key factor in addressing some of the issues resulting from the increase in FLW. Actions that can be taken in low-income countries to reduce FLW include investment in infrastructure (such as cold storage) and transportation, while in high-income countries, consumer education to change consumption patterns is key to reducing FLW. Additionally, utilizing processes for composting and energy recovery can be beneficial in managing food waste. In recent years, significant efforts have been made by some countries and local governments to reduce FLW or manage food waste more effectively. Examples include Italy (which passed a law in 2016 aimed at strengthening collaboration among key players in the food

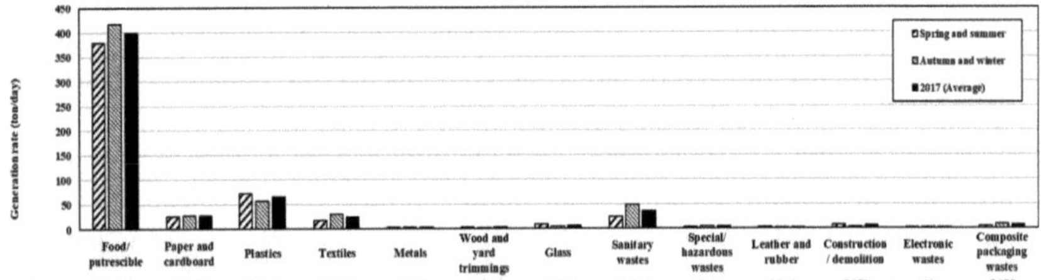

	Food/ putrescible	Paper and cardboard	Plastics	Textiles	Metals	Wood and yard trimmings	Glass	Sanitary wastes	Special/ hazardous wastes	Leather and rubber	Construction / demolition	Electronic wastes	Composite packaging wastes
Spring and summer	379,697	25,686	72,967	18,087	2,940	3,654	9,508	23,916	4,146	2,946	7,350	10	5,175
Autumn and winter	417,876	27,785	57,007	29,864	3,337	2,687	4,341	48,815	4,550	804	4,072	47	9,402
2017 (Average)	398,786	26,735	64,987	23,975	3,138	3,170	6,925	36,365	4,348	1,875	5,711	29	7,288

Fig. 2.11 Weight composition of solid waste produced in Urmia City in 2017

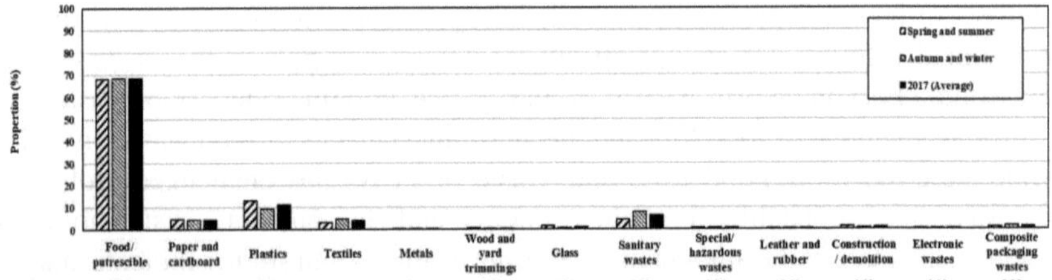

	Food/ putrescible	Paper and cardboard	Plastics	Textiles	Metals	Wood and yard trimmings	Glass	Sanitary wastes	Special/ hazardous wastes	Leather and rubber	Construction / demolition	Electronic wastes	Composite packaging wastes
Spring and summer	68.28	4.62	13.12	3.25	0.53	0.66	1.71	4.30	0.75	0.53	1.32	0.00	0.93
Autumn and winter	68.44	4.55	9.34	4.89	0.55	0.44	0.71	7.99	0.75	0.13	0.67	0.01	1.54
2017 (Average)	68.36	4.58	11.14	4.11	0.54	0.54	1.19	6.23	0.75	0.32	0.98	0.00	1.25

Fig. 2.12 Percentage presence of components in the solid waste stream in Urmia City 2017

supply chain, public education, encouraging food donations, and promoting reusable and recyclable packaging), France (which banned supermarkets from discarding or destroying unsold food and instead required them to donate unsold food to charities and food banks), San Francisco in the United States (which passed a law in 2009 mandating composting of food waste produced by residents and tourists), Ningbo City in China (which directs food waste from apartment buildings to an anaerobic digestion facility), and several cities in Sweden (where biogas produced from food waste is used as an energy source for electric vehicles and heat production).

Valuable recyclable items (including paper and cardboard, various types of plastics, metals, and glass) account for between 16% (in low-income countries) to 50% (in high-income countries) of the waste stream (Kaza et al. 2018). Referring to Fig. 2.12, these items make up approximately 17.5% of the municipal waste in Urmia.

2.4.3 Density or Specific Weight of Solid Waste

While measuring the weight of waste at transfer stations or solid waste processing facilities is straightforward, what is of high importance in waste disposal facilities is the measurement of volume. Density, in relation to the weight or mass of waste (either wet or as received), divided by the volume it occupies, is used to characterize waste and is defined by Eq. 2.5.

$$\text{Density}, \left(\text{kg/m}^3\right) = \frac{\text{mass of material}(\text{kg})}{\text{volume occupied}\left(\text{m}^3\right)} \tag{2.5}$$

The amount of compaction applied to waste, the moisture content, and the type and percentage presence of various components in the waste mass are major factors affecting waste density. Among these, the degree of mechanical compaction applied to the waste is of high importance from an environmental engineering perspective. Waste density can vary in different situations and is expressed in various forms: Loose, In compactor truck, Baled, or In landfill. Loose density refers to the mass per unit volume of waste in storage containers where no mechanical compaction has been applied, and it is used to calculate the number of storage containers required. Due to the mechanical compaction in waste collection vehicles, waste density in the vehicle is higher than in the loose state. This type of density can be utilized in calculations related to waste collection and transport costs. The baling of waste, due to its high density, can extend the useful life of a landfill by more than 50%. Waste baling can also reduce transportation costs, minimize the scattering of lightweight waste and odors, and reduce problems related to disease-carrying insects and uneven settling of the buried waste mass. Finally, waste density in a landfill, with the application of compaction by heavy machinery, will have the highest value among other density states. Apparent density increases with reduced physical disorder of waste components, and compaction primarily reduces physical irregularity. Some compaction occurs due to the weight of the overlying waste, so with an increase in the burial height, compaction, and consequently waste density in lower layers, increases.

There are other classifications for expressing waste density, including bulk density versus material density, and dry density versus wet density. Bulk density is highly sensitive to the level of compaction applied, increasing with compaction (i.e., the reduction in the volume or filling of empty spaces between waste particles). Material density is the density without void spaces. In waste management discussions, density refers to wet density or the density of the waste as received unless dry density is specified. To determine dry density, the moist weight of the waste and the volume of the container containing it are measured first, and then, through moisture evaporation and calculating the dry weight, the dry density is obtained by dividing the dry weight by the volume in the moist state.

MSW placed by households in garbage bins typically has an apparent density ranging from 90 to 150 kg/m^3. Compaction of waste inside garbage bins may increase the apparent density to around 180 kg/m^3. In collection trucks where waste is mechanically compacted, the apparent density of waste is usually between 350 and 420 kg/m^3. When waste in a landfill is subjected to higher compaction, the apparent density can reach approximately 700 kg/m^3 (Christensen 2011; Worrell et al. 2016). In Table 2.7, the specific weight of un-compacted and landfilled waste for some common municipal waste components is provided.

An ASTM standard test method, ASTM Method E1109-86, 2009, has been developed for determining the apparent density of different components of municipal solid waste. According to this standard, a sample of municipal solid waste is placed inside

Table 2.7 Specific weight of un-compacted and landfilled materials for some common components of municipal solid waste (kg/m^3) (Rhyner et al. 1995)

Waste components	Bulky specific weight			Specific weight at landfill (average)
	Range		Average	
	Min	Max		
Paper and cardboard	32	128	82	465
Yard trimmings	64	224	104	890
Food waste	130	480	288	1186
Glass	160	480	194	1345
Ferrous metals	130	1120	320	332
Tin cans (steel)	48	160	88	332
Non-ferrous metals	64	240	160	217
Plastics	32	128	64	213
Wood	128	320	240	498
Leather and rubber	96	256	128	205
Textiles	32	96	64	237
Others	320	960	480	1186

a cubic container with dimensions of 60 cm (whose inner surfaces are moisture-absorbent), and the apparent density of the container's contents is determined by weighing it. It is essential in this test that no particle of the waste is larger than two-thirds of the container's dimensions, and if such particles are observed, they must be removed from the sample. During filling the container with waste, any manual compaction or pouring of the waste should be avoided (especially to ensure that fine particles do not escape). After filling the container, it is lifted to a height of approximately 6 cm above the ground and dropped fully three times to achieve compaction. After this compaction process, the contents of the container are leveled using a straight rod. It should be noted that in cases where the contents of the container are found to be under the upper edge of the container after compaction, the container should be emptied, and the test should be repeated using a different sample.

Example 1.3 The composition of un-compacted MSW for an assumed region is according to table below. Using the values of the specific weight of the landfilled waste from Table 2.7, calculate the volumetric composition of the waste in the burial center? What is the specific weight of this waste sample? Discuss your findings by calculating the ratio of volume percentage to weight percentage of each component.

Components	Mean percentage (% by weight)
Paper and cardboard	32.3
Yard trimmings	19.0
Food waste	8.1
Glass	6.5
Ferrous metals	6.4
Non-ferrous metals	1.2
Plastics	9.8
Wood	7.3
Leather and rubber	2.7
Textiles	3.3
Others	3.4

Solution: Assuming a conventional sample of waste, for example, 100 kg, the weight of each of its components will be in accordance with the percentage presence. The volume of each component is obtained by dividing the weight by the density of that component. The total volume is also obtained by summing the volume of each component. Finally, the percentage of presence of each component in the total volume can be calculated.

According to the calculations in below table, the specific weight of this waste sample is equal to 461 (kg/m^3). In the last column of the table, a ratio of one in the case of paper means that the ratio of the space occupied by paper in the landfill is similar to the weight ratio of paper to the total available waste. Types of plastic, rubber and leather, textiles, and non-ferrous metals have ratios close to two or more, which shows that these materials occupy more space in the landfill compared to their weight percentage in the waste stream. Yard trimmings, food waste, and glass will have numerical ratios smaller than one and will occupy less space in the landfill in proportion to their weight percentage in the waste.

Components	Weight (kg)	Specific weight at landfill	Volume (m^3)	Volume percentage (%)	Ratio vol%/ mass%
Paper and cardboard	32.3	465	0.069	31.8	1.0
Yard trimmings	19.0	890	0.021	9.7	0.5
Food waste	8.1	1186	0.007	3.2	0.4
Glass	6.5	1345	0.005	2.3	0.3
Ferrous metals	6.4	332	0.019	8.7	1.4
Non-ferrous metals	1.2	217	0.005	2.3	1.9
Plastics	9.8	213	0.046	21.2	2.2
Wood	7.3	498	0.015	6.9	0.9
Leather and rubber	2.7	205	0.013	6.0	2.2
Textiles	3.3	237	0.014	6.4	1.9
Others	3.4	1186	0.003	1.5	0.4
Total	100.0		0.217	100.0	

It should be noted that in a landfill, the actual volume occupied by a non-homogeneous mixture such as waste will be less than the volume resulting from the algebraic sum of its components because some of the materials with different sizes fill the voids and holes of the waste in the landfill. Therefore, the actual specific weight is always higher than the specific weight calculated here.

2.4.4 Particle Size Distribution of Waste

Knowing the particle size distribution of waste (sieve analysis) is crucial for calculating and designing separation equipment at the beginning of waste processing lines (i.e. trommels). In this test, which is carried out concurrently with the physical analysis of waste, the diameter of various components that make up the waste stream is measured. The particle size of waste can be considered equivalent to the largest length (L) of the particle or as a function of the dimensions length (L), width (W), and height (H) of the particles (Eq. 2.6).

$$\text{Size} = \begin{cases} \text{L} \\ \frac{\text{L}+\text{W}}{2} \text{ or } \sqrt{\text{L} \times \text{W}} \\ \frac{\text{L}+\text{W}+\text{H}}{3} \text{ or } \sqrt[3]{\text{L} \times \text{W} \times \text{H}} \end{cases} \tag{2.6}$$

According to studies conducted in Italy (Arena et al. 2003), it was concluded that waste with dimensions larger than 120 mm mainly consists of materials such as paper and cardboard, wood, plastics, and ferrous fines, making them highly recyclable. On the other hand, waste larger than 60 mm and smaller than 120 mm contains valuable combustible materials, and they can be used to produce refuse-derived fuel (RDF). Additionally, waste smaller than 60 mm has a high moisture content, and the best way to manage it is through biological processes.

Particle size measurement can be performed using a set of sieves with specific dimensions, either through mechanical devices that create uniform motion or with human force. When using mechanical shaking equipment, the sieves are stacked in series and, by shaking the assembly on a vibrating table, the waste components will be classified based on their nominal size.

In Figs. 2.13 and 2.14, the particle size distribution of components in the municipal waste stream of Urmia city is presented, along with the size distribution of different components within it in the year 2017. As Fig. 2.13 indicates, over 50% of the generated waste in this city falls into two categories: larger than 150 mm and smaller than 30 mm. Food waste is predominant in all categories, and the diversity of components in the two middle categories, between 50 and 100 mm, is higher than in the other categories.

It appears that using trommel units with 150 mm mesh sizes at the beginning of the separation line can potentially separate about 10% of the waste stream as valuable items for recycling (meaning over 65% of the total recyclable waste can be separated and recycled in this way). The waste stream with sizes less than 80 mm is mainly composed of household/organic waste (meaning over 90% of waste smaller than this size consists of organic waste). Therefore, installing a second sieve at the end of the separation line with an 80 mm mesh size can separate and direct biologically decomposable waste from the waste stream towards compost production (the under screen trommel stream constitutes more than 50% of the city's organic waste).

Considering the variety of sizes of plastic waste and composite packaging materials in the Urmia city waste stream (Fig. 2.14), it seems that these categories could be considered as targets for implementing source separation policies in Urmia city, given their value in recycling markets and considering their share in the total waste stream (approximately 6% based on Fig. 2.14).

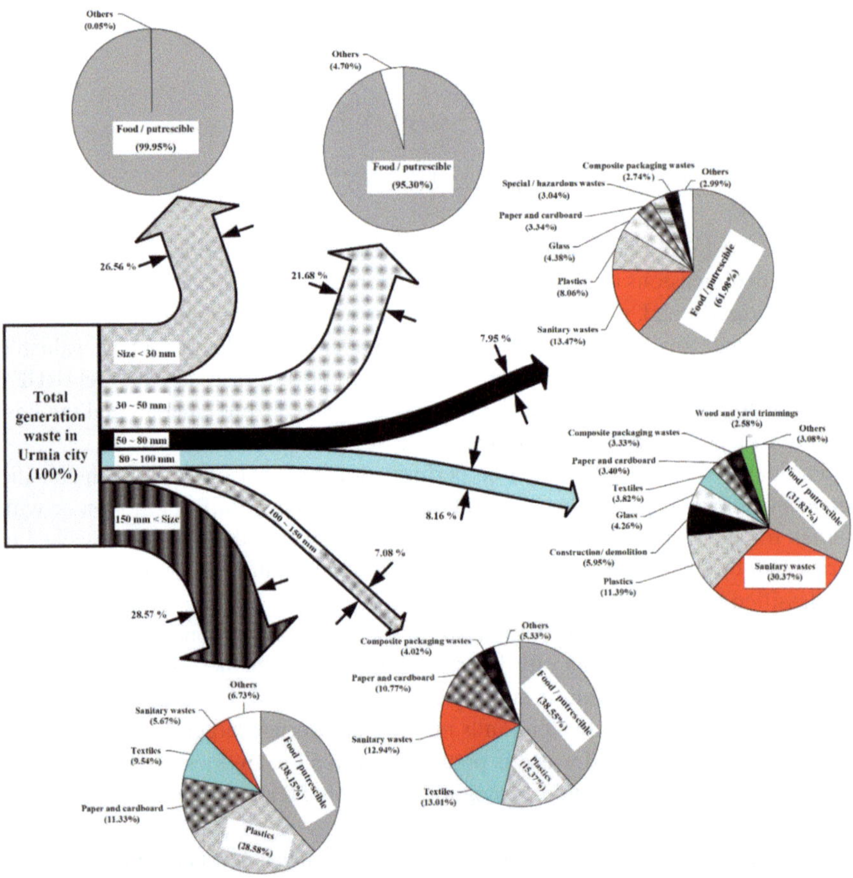

Fig. 2.13 Particle size distribution of components in the Urmia City municipal waste stream (2017)

2.4.5 Heating Value

Among the basic parameters in the selection or design of waste incineration plants, we can mention the calorific or heating value of the materials that are supposed to be burned. MSW contains large amounts of plastic, paper, wood, and other combustible items, and for this reason, it is considered as one of the cheap and available sources of renewable energy in human societies. The most common method of converting waste into energy is using waste incineration plants. The use of waste incineration in waste management has benefits such as reducing the volume of waste, providing mitigations in greenhouse gas production calculations, increasing the useful life of the landfill and reducing the need for fossil fuels in urban energy supply; Nevertheless, the exhaust gases from the chimney of waste incinerators and the management of wastewater from gas treatment units are among the challenges of operating these units. The basis of waste incineration plant design is knowing the quantitative and qualitative characteristics of waste, such as its heating value and moisture content. The heating value of generated waste varies in different countries of the world and even in different cities of the same country (Komilis et al. 2012), and the reason for that is the difference in the composition of municipal waste.

The heating value of a fuel (which can also be municipal waste) is often described as gross heat value or higher heat value (HHV) and net heat value, sometimes called lower heat value (LHV). The HHV is determined by the complete combustion of a sample of a certain weight, in the presence of oxygen and in a bomb calorimeter, by measuring the released heat (which increases the temperature of the surrounding water). The ASTM Institute has provided guidelines for measuring the gross heat value of refuse-derived fuels (RDF) (ASTM E 0711-87, 1992). The heating value of waste is generally expressed in units of J/g, British thermal unit per pound (Btu/lb), or calories per gram (cal/g). A Btu here is the amount of heat required to raise the temperature of one pound of liquid water at an initial temperature of approximately

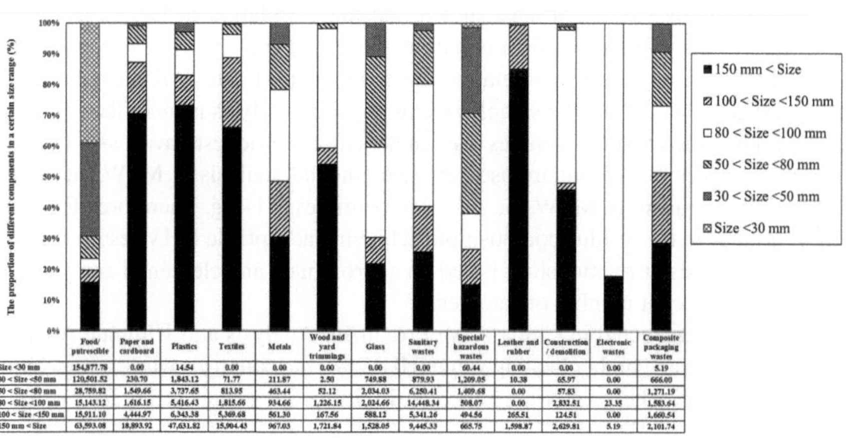

	Food/ putrescible	Paper and cardboard	Plastics	Textiles	Metals	Wood and yard trimmings	Glass	Sanitary wastes	Special/ hazardous wastes	Leather and rubber	Construction / demolition	Electronic wastes	Composite packaging wastes
Size <30 mm	154,877.78	0.00	14.54	0.00	0.00	0.00	0.00	0.00	60.44	0.00	0.00	0.00	5.19
30 < Size <50 mm	120,501.52	230.70	1,843.12	71.77	211.87	2.50	749.88	879.93	1,209.05	10.38	65.97	0.00	666.00
50 < Size <80 mm	28,759.82	1,549.66	3,737.65	813.95	463.44	52.12	2,034.03	6,250.41	1,409.68	0.00	57.83	0.00	1,271.19
80 < Size <100 mm	15,143.12	1,616.15	5,416.43	1,815.66	934.66	1,226.15	2,024.66	14,448.34	508.07	0.00	2,832.51	23.35	1,583.64
100 < Size <150 mm	15,911.10	4,444.97	6,343.38	5,369.68	561.30	167.56	588.12	5,341.26	494.56	265.51	124.51	0.00	1,660.54
150 mm < Size	63,593.08	18,893.92	47,631.82	15,904.43	967.03	1,721.84	1,528.05	9,445.33	665.75	1,598.87	2,629.81	5.19	2,101.74

Fig. 2.14 Size distribution of various components in the Urmia City municipal waste stream (2017)

39°F by one degree Fahrenheit. According to international agreements (International Table), one Btu is considered equal to 1055.06 J and one calorie is considered equal to 4.1868 J (1 Btu/lb. = 2.326 J/g and 1.8 Btu/lb. = 1.0 cal/g).

During the combustion of the waste in the calorimeter, all the water in the sample and the water formed from the oxidation of hydrogen content of waste will evaporate. Then, the resulting vapors are condensed in the calorimeter and are included in the calculation of the amount of heat released. Therefore, in determining the HHV of waste, the latent heat of water evaporation is considered. Conversely, in calculating the LHV, the latent heat of vaporization of water is subtracted from the previous result (i.e. HHV) (Eq. 2.7; at about 25 or 77°F). In addition to net and gross heat value, there are other definitions such as heat value of waste "as received" or wet waste heating value (note that both HHV and LHV can be expressed "as received"; this issue is also true for subsequent definitions); net heat value of waste in dry form (or without moisture -Moisture Free-); and the net heat value of the combustible part of dry waste (or without moisture and ash—Moisture and Ash Free = MAF-).

$$\begin{aligned} \mathrm{LHV} &= \mathrm{HHV}(\text{in MJ/kg}) - 0.0244(\mathrm{W} + 9\,\mathrm{H})\mathrm{MJ/kg} \\ \mathrm{LHV} &= \mathrm{HHV}(\text{in BTU/lb.}) - 10.50(\mathrm{W} + 9\,\mathrm{H})\mathrm{Btu/lb.} \end{aligned} \tag{2.7}$$

where W is the mass percentage of moisture in the waste (%) and H is the weight (or mass) percentage of hydrogen in the dry waste (%).

In addition to using the calorimetric test in determining the heat value of waste, so far, several mathematical equations based on the data of physical composition, proximate analysis or elemental analysis of waste have been presented by different researchers in estimating the calorific value. Nevertheless, there are uncertainties regarding the scope of application of these equations and the accuracy of the results (due to the size of different samples used in determining the independent data of physical composition, approximate or elemental analysis of MSW). Table 2.8 mentions a number of relationships in the technical literature (Kathiravale et al. 2003).

The results of a research have shown that the correlation of laboratory results with the results of mathematical equations based on waste physical composition data is higher than the results of relations based on approximate or elemental analysis (Kathiravale et al. 2003) (the sample size in physical analysis is more than 100 kg). It should be noted that the samples used in the calorimetric test have a size of about 1 g, the samples of 1–5 mg are used in the elemental analysis of MSW, and in the approximate analysis of MSW, the size of the samples is 1–5 g. Therefore, due to the high variance in the residue composition, achieving acceptable HHV results through calorimetric tests or relationships based on approximate and elemental analysis will require a significant number of samples.

In a study conducted in China using Proximate Analysis and Ultimate Analysis models, the average higher heating value (HHV) for the country's waste composition was determined. The results are reflected in Table 2.9 (Zhou et al. 2014). As in

Table 2.8 Computational models for heating value

Name	Equation	Units	Remarks	Application
1. Models based on ultimate analysis				
Dulong	HHV = 81 C + 342.5 (H − O/8) + 22.5 S − 6 (9 H − W)	kcal/kg	Modified (wt%)	MSW/coal
Steuer	HHV = 81 (C − 3 × O/8) + 57 × 3 × O/8 + 345 (H − O/10) + 25 S − 6 (9 H + W)	kcal/kg	(wt%)	MSW
Scheurer—Kestner	HHV = 81 (C − 3 × O/4) + 342.5 H + 22.5 S + 57 × 3 × O/4 − 6 (9 H + W)	kcal/kg	(wt%)	MSW
Chang	HHV = 8561.11 + 179.72 H − 63.89 S − 111.17 O − 91.11 Cl − 66.94 N	kcal/kg	(wt%)	MSW
Boie	HHV = 83.22 C + 274.3 H − 25.8 O + 15 N + 9.4 Cl + 65 P	kcal/kg	(wt%)	Refuse
Vondracek	HHV = C (89.17 − 0.0622 C_1) + 270 (H − O/10) + 25 S (C_1 − carbon content on moisture and ash free basis)	kcal/kg	(wt%)	Refuse
Wilson	HHV = 7831 C_{org} + 35.932 (H − O/8) + 2212 S − 3545 C_{inorg} + 1187 O + 578 N	kcal/kg	(wt fraction)	MSW
Mott and Spooner	HHV = 0.336 C + 1.418 H − 0.0145 O + 0.0941 S	MJ/kg	(wt%)	Coal/refuse
Institute for Gas Technology, USA	HHV = 0.3417 C + 1.3221 H + 0.1232 S − 0.1198 (O + N) − 0.0153 A	MJ/kg	(wt%)	Coal/refuse
Kathiravale et al	HHV = 416.638 C − 570.017 H + 259.031 O + 598.955 N − 5829.078	kJ/kg	Dry (wt%)	MSW
2. Models based on proximate analysis				
Goutal	HHV = 147.6 × FC + K × VM (K is a constant that varies with the value of VM)	Btu/lb	(wt%)	Coal/refuse

(continued)

Table 2.8 (continued)

Name	Equation	Units	Remarks	Application
Benton	$HHV = 44.75 \times VM - 5.85 \times W + 21.2$	kcal/kg	(wt%)	Refuse
Traditional	$HHV = 45 \times VM - 6 \times W$	kcal/kg	(wt%)	Refuse
Kathiravale et al	$HHV = 356.047 \times VM - 118.035 \times FC - 5600.613$	kJ/kg	Dry (wt%)	MSW
3. Models based on physical composition				
Conventional	$HHV = 88.2\ Pl + 40.5\ (Ga + Pa) - 6\ W$	kcal/kg	(wt%)	Refuse
Tokyo	$HHV = [(100\ W)/100]\ \{38.8\ (Pa + Ga + T + Oc) + 50.9\ (Te + Ru) + 73.7\ Pl\} - 6\ W$	kcal/kg	(wt%)	MSW
Ali Khan	$HHV = [23\ (Ga + 3.6 \times Pa)] + [160\ (Pl + Ru)]$	Btu/lb	(wt%)	MSW
Kathiravale et al	$HHV = 112.157\ Ga + 183.386\ Pa + 288.737\ Pl + 5064.701$ $HHV = 112.815\ Ga + 184.366\ Pa + 298.343\ Pl - 1.920\ W + 5130.380$	kJ/kg	(wt%)	MSW

Table 2.9 Average heating values of China's waste composition

Components	Average of HHV (kJ/kg)
Chlorine-free plastics*	43,448
Rubber	29,789
PVC	21,172
Textiles	20,162
Wood waste	19,461
Paper	15,894
Food residues	15,386

* Polyethylene, Polypropylene, and Polystyrene

Table 2.9, among the waste composition in China, plastics have the highest heating value, while food waste has the lowest one.

Example 2.4 The chemical formula of the combustible part of a waste sample is given as $C_{760}H_{1980}O_{874.7}N_{12.7}S$. Estimate the energy content (HHV) of this sample using the relations provided by Boie and Kathiravale et al. in Table 2.8.

Solution: First, it is necessary to calculate the percentage of elements that make up the waste sample (table on the next page).

Based on the Boie relation

Note that (%wt) in Table 2.8 means the weight percentage of the elements in the "as received" definition.

$$\begin{aligned}\mathrm{HHV(kcal/kg)} &= [83.22 \times (36.04)] + [274.3 \times (7.82)] - [25.8 \times (55.31)] \\ &\quad + [15 \times (0.70)] + [9.4 \times (0)] + [65 \times (0)] = 3727.78 \\ &= 3727.78\left(\frac{\mathrm{kcal}}{\mathrm{kg}}\right) \times \frac{4.1868\mathrm{J}}{1\mathrm{cal}} = 15607.47(\mathrm{kJ/kg})\end{aligned}$$

Element	Number of moles	Atomic weight (g/mol)	Weight (g)	Percent
Carbon	760	12	760 × 12 = 9120	36.04
Hydrogen	1980	1	1980	7.82
Oxygen	874.7	16	13,995.2	55.31
Nitrogen	12.7	14	177.8	0.70
Sulfur	1	32	32	0.13
Total	–	–	25,305	100

Based on the relationship of Kathiravale et al. (Given that the chemical formula is generally calculated and presented on a dry basis, therefore, the numbers in the previous table will be used):

$$\begin{aligned} HHV(kJ/kg) = & [416.638 \times (36.04)] - [570.017 \times (7.82)] \\ & + [259.031 \times (55.31)] + [598.955 \times (0.70)] \\ & - 5829.078 = 19475.30 \end{aligned}$$

The value obtained from the relationship of Kathiravale et al. is less than 25% different from the result obtained from Boie's relation, the reasons for which were mentioned in this section (the difference in the characteristics of the residuals used in deriving empirical relations).

In this sub-section, the approximate and elemental analysis of waste was discussed. Waste analysis is called approximate analysis from the point of view of determining calorific value, amount of moisture and ash percentages. The flammable part of the waste can be divided into two parts: volatile materials and fixed carbon or remaining materials. Determining the percentage of the mentioned components is done by heating wastes in the absence of air and under standard conditions. In this analysis, some of materials will be vaporized and residues like coal will remain. This remaining part is referred to as fixed or residual carbon. The presence of volatile substances in this test is also directly related to the presence of flame during the process. An elemental or final analysis determines the chemical composition of the combustible part of the waste as the amount of ash and chemical elements such as carbon, hydrogen, oxygen, nitrogen, sulfur and chlorine.

2.4.6 *Moisture Content*

Moisture content in the waste can be expressed in two ways: (a) dry basis moisture content, which is the ratio of the weight of moisture content in the waste to its dry weight, and (b) wet basis moisture content, which is the ratio of the weight of moisture content in the waste to its wet weight. Moisture content is typically calculated as a field test using Eq. 2.8:

$$M = \frac{(W - d) \times 100}{W} \tag{2.8}$$

where M is the percentage of moisture content on a wet basis, W is the initial weight of the sample (kg), d is the weight of the sample after drying at 105 °C (kg). Dry basis moisture content can also be calculated using Eq. 2.8:

$$M_d = \frac{(W - d) \times 100}{d} \tag{2.9}$$

where M_d is the percentage of moisture content on a dry basis. Table 2.10 provides information on the specific weight and moisture content of various types of wastes.

Moisture content is determined by weighing the waste before and after drying. Drying is typically done at temperatures of 77 or 105 °C until a constant weight is reached. At 105 °C, not only water but also volatile fatty acids, organic matter, and mercury can evaporate.

Moisture is a volatile property and can change rapidly, so it should be considered when determining specifications. If different materials are placed together in a waste bin or bag, moisture can be transferred from one material to another, such as from moist food waste to dry paper. Vassiljev et al. (2002) reported that newspaper in a waste bin has about 7% moisture content, but the average moisture content of newspaper collected from waste trucks often exceeds 20%. Table 2.11 provides the moisture content percentage of waste components before and after mixing (Chandler et al. 1997).

2.4.7 Total Organic Carbon (TOC)

Knowing the amount of organic matter in the waste stream and its availability for biological processes is of particular importance. For example, mentioned information are used in calculating the amount of air required in the aerobic composting process; or in determining the potential of methane gas production in anaerobic digesters and landfills.

The organic content of waste can be determined through several different approaches. Volatile solids (VS), defined as the amount of weight loss of dry matter after combustion of the sample; Note that here it is assumed that all volatile substances are organic. A common method is to dry the sample at a temperature of 105 °C and then put it in a muffle oven at a temperature of 550 °C in the vicinity of air for two hours. The remaining part after ignition, the ash content and the evaporated part, form the volatile solids in the waste mass. Total organic carbon (TOC) is determined by measuring the amount of CO_2 produced by the thermal degradation of the sample. Please note that in determining TOC, CO_2 caused by heating inorganic carbon (e.g. carbonates) should be deducted from the total amount. In addition, the organic content of the waste can also be determined by chemical oxidation of the sample by (K_2CrO_4) (commonly expressed as chemical oxygen demand or COD). Certain organic compounds, such as volatile fatty acids (VFA such as acetic acid and propionic acid); protein; carbohydrate; fat; cellulose; hemicellulose; and lignin; are also measured in some cases.

Since microorganisms can metabolize paper, yard waste, food and wood, these wastes are classified as biodegradable wastes. Disposable diapers and their contents and cotton and wool textiles are also highly biodegradable. In general, some solid

Table 2.10 Specific weight and moisture percentage of waste types (Tchobanoglous et al. 1993)

Type of waste	Specific weight (lb/yb^3)		Moisture content (% by weight)	
	Range	Typical	Range	Typical
Residential(un-compacted)				
Food waste (mixed)	220–810	490	50–80	70
Paper	70–220	150	4–10	6
Cardboard	70–135	85	4–8	5
Plastics	70–220	110	1–4	2
Textiles	70–170	110	6–15	10
Rubber	170–340	220	1–4	2
Leather	170–440	270	8–12	10
Yard wastes	100–380	170	30–80	60
Wood	220–540	400	15–40	20
Glass	270–810	330	1–4	2
Tin case	85–270	150	2–4	3
Aluminum	110–405	270	2–4	2
Other metals	220–1940	540	2–4	3
Dirt, ash, etc.	540–1685	810	6–12	8
Ashes	1095–1400	1255	6–12	6
Rubbish	150–305	220	5–20	15
Residential yard waste				
Leaves (loose and dry)	50–250	130	20–40	30
Green grass (loose and moist)	350–500	400	40–80	60
Green grass (wet and compacted)	1000–1400	1000	50–90	80
Yard waste	450–600	500	20–70	50
Yard waste (composted)	450–650	550	40–60	50
Municipal				
In compactor truck	300–760	500	15–40	20
In landfill				
Normally compacted	610–840	760	15–40	25
Well compacted	995–1250	1010	15–40	25
Commercial				
Food wastes (wet)	800–1600	910	50–80	70
Appliances	250–340	305	0–2	1
Wooden crates	185–270	185	10–30	20
Tree trimmings	170–305	250	20–80	5
Rubbish (combustible)	85–305	200	10–30	15

(continued)

Table 2.10 (continued)

Type of waste	Specific weight (lb/yb^3)		Moisture content (% by weight)	
	Range	Typical	Range	Typical
Rubbish (noncombustible)	305–610	505	5–15	10
Rubbish (mixed)	235–305	270	10–25	15
Construction and demolition				
Mixed demo. (noncombustible)	1685–2695	2395	2–10	4
Mixed demo. (combustible)	505–675	605	4–15	8
Mixed const. (combustible)	305–605	440	4–15	8
Broken concrete	2020–3035	2595	0–5	–
Industrial				
Chemical sludges (wet)	1350–1855	1685	75–99	80
Fly ash	1180–1515	1350	2–10	4
Leather scraps	170–420	270	6–15	10
Metal scrap (heavy)	2530–3370	3000	0–5	–
Metal scrap (light)	840–1515	1245	0–5	–
Metal scrap (mixed)	1180–2530	1515	0–5	–
Oils, tars, asphalts	1350–1685	1600	0–5	2
Sawdust	170–590	490	10–40	20
Textile wastes	170–370	305	6–15	10
Wood (mixed)	675–1140	840	30–60	25
Agricultural (mixed)	675–1265	945	40–80	50
Dead animals	340–840	605	–	–
Fruit wastes (mixed)	420–1265	605	60–90	75
Manure (wet)	1515–1770	1685	75–96	94
Vegetable wastes (mixed)	340–1180	605	60–90	75

waste materials are easier to biodegrade (have high bioavailability) than others. Generally, these materials have a high nitrogen and moisture content, and examples of them include food scraps, lawn clippings, and other green and horticultural waste. Tree leaf waste (brown debris) generally has moderate bioavailability. Although wood, cotton and wool are biodegradable, they have relatively low bioavailability and are known as non-compostable materials in waste management.

In the discussions of biological waste management, the carbon content of wastes is divided into two categories: degradable organic carbon (DOC) and fossil carbon. Food scraps, horticultural waste, paper and cardboard, wood, textiles and disposable

Table 2.11 Moisture percentage of waste components before and after mixing (Chandler et al. 1997)

Category	Before mixing	After mixing
Food waste	63.6	70
Yard waste	39.7	55
Glass	3	2
Metal	6.6	2
Paper	24.3	8
Plastic	13.8	2
Leather, rubber	13.8	2
Textile	23.8	10
Wood	15.4	15
Other	3	2

diapers have the largest share in the total DOC content of municipal waste. Ash, dust (for example, the contents of a vacuum cleaner bag), rubber and leather contain certain amounts of non-fossil carbon, but this carbon is difficult to break down. Some textiles, plastics (including plastic used in disposable diapers), rubber, and e-waste make up the majority of fossil carbon in the waste stream. Paper (coated) and leather (synthetic) can also contain small amounts of fossil carbon.

Bibliography

Abubakar IR, Maniruzzaman KM, Dano UL, AlShihri FS, AlShammari MS, Ahmed SMS, Al-Gehlani WAG, Alrawaf TI (2022) Environmental sustainability impacts of solid waste management practices in the global South. Int J Environ Res Public Health 19(19):12717. https://doi.org/10.3390/ijerph191912717

Adnan M, Jha A, Kumar S (2020) Municipal solid waste management and its impact: a review. Int J Adv Res Eng Technol (IJARET) 11(5):685–693. https://doi.org/10.34218/IJARET.11.5.2020.071

Akmal T, Jamil F (2021) Health impact of solid waste management practices on household: the case of metropolitans of Islamabad-Rawalpindi, Pakistan. Heliyon 7:e07327. https://doi.org/10.1016/j.heliyon.2021.e07327

Arena U, Mastellone ML, Perugini F (2003) The environmental performance of alternative solid waste management options: a life cycle assessment study. Chem Eng J 96(1–3):207–222. https://doi.org/10.1016/j.cej.2003.08.019

Chandler AJ, Eighmy TT, Hjelmar O, Kosson DS, Sawell SE, Vehlow J, Van der Sloot HA, Hartlén J (1997) Municipal solid waste incinerator residues. Elsevier

Christensen T (ed) (2011) Solid waste technology and management. John Wiley & Sons

Danthurebandara M, Van Passel S, Nelen D, Tielemans Y, Van Acker K (2012) Environmental and socio-economic impacts of landfills. Linnaeus Eco-Tech 2012:40–52

Eggleston HS, Buendia L, Miwa K, Ngara T, Tanabe K (2006) IPCC guidelines for national greenhouse gas inventories

Environmental Protection Agency (EPA) (2022) A town, a flood, and superfund: looking back at the times beach disaster nearly 40 years later. Available at: https://www.epa.gov/mo/town-flood-and-superfund-looking-back-times-beach-disaster-nearly-40-years-later

Ghadban S, Shames M, Abou Mayaleh H (2017) Trash crisis and solid waste management in Lebanon-Analyzing Hotels' commitment and guests' preferences. J Tourism Res Hospitality 6(3):1000169. https://doi.org/10.4172/2324-8807.1000171

Ghaffariraad M, Ghanbarzadeh LM (2021) Landfill leachate treatment through coagulation-flocculation with lime and bio-sorption by Walnut-shell. Environ Manage 68(2):226–239. https://doi.org/10.1007/s00267-021-01489-4

Ghanbarzadeh LM, Sabour MR, Amiri A, Rabbani O (2012) Application of quadratic regression model for Fenton treatment of municipal landfill leachate. Waste Manage 32(10):1895–1902. https://doi.org/10.1016/j.wasman.2012.05.020

Ghanbarzadeh LM, Sabour MR, Ghafari E, Amiri A (2018) Energy consumption and relative efficiency improvement of Photo-Fenton—optimization by RSM for landfill leachate treatment, a case study. Waste Manage 79:58–70. https://doi.org/10.1016/j.wasman.2018.07.029

Henry JG, Heinke GW (1989) Solid wastes. In: Henry JG, Heinke GW (eds) Environmental science and engineering 1st Edn. Prentice-Hall International (UK) Limited, London

Hirpe L, Yeom C (2021) Municipal solid waste management policies, practices, and challenges in Ethiopia: a systematic review. Sustainability 13(20):11241. https://doi.org/10.3390/su132011241

Hurd DJ (1992) Feasibility study for the implementation of consumer dry cell battery recycling as an alternative to disposal: "getting a charge out of the Wastestream": Council of State Governments

Johnson R, Hirth C (1990) Myths and realities: collecting household batteries. Waste Age 21(6):48–48

Kathiravale S, Yunus MNM, Sopian K, Samsuddin AH, Rahman RA (2003) Modeling the calorific value of Municipal Solid Waste☆. Fuel 82(9):1119–1125. https://doi.org/10.1016/S0016-2361(03)00009-7

Kaza S, Yao L, Bhada-Tata P, Van Woerden F (2018) What a waste 2.0: a global snapshot of solid waste management to 2050. World Bank Publications

Kim KH (1998) An analysis of optimum number of response categories for Korean consumers. J Glob Acad Mark Sci 1(1):61–86. https://doi.org/10.1080/12297119.1998.9707386

Komilis D, Evangelou A, Giannakis G, Lymperis C (2012) Revisiting the elemental composition and the heating value of the organic fraction of municipal solid wastes. Waste Manage 32(3):372–381. https://doi.org/10.1016/j.wasman.2011.10.034

Kwatra S, Pandey S, Sharma S (2014) Understanding public knowledge and awareness on e-waste in an urban setting in India: a case study for Delhi. Manag Environ Q Int J 25(6):752–765. https://doi.org/10.1108/MEQ-12-2013-0139

Lavigne F, Wassmer P, Gomez C, Davies TA, Sri Hadmoko D, Iskandarsyah TYW, Gaillard JC, Fort M, Texier P, Heng MB, Pratomo I (2014) The 21 February 2005, catastrophic waste avalanche at Leuwigajah dumpsite, Bandung, Indonesia. Geoenviron Disasters 1(1):1–12. https://doi.org/10.1186/s40677-014-0010-5

Leonard B (ed) (2003) Solid waste management and greenhouse gases: a life-cycle assessment of emissions and sinks. DIANE Publishing

Matsumoto S (2011) Waste separation at home: are Japanese municipal curbside recycling policies efficient? Resour Conserv Recycl 55(3):325–334. https://doi.org/10.1016/j.resconrec.2010.10.005

Moh YC, Abd Manaf L (2014) Overview of household solid waste recycling policy status and challenges in Malaysia. Resour Conserv Recycl 82:50–61. https://doi.org/10.1016/j.resconrec.2013.11.004

Nguyen TTP, Zhu D, Le NP (2015) Factors influencing waste separation intention of residential households in a developing country: evidence from Hanoi, Vietnam. Habitat Int 48:169–176. https://doi.org/10.1016/j.habitatint.2015.03.013

Rhyner CR, Schwartz LJ, Wenger RB, Kohrell MG (1995) Waste management and resource recovery, 1st ed. CRC-Press

Rousta K, Bolton K, Lundin M, Dahlén L (2015) Quantitative assessment of distance to collection point and improved sorting information on source separation of household waste. Waste Manage 40:22–30. https://doi.org/10.1016/j.wasman.2015.03.005

Sabour MR, Ghanbarzadeh LM, Rabbani O (2010) Evaluation of the main parameters affecting Fenton oxidation process in municipal landfill leachate treatment. Waste Manage Res 29(4):397–405. https://doi.org/10.1177/0734242X10375332

Saffron L, Giusti L, Pheby D (2003) The human health impact of waste management practices: a review of the literature and an evaluation of the evidence. Manag Environ Q Int J 14(2):191–213. https://doi.org/10.1108/14777830310470422

State of New York, Department of Health (1978, revised 2018) Love canal—public health time bomb, a special report to the governor and legislature. Available at (access in July 2023): https://www.health.ny.gov/environmental/investigations/love_canal/lctimbmb.htm

Tchobanoglous G, Theisen H, Vigil SA (1993) Integrated solid waste management: engineering principles and management issues (No Title)

US Environmental Protection Agency (2018) Advancing sustainable materials management: 2018 tables and figures, assessing trends in materials generation and management in the United States. Available at (access in July 2023): https://www.epa.gov/facts-and-figures-about-materials-waste-and-recycling/national-overview-facts-and-figures-materials#Trends1960-Today.

Worrell WA, Vesilind PA, Ludwig C (2016) Solid waste engineering: a global perspective. Cengage Learning

Worthley JA, Torkelson R (1981) Managing the toxic waste problem: lessons from the love canal. Adm Soc 13(2):145–160. https://doi.org/10.1177/009539978101300202

Yean Yng Ling F, Song Anh Nguyen D (2013) Strategies for construction waste management in Ho Chi Minh city, Vietnam. Built Environ Project Asset Manag 3(1):141–156. https://doi.org/10.1108/BEPAM-08-2012-0045

Zhou H, Meng A, Long Y, Li Q, Zhang Y (2014) An overview of characteristics of municipal solid waste fuel in China: physical, chemical composition and heating value. Renew Sustain Energy Rev 36:107–122. https://doi.org/10.1016/j.rser.2014.04.024

Chapter 3
Characteristics and Impact Assessment of Municipal Solid Waste (MSW)

Mohammed Zari

Abstract Landfilling and open dumping have been the most widely used municipal solid waste (MSW) management strategies over the past decades due to lower costs and less treatment efforts. However, landfilling continues to be an acceptable final disposal strategy for MSW. Non-sanitary landfill sites (lacking modern environmental technology) that rely on direct landfilling without waste pre-treatment pose a significant threat to the entire ecosystem. Many of these sites are situated in environments adjacent to residential districts or water bodies, affecting the local environment and population health. The assessment of the environmental impact of landfills is thus a key issue in the literature that has recently received more attention as a result of growing environmental concerns. This chapters reviews the major impacts due to waste mismanagement worldwide, particularly in developing countries where the unsustainable management of solid waste is common. Potential analytes and composition found in MSW landfills and their related risk assessment are also highlighted in this book chapter with a focus on heavy metals accumulation. This narrative literature review assessed global issues due to improper waste management and showing how potential contaminants found within landfills waste can affect the environment and human health. It can also be of reference for scholars and stakeholders to evaluate and estimate the potential risk posed by poor MSW management in order to improve sustainability at a global level.

Keywords Municipal solid waste · Heavy metals · Environmental impact assessment · Health impact assessment

M. Zari
Chemical and Environmental Engineering Department, Faculty of Engineering, University of Nottingham, Coates Building, University Park, Nottingham NG7 2RD, UK

M. Zari (✉)
Department of Environment, Faculty of Environmental Sciences, King Abdulaziz University, Jeddah 21589, Saudi Arabia
e-mail: mtzari@kau.edu.sa

A. Anouzla and S. Souabi (eds.), *Technical Landfills and Waste Management*, Springer Water, https://doi.org/10.1007/978-3-031-52633-6_3

3.1 Introduction

Waste is an unavoidable by-product of human activity, and rising living standards have led to increases in the quantity and complexity of generated waste (Rathi 2006). The rate of municipal solid waste (MSW) output will rise dramatically in the coming decades (Hoang et al. 2022; Naveenkumar et al. 2023). Handling and disposing of MSW is a serious issue, especially in developing nations where it poses significant threats to the environment and society (Varjani et al. 2022). Due to lower costs, lack of technologies, and an absence of policy framework, landfilling and open dumping are the most commonly applied strategies for MSW management for the past decades (Smart-Ground 2018; Vaverková 2019; Weng et al. 2015). Nonetheless, landfilling is still utilized and acceptable final disposal method for MSW In some countries (Jovanov et al. 2018; Qian et al. 2021). Non-engineered landfill sites that are reliant on direct landfilling without pre-treatment of waste are of major concern to the whole ecosystem (water, air, soil, and land) (Fig. 3.1) (Chandana et al. 2020; Zoungrana et al. 2022). These landfill sites are the primary sources of long-term landfill gas emissions, which contribute to global warming and acidification (Danthurebandara et al. 2015). They are also the primary source of groundwater pollution due to toxic chemical leaching (Van Passel et al. 2013). Many of these sites are located in semi-urban environments in close proximity to water bodies or residential areas, rising concern to many experts in the landfill sector (Brand et al. 2018; Maheshi 2015; Nguyen et al. 2018).

Waste management policy has evolved rapidly over recent time as a result of the significantly increased MSW waste volume (Ferronato and Torretta 2019). The Landfill Directive (European Council 1999) and, later the Waste Framework Directive (2008/98/EC) (Directive 2008), are two significant EU Directives that have had a significant impact on waste management practices (Jones et al. 2013; Laner et al. 2012; Ortner et al. 2014). Reducing the negative impacts of landfill activities on human health and the environment is a major emphasis of the both Directives (Smart-Ground 2018). They encourage a shift away from waste disposal and toward more environmentally friendly waste and resource management to be in line with the circular economy (Zhang et al. 2023).

local authorities in countries with good environmental performance encourage closure of these landfills to reduce risks and construct new sanitary landfills with modern design of engineering systems (Burlakovs et al. 2017). The evaluation of the environmental impact of landfills is therefore a critical topic in the literature that has recently garnered more attention due to growing environmental concerns (Havukainen et al. 2017). Hence, extensive risk assessments are required to inform the implementation of intervention plans for the protection of workers' health and adjacent residents.

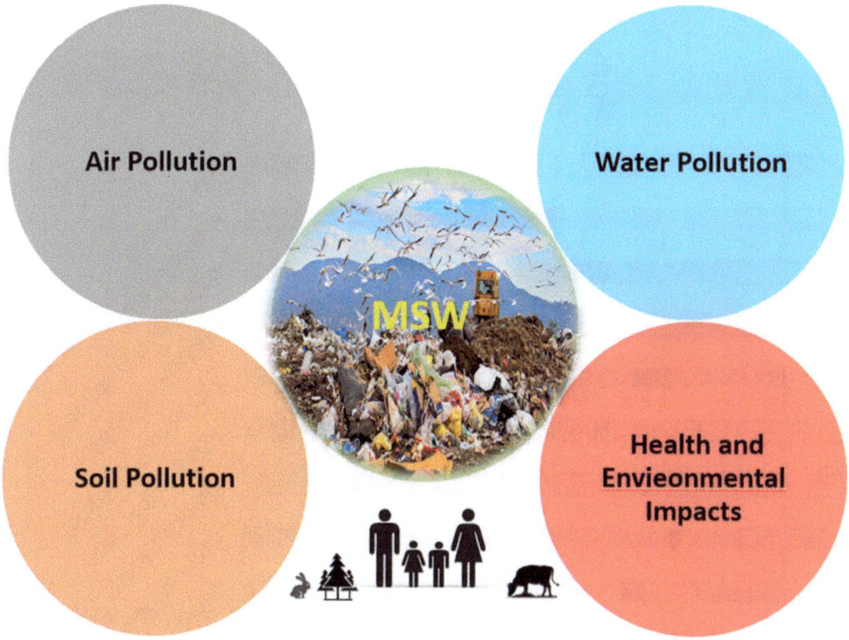

Fig. 3.1 Potential impact of landfills on the environment

3.2 Characteristics of MSW

MSW refers to waste generated by municipalities and local authorities. Waste that falls in this category stems from both residential and commercial activities. MSW contains a wide range of materials, including, papers, organics, plastics, glass, metals, and other miscellaneous items (Kaza et al. 2018). Because of their high organic content (between 40 and 60%), MSWs offer an excellent opportunity for energy recovery when used in conjunction with the right waste-to-energy technology (Varjani et al. 2022), It should be highlighted, nonetheless, if the waste was not appropriately handled, high organic contents could result in severe environmental pollution (Varghese et al. 2020). The composition of MSW might differ from one community to the next based on the country's waste consumption and factors, such as geography, economic status, and culture (Naveenkumar et al. 2023; Zhang et al. 2022). Waste consumption of different regions is shown in Fig. 3.2. It can be clearly seen that the wealthier and more developed a city is, the more waste it generates. Richer cities typically have more complicated waste compositions, while poorer countries generally have more organic fractions within their waste.

MSW can also include industrial waste, construction and demolition waste, and agricultural waste (Mukherjee et al. 2020). According to Varjani et al. (2022). Global MSW generation is anticipated to rise from 2.02 billion metric tons in 2016 to 3.4 billion metric tons in 2050. The globally distribution of MSW production is shown

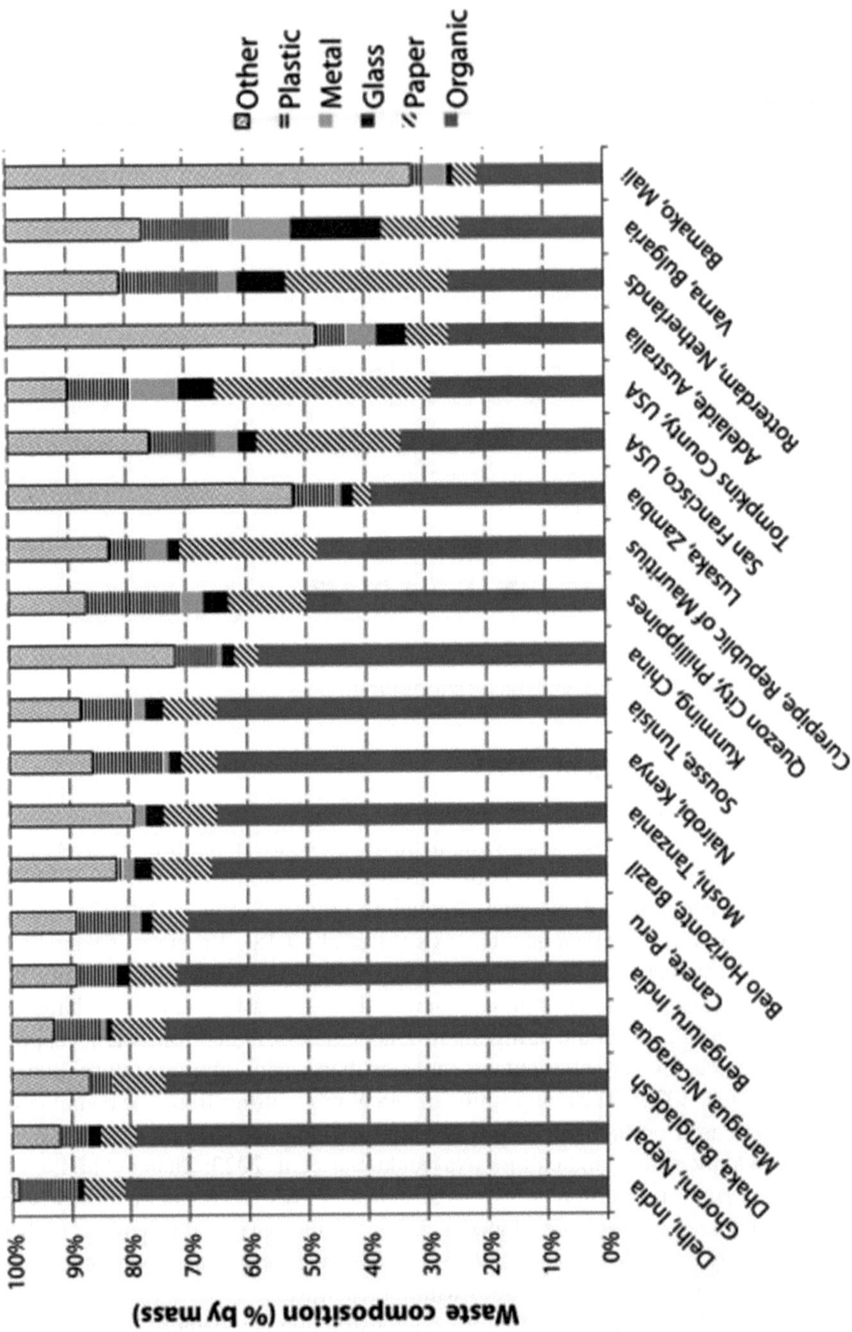

Fig. 3.2 Waste composition for 20 different cities worldwide (Vergara and Tchobanoglous 2012)

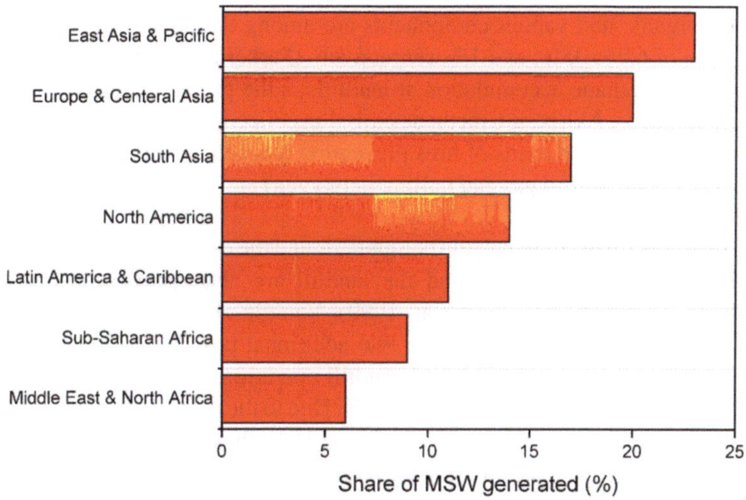

Fig. 3.3 Worldwide MSW production trend in 2016 shown by major regions (Varjani et al. 2022)

in Fig. 3.3. Therefore, constant efforts should be carried out to reduce waste as a preventive measure.

MSW can also be classified according to its physical, biological, and chemical constituents. For example, the main metrics used for the physical characterization of MSW are moisture content, density, and particle size (Naveenkumar et al. 2023). The chemical composition present in MSW includes phosphorous, pH, electrical conductivity, potassium, volatile solid, nitrogen/carbon-to-nitrogen ratio, and elemental/heavy metals analysis (Naveenkumar et al. 2023; Roy et al. 2022). Metrics known for leachate quality are represented in the biological parameters (also classified as physiochemical parameter), such as BOD, COD, etc. (Mor et al. 2006; Naveen et al. 2017).

It is vital to understand the quantity, composition, and characteristics of MSW in order to properly handle and manage these wastes (Roy et al. 2022; Zhu et al. 2021).

3.3 Potential MSW Contaminants

MSW landfills are considered an environmental threat because of the production of landfill gas and leachate (Singh and Chandel 2020). Although a landfill site comprises of many components, landfill gas is the most significant emission (Macklin et al. 2011). When waste materials are dumped in a landfill, the biodegradable fractions begin to break down through aerobic and anaerobic processes. These processes involve intricate physical, chemical, and biological transformations, and as a result, landfill gas and leachate are released (Danthurebandara 2015). Landfill gas contains several gases depending on the type of waste. Methane, carbon dioxide, and trace organic and vapour components are among the bulk gases, with relative proportions of 65%, 35%, and 1%, respectively (Parker et al. 2002). Because of its flammability, methane accumulation at landfills is the primary cause of explosions (Weng et al. 2015). As a result, methane emission collection and reduction are important for public health in terms of fires prevention (caused by waste streams or often over extraction of landfill gas and pulling in oxygen) and global warming mitigation (Frändegård et al. 2013).

The evaporation of some volatile compounds and the exchange of gaseous components between the atmosphere and the landfill are also contributors to varying landfill gas composition (Fisher et al. 1999). Volatile organic compounds (VOCs) and hydrogen sulphide (H_2S) are also two additional possible gases that might be discharged into the atmosphere during landfill operations (e.g., landfill mining) and should not be overlooked (Ziyang et al. 2015). Due to the potentially explosive nature of methane and the toxic effects of VOCs and H_2S, these landfill gases can have a substantial negative impact on the environment and human health (Smart-Ground 2017).

High quantities of trace metal (loid)s, such as, Cd, Cr, Cu, Hg, Pb, Sb, and Zn, as well as inorganic salts (such as chlorides and sulphates), are frequently found in solid waste (Assi et al. 2020). These metals have a toxic impact on human health with a tendency for bioaccumulation (Singh and Chandel 2022). Some researchers believe that heavy metals contamination in landfills can reach 51%. This large weighting factor highlights the potential environmental harm posed by heavy metals contamination (Abu-Daabes et al. 2013).

In additions, minerals and natural inorganic fibers can also be found within the MSW, resulting in localized environmental and health impacts when released (Warren and Read 2014), Some of these materials, such as asbestos, crystalline silica, and quartz dust, can be harmful to human health when inhaled in high doses (Smart-Ground 2017).

Persistent organic pollutants (POPs) are a diverse group of chemicals generated by industrial processes that can be disposed of in MSW landfills (Sun et al. 2020). Polybrominated diphenyl ethers, polybrominated biphenyls, and polychlorinated biphenyls are a few examples. Even at extremely low concentrations, POPs in water pose a significant harm to the environment and human health (Nomngongo et al. 2012). These contaminants exhibit high resistance to chemical or biological degradation, high environmental mobility, and a strong bioaccumulation in the food chain (Weber et al. 2011).

Moreover, according to (Nair 2021), MSW landfills consists of nutrients for the growth of harmful microorganisms, which are aerosolized into the atmosphere as a result of local meteorology and different waste disposal activities. Bioaerosols emitted from landfills can cause negative health effects for workers and residents near landfill sites.

3.4 Conceptual Site Model

Prior to conducting a risk assessment, it is critical to develop a conceptual model that assists in the identification of potential pollutants linkage (Defra 2012). Environmental risk assessment is based on three critical elements that form a pollutant linkage. These key elements are:

- **Source**: this refers to the contaminant;
- **Pathway**: the route through which the contaminant can migrate;
- **Receptor**: this refers to any organism, including humans, plants, animals, and properties or controlled water, that is likely to be affected by the contaminants.

There is no risk considered on any site if any of these elements is missing (Defra 2012). An example of a generic conceptual site model is shown in Fig. 3.4. The conceptual site model depicted the region of concern, contamination sources, potentially affected environmental media, and the processes that regulate the transport of contaminants to possible receptors.

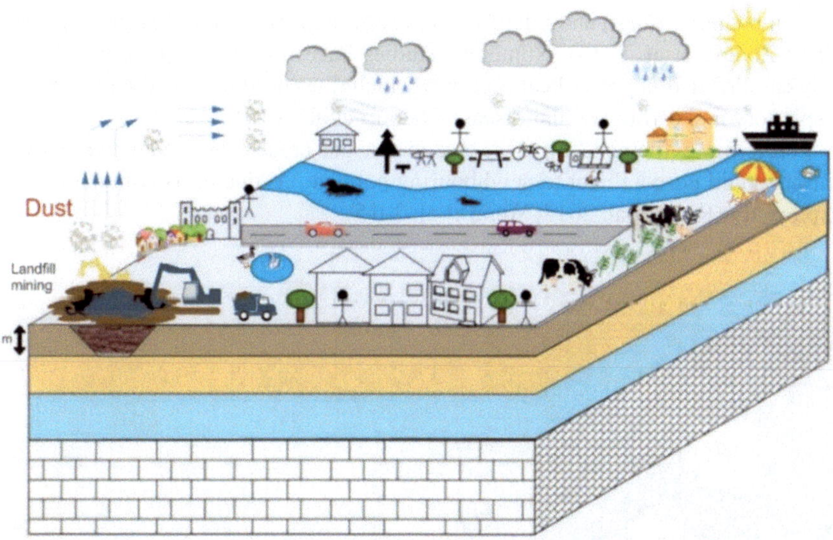

Fig. 3.4 Generic block diagram of the conceptual site model. Produced by using CorelDRAW Software

3.5 Impacts on Air

Landfills and its associated activities represented in landfill mining (LFM) can greatly increase air pollution If not properly managed. Due to anthropogenic activities or natural events, dust particles can become airborne (Smart-Ground 2017). Dust particles from landfills are likely to be air dispersed from activities, such as vehicle movements on site, mechanical recycling and composting, the action of tipping waste particularly on elevated ground, and waste unloading and sorting as shown in Fig. 3.5 (Ilse et al. 2018).

The most serious issue associated with landfill sites is the formation of fine dust, because fine fractions of soil-like materials ranging in size from < 10 mm to > 2 mm can account for up to 40–80 wt.% of total waste excavated (Chandana et al. 2020; Jia et al. 2013; Somani et al. 2018). This was demonstrated by a recent analysis of nine landfill sites in the UK, which revealed that fine soil-like material accounted for 30–74% (w/w) of the total waste recovered (Wagland et al. 2019). Heavy metals such as lead, cobalt, arsenic, manganese, cadmium, chromium, and copper may be present in particulate matter (PM) in the form of dust emitted from landfill sites (Hogland et al. 2014; Rodriguez et al. 2018). Furthermore, landfill waste dust can contain bacteria, fungi, and microbial toxins in the form of bioaerosols (Kim et al. 2018; Ziyang et al. 2015). Numerous studies have revealed correlations between airborne PM and negative health impacts, including morbidity and mortality from respiratory and cardiovascular disease (Tsiouri et al. 2015).

Furthermore, methane, a potent greenhouse gas created by organic waste degradation, could pose a threat to neighbouring populations due to its flammability and explosivity (Rong et al. 2017; Weng et al. 2015).

Fig. 3.5 Potential release of Greenhouse gases and fine dust containing potentially toxic elements

Because some airborne PM can be detrimental to human health in high concentrations (e.g., asbestos fibers, heavy metals, and quartz dusts), their likely existence and release in air must be carefully evaluated before initiating LFM activities. In fact, if contamination is found, more thorough studies must be carried out, and mining activities must be controlled to reduce the risk of exposure (Smart-Ground 2017). Due to the heterogeneity of landfilled waste, serious environmental and health issues could result from the dust emissions (Bastian et al. 2018). Large volumes of these emissions have significant negative health impacts on workers and people living in close proximity (Douglas et al. 2017).

3.6 Impacts on Soil

A number of sites in Europe have been highlighted in recent years as requiring further attention or intervention mainly because the high potential risk associated with different mining activities (Smart-Ground 2017). Additionally, several investigations found that the soils near MSW landfills were contaminated by trace metals, posing an ecotoxicological risk (de Souza et al. 2023). Heavy metals stand out among MSW pollutants in MSW owing to their harmful and bioaccumulative potential. Aside from endangering water and soil, they can jeopardize food safety, impact crops in the area, and pose a health concern, particularly for individuals living near a waste disposal site (Ma et al. 2018). Figure 3.6 illustrates the potential human health impact of soil heavy metals pollution in different land use types.

Due to the interaction of mineral particles with waste and leachate, landfill soils typically include excessive amounts of salts, nutrients, and heavy metals (Adelopo et al. 2018). Thus, the accumulation of these contaminants in landfill soil, particularly heavy metals, can be transferred to crops, surface and groundwater ecosystems, and

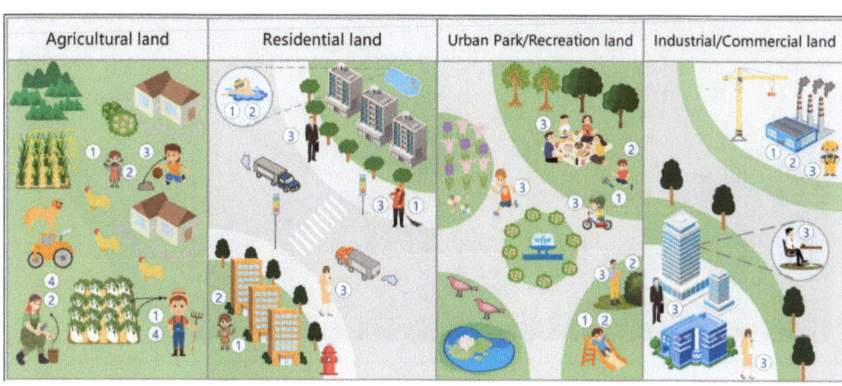

Fig. 3.6 Health risk potential impact under different land use types (1) ingestion of soil; (2) dermal contact of soil; (3) inhalation of soil; (4) ingestion of homegrown food) (Yang et al. 2022)

eventually the food chain, harming the health of people and ecosystems (Sagbara et al. 2020).

As a result, heavy metal concentrations in the soil surrounding a landfill must be regularly monitored, even after the landfill has been inactivated, because leachate is produced and has an impact on the ecosystem for a lengthy period of time (Marinho et al. 2022).

3.7 Impacts on Water

MSW is often related to water resources pollution, if not well managed (Siddiqua et al. 2022). Although wastes are an unavoidable part of daily life, they may pose long-term concerns to groundwater and surface waterways that are hydrologically linked. European regulations were put into place to maintain the purity of groundwater, and some landfills were forced to employ liners to collect and treat leachate (the water solution that is produced when water percolates through a landfill) (Laner et al. 2012). MSW landfills are also bordered by ditches, which are designed to restrict surface water transfer between landfills and the surrounding areas. However, since many disposal sites are old and hence exempted from these regulations, the major impact on water resources remains leachate formation (de Souza et al. 2023) (Fig. 3.7).

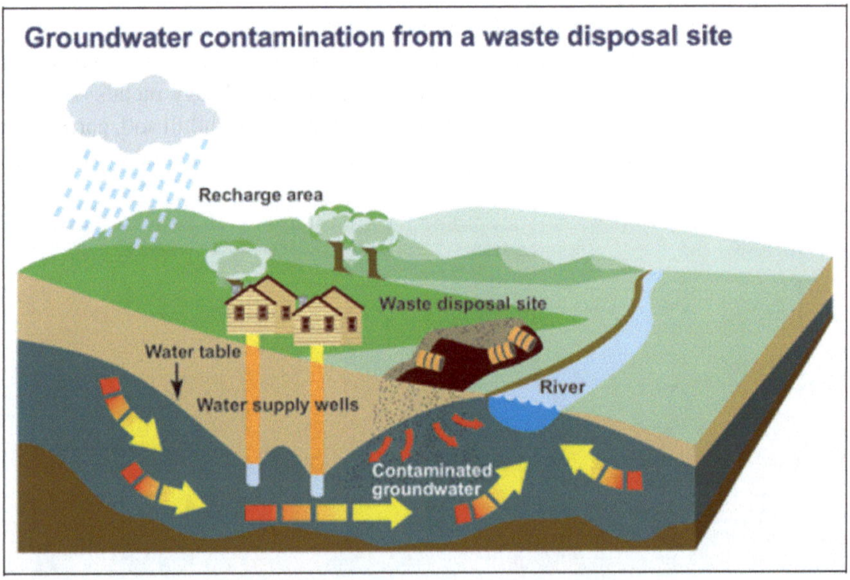

Fig. 3.7 Simplified scheme of groundwater and surface water contamination from a waste disposal site (Smart-Ground 2017)

The potential of groundwater pollution is likely the most serious environmental effect from landfills since the majority of landfills were historically built without designed liners and leachate collection systems (Przydatek and Kanownik 2019). Regulations in a number of countries have more recently enforced the installation of liners, leachate collection systems, and a leachate treatment plan (Kjeldsen et al. 2002). An impermeable geotextile lining for landfill pits has been required by the EU Water Quality Directives for groundwater protection and disposal since 1993 (Smart-Ground 2017).

Nitrate, ammonia, certain metals, chlorinated solvents, are some examples of potential contaminants typically linked to the impact on water in MSW. These compounds are also most likely to be released during landfill excavation as illustrated previously in Fig. 3.4 in a form of dust, and there is a potential risk that these chemicals will drift into natural waters in the landfill's vicinity through atmospheric deposition (Ziyang et al. 2015). As a result, concentrations of water contamination should be closely compared to relevant standards and legislation.

3.8 Heavy Metal Pollution Assessment

In general, pollution indicators are the most effective and suitable tools for assessing heavy metal pollution in soil, sediment, and waste (Doležalová Weissmannová et al. 2019). There are many pollution indicators (also called pollution indices) that can be used in the heavy metal pollution assessment, the most commons indices are: geo-accumulation index (I_{geo}), contamination factor (CF), enrichment factor (EF), ecological risk factor (Er), and pollution load index (PLI). The use of these indices is dependent on the aim of a study and the suitability of data input. A detailed description of the role of each indicator is provided below.

3.8.1 Geo-Accumulation Index

I_{geo} is used to examine the contamination level of a sample affected by heavy metals. It is a geochemical criterion (unitless) defined by (Muller 1969) and has been frequently used in European trace metal studies (Li et al. 2014). This index compares the current state of metal contamination to its background concentration in samples (Gujre et al. 2021). It is calculated using the following Eq. (3.1):

$$I_{geo} = \log_2\left(\frac{C_n}{1.5B_n}\right) \tag{3.1}$$

where Cn is the concentration of a selected heavy metal while, Bn is the background concentration of that element. A constant of 1.5 is introduced to minimise potential differences in background values, referred to as lithogenic variations. I_{geo} values can be classified into seven different classes: no pollution ($I_{geo} \leq 0$), no to moderate pollution ($0 < I_{geo} \leq 1$), moderate pollution ($1 < I_{geo} \leq 2$), moderate to heavy pollution ($2 < I_{geo} \leq 3$), heavy pollution ($3 < I_{geo} \leq 4$), heavy to extreme pollution ($4 < I_{geo} \leq 5$), or extreme pollution ($I_{geo} > 5$) (Thongyuan et al. 2020).

3.8.2 Contamination Factor

CF is a single element index indicator used to estimate the waste sample contamination by heavy metals. The CF is obtained by dividing the concentration of heavy metals in a sample by their background concentrations (Chen et al. 2015). The CF is (unitless) and calculated according to Eq. (3.2) (Hakanson 1980):

$$CF = \frac{C_i}{B_i} \tag{3.2}$$

where C_i is the concentration of the analysed heavy metal and B_i is the geochemical background value of that metal. The pollution levels of the CF are divided into the following classes: CF < 1: low degree of pollution; 1 ≤ CF < 3: moderate degree of pollution, 3 ≤ CF < 6: considerable degree of pollution, and CF ≥ 6: very high degree of pollution (Thongyuan et al. 2020).

3.8.3 Enrichment Factor

EF is also a single element index used to study the effect of anthropogenic activities on heavy metals contamination. It also takes into account the lithogenic input by normalizing the metal concentration with a reference metal such as iron (Fe), manganese (Mn), titanium (Ti), and vanadium (V) (Brady et al. 2014; Gujre et al. 2021). Equation (3.3) is used for the calculation of EF which is as follows:

$$EF = \left(\frac{\frac{Ci_{sample}}{Cr^{ref}_{sample}}}{\frac{Bi_{crust}}{Br^{ref}_{crust}}} \right) \tag{3.3}$$

where $(Ci/C\ ref)_{sample}$ and $(Bi/Br\ ref)_{crust}$ denote the concentration levels of i metal and the iron metal in the sample and crust. The most common reference metals are Sc, Mn, Ti, Al, and Fe (Siddque et al. 2023). EF values are classified into five levels based on the influence of human activities (Faragó et al. 2023; Sakunkoo et al. 2022) as follows: (EF < 1) not influenced by human activities which indicates no enrichment, (1 ≤ EF < 2) = deficiency to minimal enrichment, (2 ≤ EF < 5) =

moderate enrichment, (5 ≤ EF < 20) = significant enrichment, (20 ≤ EF < 40) = very high enrichment, (EF ≥ 40) = extremely high enrichment.

3.8.4 Ecological Risk Assessment

Er is proposed by Swedish scientist Hakanson (1980) and can be used to evaluate the potential negative impact on the ecosystem of a study area. According to (Singh and Chandel 2022), the primary function of the risk index is to indicate the most contaminated heavy metals and prioritize the contaminated sites for remediation measures. Ecological risk assessment can be calculated using Eqs. (3.4) and (3.5) which are as follows:

$$E_r^i = T_r^i \times CF \tag{3.4}$$

$$RI = \sum E_r^i \tag{3.5}$$

where E_r^i is the potential ecological risk factor of a single element. T_r^i is the toxic response factor for the given element, indicated as follows: Mn = Zn = 1, Cr = 2, Co = Cu = Ni = Pb = 5, Sb = 7, As = 10, and Cd = 30 (Faragó et al. 2023; Hakanson 1980; Wang et al. 2018). CF represents the single index of contamination defined as the ratio of the measured concentration of the metal (loid) in the soil (Ci) to its background concentration (Bi), explained previously in Eq. (3.2). The potential ecological risk index (RI) is the sum of the ecological risk factor (Er) for all heavy metals. The pollution degree of E_r^i can be graded as follows: low pollution ($E_r^i < 40$), moderate pollution ($40 \leq E_r^i < 80$), considerable pollution ($80 \leq E_r^i < 160$), high pollution ($160 \leq E_r^i < 320$), very high pollution ($E_r^i \geq 320$). Similarly, RI is classified into 4 grades: low risk (RI < 150), moderate risk (150 ≤ RI < 300), high risk (300 ≤ RI < 600) and very high risk (RI ≥ 600) (Gujre et al. 2021).

3.8.5 Pollution Load Index

To assess the overall pollution, the PLI provides an established approach for calculating the accumulation of heavy metals in samples (Wang et al. 2020). The PLI (unitless) can be obtained by calculating the geometric mean of the CFs of each element analysed (Zari et al. 2022). The PLI of heavy metals is calculated using Eq. (3.6):

$$PLI = (CF_1XCF_2XCF_3X\ldots XCF_n)^{1/n} \tag{3.6}$$

where n is the number of analysed heavy metals and CF is the contamination factor of each metal. According to contamination degree, the PLI is classified as unpolluted (PLI < 1), unpolluted to moderately polluted ($1 \leq$ PLI ≤ 2), moderately polluted ($2 \leq$ PLI ≤ 3), moderately to highly polluted ($3 \leq$ PLI ≤ 4), highly polluted ($4 \leq$ PLI ≤ 5), or very highly polluted (PLI > 5) (Qing et al. 2015).

3.9 Health Risk Assessment

A human health risk assessment is used to evaluate the potential health effects of chemical exposure in polluted environmental media (Yang et al. 2022). It is often used to estimate the health impacts of heavy metals as a result of chemical exposure (Chen et al. 2022). In human health risk assessments, heavy metal quantification has been classified as non-carcinogenic or carcinogenic by the (USEPA 2002; Wang et al. 2020). According to the recommendations and methodology of the (USEPA 2002), exposure of humans to heavy metals from soil is estimated through three main exposure routes: ingestion of substrate dust particles, inhalation of suspended dust particles through mouth/nose, and dermal contact/absorption of heavy metals in particles adhered to exposed skin (Yang et al. 2019). The hazard quotient (HQ) is often used to characterize the non-carcinogenic risk effect of a chemical, which is defined as the ratio of the average daily intake to the toxicity threshold value (also known as the reference dose) for the same exposure (Yan et al. 2022). It is further defined by the hazard index (HI), which calculates the overall risk of non-carcinogenic implications (Doležalová Weissmannová et al. 2019). The HQ and HI are both unitless, expressing an individual's potential for encountering negative impacts. The HQ of each chemical is the ratio of the ADI of an element to its reference dose (RfD) for the same exposure pathway to yield a non-carcinogenic HQ, which is determined using Eq. (3.7). HI is the sum of the hazard quotient and indicates the cumulative noncarcinogenic risk and is determined according to Eq. (3.8):

$$HQ = \frac{ADI}{RfD} \tag{3.7}$$

$$HI = \sum HQ_i \tag{3.8}$$

where ADI (mg kg/ day) is the average daily intake through each exposure pathway to soil: ingestion (ADI_{ing}), inhalation (ADI_{inh}), and dermal contact (ADI_{dermal}). RfD is the corresponding reference toxicity threshold dose (mg kg/ day). If HQ or HI < 1, there is no non-carcinogenic risk, whereas HQ or HI $\geq$ 1, adverse health effects may occur, and the likelihood of effects increases as the HQ/HI value increases. The formula for the daily exposure to these three pathways can be calculated according to Eqs. (3.9), (3.10), and (3.11) for exposure assessment:

$$ADI_{ing} = \frac{C \times IngR \times EF \times ED \times CF}{BW \times AT} \tag{3.9}$$

$$ADI_{inh} = \frac{C \times InhR \times EF \times ED}{BW \times AT \times PEF} \tag{3.10}$$

$$ADI_{dermal} = \frac{C \times SA \times SL \times ABS \times EF \times ED \times CF}{BW \times AT} \tag{3.11}$$

where C represents the concentration of heavy metals (mg/kg), IngR is the ingestion rate (mg/day), EF is the exposure frequency (day/year), ED is the exposure duration (years), CF is the conversion coefficient 1.0×10^{-6} (kg/mg), BW is the standard body weight (kg), AT is the average exposure time (day), InhR is the inhalation rate (m^3/day), PEF is the particulate emission factor (m^3/kg), SA is the skin exposed area (cm^2/day), and SL is the skin adherence factor (mg/cm^2). ABS is the skin absorption factor.

Carcinogenic risk (CR) is estimated by calculating the incremental probability of an individual developing cancer over a lifetime due to exposure to the potential carcinogen (Li et al. 2014). The CR and lifetime cancer risk (LCR) are calculated using Eqs. (3.12) and (3.13), respectively:

$$CR = ADI \times SF \tag{3.12}$$

$$LCR = \sum CR \tag{3.13}$$

where CR is the unitless probability of an individual developing cancer over a lifetime. ADI indicates average daily intake (mg/kg/day) of each element, and SF indicates the cancer slope factor (mg/kg/day). The slope factor converts the estimated daily intake of a toxin averaged over a lifetime of exposure directly to the incremental risk of an individual developing cancer. The LCR is the total of the CR values and represents the overall risk. Risks surpassing 1×10^{-4} are viewed as unacceptable, risks below 1×10^{-6} are not considered to pose significant health effects, and risks lying between 1×10^{-4} and 1×10^{-6} are generally considered an acceptable range, depending on the situation and circumstances of exposure (Wang et al. 2022). Yang et al. (2022) provide detailed information of RfD and SF parameters of various heavy metals.

In some countries such as, USA and UK that have developed their own soil screening level and soil guideline values (SGVs), respectively, can directly use their soil threshold values in the health risk model assessment. For example, (Zari et al. 2022) has integrated the UK SGVs in the health risk exposure assessment from a MSW landfill in the view of LFM. Many countries, however, have not established their own health risk assessment (HRA) framework, and the majority of existing research have directly related to the USEPA risk assessment model and parameters. For those countries who do not present an original HRA framework, learning from the experience of developed countries is critical for advancing their own HRA system because the USEPA calculation (exposure factors and parameters used for health risk assessment of heavy metals) might not be suitable for other countries' practical situation and thereby influenced by several of uncertainty factors.

The exposure dose of receptors for different exposure pathways is estimated using the concentration of metals in soil (C). C had the greatest impact on the risk output, according to previous sensitive analysis (Yang et al. 2019). A significant amount of sampling data is required to provide a comprehensive insight of the national and regional health risks of soil metals, given that metal concentrations can vary by many orders of magnitude between different regions (Yang et al. 2022).

3.10 Conclusion

The article presented a narrative review about the potential impacts that MSW mismanagement can have on the environment and human health. Potential contaminants and compositions that exist within MSW landfills were highlighted and discussed. Additionally, physiochemical parameters associated with MSW characterisation were identified from several studies. Furthermore, pollutant linkages of possible contaminants were thoroughly addressed in different environmental compartments, ultimately impacting human health and the environment. Moreover, an overview of heavy metals pollution and health risk assessment were extensively discussed from multiple perspectives. Stakeholders and local authorities need to be aware of the complex system of solid waste management which involve environmental and social concerns that should be comprehensively evaluated in order to optimize the life cycle of waste and reduce air, water, and soil contamination caused by open dumping and/or landfilling, both of which are common practices around the world.

References

Abu-Daabes M, Qdais HA, Alsyouri H (2013) Assessment of heavy metals and organics in municipal solid waste leachates from landfills with different ages in Jordan. J Environ Prot 4:344

Adelopo A, Haris PI, Alo B, Huddersman K, Jenkins R (2018) Multivariate analysis of the effects of age, particle size and landfill depth on heavy metals pollution content of closed and active landfill precursors. Waste Manage 78:227–237

Assi A, Bilo F, Zanoletti A, Ponti J, Valsesia A, La Spina R et al (2020) Zero-waste approach in municipal solid waste incineration: Reuse of bottom ash to stabilize fly ash. J Clean Prod 245:118779

Bastian K, Alexander M, Daniel H, Robert R (2018) Landfill mining of a mixed municipal solid waste and commercial waste landfill: application of existing processing technology—opportunities and limitations In: Jones PT, Machiels L (eds) 4th International Symposium On Enhanced Landfill Mining, Mechelen, Belgium

Brady JP, Ayoko GA, Martens WN, Goonetilleke A (2014) Enrichment, distribution and sources of heavy metals in the sediments of Deception Bay, Queensland, Australia. Marine Pollut Bull 81:248–255

Brand JH, Spencer KL, O'shea FT, Lindsay JE (2018) Potential pollution risks of historic landfills on low-lying coasts and estuaries. Wiley Interdiscip Rev Water 5:e1264

Burlakovs J, Kriipsalu M, Klavins M, Bhatnagar A, Vincevica-Gaile Z, Stenis J et al (2017) Paradigms on landfill mining: from dump site scavenging to ecosystem services revitalization. Resour Conserv Recycl 123:73–84

Chandana N, Goli VSNS, Mohammad A, Singh DN (2020) Characterization and utilization of landfill-mined-soil-like-fractions (LFMSF) for sustainable development: a critical appraisal. Waste Biomass Valorization, 1–22

Chen H, Teng Y, Lu S, Wang Y, Wang J (2015) Contamination features and health risk of soil heavy metals in China. Sci Total Environ 512:143–153

Chen L, Zhou M, Wang J, Zhang Z, Duan C, Wang X et al (2022) A global meta-analysis of heavy metal (loid) s pollution in soils near copper mines: evaluation of pollution level and probabilistic health risks. Sci Total Environ 835:155441

Danthurebandara M (2015) Environmental and economic performance of enhanced landfill mining. KU Leuven

Danthurebandara M, Van Passel S, Vanderreydt I, Van Acker K (2015) Environmental and economic performance of plasma gasification in enhanced landfill mining. Waste Manage 45:458–467

de Souza VB, Hollas CE, Bortoli M, Manosso FC, de Souza DZ (2023) Heavy metal contamination in soils of a decommissioned landfill, southern Brazil, ecological and health risk assessment. Chemosphere, 139689

Defra (2010) Environmental protection act 1990: part 2A. Contaminated land statutory guidance. Department of Environment, Food and Rural Affairs London

Directive E (2008) Directive 2008/98/EC of the European Parliament and of the Council of 19 November 2008 on waste and repealing certain Directives. Official Journal of the European Union L 2008, 312

Doležalová Weissmannová H, Mihočová S, Chovanec P, Pavlovský J (2019) Potential ecological risk and human health risk assessment of heavy metal pollution in industrial affected soils by coal mining and metallurgy in Ostrava, Czech Republic. Int J Environ Res Public Health 16:4495

Douglas P, Hayes ET, Williams W, Tyrrel SF, Kinnersley R, Walsh K et al (2017) Use of dispersion modelling for environmental impact assessment of biological air pollution from composting: progress, problems and prospects. Waste Manage 70:22–29

European Council (1999) Council directive 1999/31/EC of 26 April 1999 on the landfill of waste. Off J Eur Communities 16:1999

Faragó T, Špirová V, Blažeková P, Lalinská-Voleková B, Macek J, Jurkovič Ľ et al Environmental and health impacts assessment of long-term naturally-weathered municipal solid waste incineration ashes deposited in soil—old burden in Bratislava city, Slovakia. Heliyon 2023; 9.

Ferronato N, Torretta V (2019) Waste mismanagement in developing countries: a review of global issues. Int J Environ Res Public Health 16:1060

Fisher C, Maurice C, Lagerkvist A (1999) Gas emission from landfills—an overview of issues and research needs. Swedish Environmental Protection Agency, Stockholm (1999)

Frändegård P, Krook J, Svensson N, Eklund M (2013) Resource and climate implications of landfill mining: a case study of Sweden. J Ind Ecol 17:742–755

Gujre N, Rangan L, Mitra S (2021) Occurrence, geochemical fraction, ecological and health risk assessment of cadmium, copper and nickel in soils contaminated with municipal solid wastes. Chemosphere 271:129573

Hakanson L (1980) An ecological risk index for aquatic pollution control. A sedimentological approach. Water Res 14:975–1001

Havukainen J, Zhan M, Dong J, Liikanen M, Deviatkin I, Li X et al (2017) Environmental impact assessment of municipal solid waste management incorporating mechanical treatment of waste and incineration in Hangzhou, China. J Cleaner Prod 141:453–461

Hoang AT, Varbanov PS, Nižetić S, Sirohi R, Pandey A, Luque R et al (2022) Perspective review on Municipal Solid Waste-to-energy route: Characteristics, management strategy, and role in circular economy. J Clean Prod 359:131897

Hogland M, Hogland W, Jani Y, Kaczala F, de Sá Salomão AL, Kriipsalu M, et al (2014) Experiences of three landfill mining projects in the baltic sea area: with focus on machinery for material recovery. Linnaeus Eco-Tech

Ilse B, Patrick B, Jan P, Yves T (2018) Development of an early warning system for the closing the circle project at the remo landfill site. In: 4th International Symposium On Enhanced Landfill Mining, Mechelen, Belgium

Jia Q, Huang Y, Al-Ansari N, Knutsson S (2013) Dust emissions from landfill deposition: a case study in Malmberget mine, Sweden. J Earth Sci Geotech Eng 3:25–34

Jones PT, Geysen D, Tielemans Y, Van Passel S, Pontikes Y, Blanpain B et al (2013) Enhanced Landfill Mining in view of multiple resource recovery: a critical review. J Clean Prod 55:45–55

Jovanov D, Vujić B, Vujić G (2018) Optimization of the monitoring of landfill gas and leachate in closed methanogenic landfills. J Environ Manage 216:32–40

Kaza S, Yao L, Bhada-Tata P, Van Woerden F (2018) What a waste 2.0: a global snapshot of solid waste management to 2050. World Bank Publications

Kim K-H, Kabir E, Jahan SA (2018) Airborne bioaerosols and their impact on human health. J Environ Sci 67:23–35

Kjeldsen P, Barlaz MA, Rooker AP, Baun A, Ledin A, Christensen TH (2002) Present and long-term composition of MSW landfill leachate: a review. Crit Rev Environ Sci Technol 32:297–336

Laner D, Crest M, Scharff H, Morris JW, Barlaz AM (2012) A review of approaches for the long-term management of municipal solid waste landfills. Waste Manage 32:498–512

Li Z, Ma Z, van der Kuijp TJ, Yuan Z, Huang L (2014) A review of soil heavy metal pollution from mines in China: pollution and health risk assessment. Sci Total Environ 468:843–853

Ma W, Tai L, Qiao Z, Zhong L, Wang Z, Fu K et al (2018) Contamination source apportionment and health risk assessment of heavy metals in soil around municipal solid waste incinerator: a case study in North China. Sci Total Environ 631:348–357

Macklin Y, Kibble A, Pollitt F (2011) Impact on health from emissions from landfill sites: Advice from the health protection agency. Centre for radiation. Health Protection Agency UK

Maheshi D (2015) Environmental and economic assessment of 'open waste dump'mining in Sri Lanka. Resour Conserv Recycl 102:67–79

Marinho APFD, Nascimento CWAD, Cunha KPVD (2022) Soil degradation and Cu, Cr, Ni, Pb and Zn contamination in dumpsites of humid and semiarid tropical regions in northeastern Brazil. Environ Monit Assess 194:459

Mor S, Ravindra K, Dahiya R, Chandra A (2006) Leachate characterization and assessment of groundwater pollution near municipal solid waste landfill site. Environ Monit Assess 118:435–456

Mukherjee C, Denney J, Mbonimpa E, Slagley J, Bhowmik R (2020) A review on municipal solid waste-to-energy trends in the USA. Renew Sustain Energy Rev 119:109512

Muller G (1969) Index of geoaccumulation in sediments of the Rhine River. GeoJournal 2:108–118

Nair AT (2021) Bioaerosols in the landfill environment: an overview of microbial diversity and potential health hazards. Aerobiologia 37:185–203

Naveen B, Mahapatra DM, Sitharam T, Sivapullaiah P, Ramachandra T (2017) Physico-chemical and biological characterization of urban municipal landfill leachate. Environ Pollut 220:1–12

Naveenkumar R, Iyyappan J, Pravin R, Kadry S, Han J, Sindhu R, et al (2023) A strategic review on sustainable approaches in municipal solid waste management and energy recovery: role of artificial intelligence, economic stability and life cycle assessment. Bioresource Tech, 129044

Nguyen F, Ghose R, Isunza Manrique I, Robert T, Dumont G (2018) Managing past landfills for future site development: a review of the contribution of geophysical methods. In: Proceedings of the 4th International Symposium on Enhanced Landfill Mining, pp 27–36

Nomngongo PN, Ngila JC, Msagati TA, Gumbi BP, Iwuoha EI (2012) Determination of selected persistent organic pollutants in wastewater from landfill leachates, using an amperometric biosensor. Phys Chem Earth, Parts A/B/C 50:252–261

Ortner ME, Knapp J, Bockreis A (2014) Landfill mining: objectives and assessment challenges. In: Proceedings of the Institution of Civil Engineers-Waste and Resource Management. Thomas Telford Ltd., pp 51–61

Parker T, Dottridge J, Kelly S () Investigation of the composition and emissions of trace components in landfill gas. Environ Agency, R&D Technical Report P1-438/TR, Bristol

Przydatek G, Kanownik W (2019) Impact of small municipal solid waste landfill on groundwater quality. Environ Monit Assess 191:1–14

Qian G, Duanmu C, Ali N, Khan A, Malik S, Yang Y, et al (2021) Hazardous wastes, adverse impacts, and management strategies: a way forward to environmental sustainability. Environ, Develop Sustain, 1–26

Qing X, Yutong Z, Shenggao L (2015) Assessment of heavy metal pollution and human health risk in urban soils of steel industrial city (Anshan), Liaoning, Northeast China. Ecotoxicol Environ Safety 120:377–385

Rathi S (2006) Alternative approaches for better municipal solid waste management in Mumbai, India. Waste Manage 26:1192–1200

Rodriguez N, Machiels L, Jones PT, Binnemans K (2018) Towards near-zero-waste recycling of mine tailings and metallurgical process residues through a novel solvometallurgical process based on deep eutectic solvents. In: Jones PT, Machiels L (eds) Proceedings of the 4th International Symposium on Enhanced Landfill Mining (ELFM IV), Mechelen (Belgium), 5–6 February, pp 95–100

Rong L, Zhang C, Jin D, Dai Z (2017) Assessment of the potential utilization of municipal solid waste from a closed irregular landfill. J Clean Prod 142:413–419

Roy H, Alam SR, Bin-Masud R, Prantika TR, Pervez MN, Islam MS et al (2022) A review on characteristics, techniques, and waste-to-energy aspects of municipal solid waste management: Bangladesh perspective. Sustain 14:10265

Sagbara G, Zabbey N, Sam K, Nwipie GN (2020) Heavy metal concentration in soil and maize (Zea mays L.) in partially reclaimed refuse dumpsite 'borrow-pit' in Port Harcourt, Nigeria. Environ Tech Inno 18:100745

Sakunkoo P, Thonglua T, Sangkham S, Jirapornkul C, Limmongkon Y, Daduang S, et al (2022) Human health risk assessment of PM2. 5-bound heavy metal of anthropogenic sources in the Khon Kaen Province of Northeast Thailand. Heliyon, 8

Siddiqua A, Hahladakis JN, Al-Attiya WAK (2022) An overview of the environmental pollution and health effects associated with waste landfilling and open dumping. Environ Sci Pollut Res 29:58514–58536

Siddque N, Waheed S, uz Zaman Q, Aslam A, Tufail M, Nasir R (2023) Uptake of heavy metal in wheat from application of different phosphorus fertilizers. J Food Comp Anal 115:104958

Singh A, Chandel MK (2020) Effect of ageing on waste characteristics excavated from an Indian dumpsite and its potential valorisation. Process Saf Environ Prot 134:24–35

Singh A, Chandel MK (2022) Mobility and environmental fate of heavy metals in fine fraction of dumped legacy waste: Implications on reclamation and ecological risk. J Environ Manage 304:114206

Smart-Ground (2017) Report on the environmental issues

Smart-Ground (2018) Enhanced landfill mining toolkit for municipal solid waste streams, Europe

Somani M, Datta M, Ramana G, Sreekrishnan T (2018) Investigations on fine fraction of aged municipal solid waste recovered through landfill mining: case study of three dumpsites from India. Waste Manage Res 36:744–755

Sun B, Li Q, Zheng M, Su G, Lin S, Wu M et al (2020) Recent advances in the removal of persistent organic pollutants (POPs) using multifunctional materials: a review. Environ Pollut 265:114908

Thongyuan S, Khantamoon T, Aendo P, Binot A, Tulayakul P (2020) Ecological and health risk assessment, carcinogenic and non-carcinogenic effects of heavy metals contamination in the

soil from municipal solid waste landfill in Central, Thailand. Human Ecol Risk Assess: An Inter J, 1–22
Tsiouri V, Kakosimos KE, Kumar P (2015) Concentrations, sources and exposure risks associated with particulate matter in the Middle East Area—a review. Air Qual Atmos Health 8:67–80
USEPA (2002) Supplemental guidance for developing soil screening levels for superfund sites, peer review draft. US Environmental Protection Agency Office of Solid Waste and Emergency Response, OSWER 9355 9355, Washington, DC, pp 4–24
Van Passel S, Dubois M, Eyckmans J, De Gheldere S, Ang F, Jones PT et al (2013) The economics of enhanced landfill mining: private and societal performance drivers. J Clean Prod 55:92–102
Varghese D, Puranik P, Palamattam S (2020) Shaikh S. Solid waste management: Green index calculation and route optimization. In: 2020 3rd International Conference on Communication System, Computing and IT Applications (CSCITA). IEEE, pp 186–191
Varjani S, Shahbeig H, Popat K, Patel Z, Vyas S, Shah AV et al (2022) Sustainable management of municipal solid waste through waste-to-energy technologies. Biores Technol 355:127247
Vaverková MD (2019) Landfill impacts on the environment. Geosci 9:431
Vergara SE, Tchobanoglous G (2012) Municipal solid waste and the environment: a global perspective. Annu Rev Environ Resour 37:277–309
Wagland ST, Coulon F, Canopoli L (2019) Developing the case for enhanced landfill mining in the UK
Wang N, Wang A, Kong L, He M (2018) Calculation and application of Sb toxicity coefficient for potential ecological risk assessment. Sci Total Environ 610:167–174
Wang W, Chen C, Liu D, Wang M, Han Q, Zhang X et al (2022) Health risk assessment of PM2. 5 heavy metals in county units of Northern China based on Monte Carlo simulation and APCS-MLR. Science of the Total environment, 156777
Wang X, Dan Z, Cui X, Zhang R, Zhou S, Wenga T et al (2020) Contamination, ecological and health risks of trace elements in soil of landfill and geothermal sites in Tibet. Sci Total Environ 715:136639
Warren K, Read A (2014) Landfill mining: goldmine or minefield. http://www.ismenvis.nic.in/Database/Landfill_Mining-Goldmine_or_Minefield_5478.aspx (accessed 11 Octobor 2018), 2014.
Weber R, Watson A, Forter M, Oliaei F (2011) Persistent organic pollutants and landfills—a review of past experiences and future challenges. Waste Manage Res 29:107–121
Weng Y-C, Fujiwara T, Houng HJ, Sun C-H, Li W-Y, Kuo Y-W (2015) Management of landfill reclamation with regard to biodiversity preservation, global warming mitigation and landfill mining: experiences from the Asia-Pacific region. J Clean Prod 104:364–373
Yan T, Shen S-L, Zhou A (2022) Indices and models of surface water quality assessment: Review and perspectives. Environ Pollut, 119611
Yang S, Zhao J, Chang SX, Collins C, Xu J, Liu X (2019) Status assessment and probabilistic health risk modeling of metals accumulation in agriculture soils across China: a synthesis. Environ Int 128:165–174
Yang S, Sun L, Sun Y, Song K, Qin Q, Zhu Z, et al (2022) Towards an integrated health risk assessment framework of soil heavy metals pollution: theoretical basis, conceptual model, and perspectives. Environ Pollut, 120596
Zari M, Smith R, Wright C, Ferrari R (2022) Health and environmental impact assessment of landfill mining activities: a case study in Norfolk, UK. Heliyon, e11594
Zhang X, Liu C, Chen Y, Zheng G, Chen Y (2022) Source separation, transportation, pretreatment, and valorization of municipal solid waste: a critical review. Environ, Develop Sustain, 1–43
Zhang J, Yu X, Wang J, Sui Q, Zhao W (2023) Impacts of garbage classification and disposal on the occurrence of pharmaceutical and personal care products in municipal solid waste leachates: A case study in Shanghai. Sci Total Environ 874:162467
Zhu Y, Zhang Y, Luo D, Chong Z, Li E, Kong X (2021) A review of municipal solid waste in China: characteristics, compositions, influential factors and treatment technologies. Environ Dev Sustain 23:6603–6622

Ziyang L, Luochun W, Nanwen Z, Youcai Z (2015) Martial recycling from renewable landfill and associated risks: a review. Chemosphere 131:91–103

Zoungrana A, Hasnine MT, Yuan Q (2022) Landfill mining: significance, operation and global perspectives. In: Circular economy in municipal solid waste landfilling: biomining & Leachate treatment: sustainable solid waste management: waste to wealth. Springer, pp 25–45

Chapter 4
Characteristics and Impacts of Municipal Solid Waste (MSW): A Review

Zakia Gueboudji, Maher Mahmoudi, Kenza Kadi, and Kamel Nagaz

Abstract Municipal solid waste (MSW) refers to the waste generated by households, businesses, and institutions in urban areas. It is a complex mixture of materials, including food waste, plastics, paper, glass, metals, and yard waste. The characteristics of municipal solid waste can vary depending on factors such as the size and density of the population, consumption patterns, and waste management practices. The impacts of MSW are significant. It can affect the environment, public health, and the economy. Improperly managed MSW can result in air, water, and soil pollution, leading to environmental degradation and health problems. In addition, MSW can attract pests and vermin, creating public health risks. The accumulation of MSW can also cause aesthetic problems and reduce property values. Effective management of MSW can mitigate these impacts. The most common methods of MSW management include landfilling, incineration, and recycling. Landfills are the most commonly used method of MSW disposal, but they can pose environmental risks if not properly designed and maintained. Incineration can reduce the volume of MSW and generate energy, but it can also produce air pollution. Recycling can reduce the volume of MSW and conserve resources, but it requires significant infrastructure and education to be effective. In conclusion, MSW is a significant challenge for urban areas worldwide, with complex characteristics and significant impacts. Effective management of MSW requires a multifaceted approach that considers environmental, public health, and economic factors.

Z. Gueboudji (✉) · K. Kadi
Biotechnology, Water, Environment and Health Laboratory, Abbes Laghrour University of Khenchela, Khenchela, Algeria
e-mail: gueboudji.zakia@gmail.com; gueboudji.zakia@univ-khenchela.dz

M. Mahmoudi
Faculty of Sciences of Gabes, University of Gabes, Gabes, Tunisia

Laboratory of Plant, Soil and Environment Interactions (LIPSE), LR21LS01, University of Tunis El Manar, 1068 Tunis, Tunisia

K. Kadi
Faculty of Life and Nature Sciences, Abbes Laghrour University, Khenchela, Algeria

K. Nagaz
Dryland and Oases Cropping Laboratory, Institute of Arid Regions, 4119 Mednine, Tunisia

A. Anouzla and S. Souabi (eds.), *Technical Landfills and Waste Management*, Springer Water, https://doi.org/10.1007/978-3-031-52633-6_4

Keywords Municipal solid waste · Landfills · Characteristics of MSW · Impacts of MSW · Recycling · Waste management

4.1 Introduction

Municipal solid waste (MSW) pertains to the refuse produced by residences, enterprises, and establishments in urban localities. This encompasses a diverse range of materials, including food remnants, paper, plastics, glass, metals, and yard debris (Aziz et al. 2023; Tawfik et al. 2022). The attributes of MSW exhibit variations influenced by factors such as population density, socioeconomic status, and cultural norms within the community. Every year, 2 billion tones of MSW are generated globally, with that number anticipated to rise to 3.4 billion tones by 2050 (Pujara et al. 2023). As a result, the unregulated dumping of such a massive waste volume is one of the world's most serious concerns, with developing countries facing a significantly larger burden (Dastjerdi et al. 2021). In developed nations, the collection rate of municipal solid waste (MSW) stands at 90%, while in developing nations, it is at 41%. Consequently, the remaining waste in both types of countries is directly disposed of in the open environment. The significant volume of waste generated requires a larger allocation of land for waste management in both developed and developing nations (Hoang et al. 2022; Pujara et al. 2023). The primary characteristics of municipal solid waste (MSW) include its substantial quantities of organic matter and high moisture content. These factors contribute to the proliferation and multiplication of pathogenic microorganisms within the waste (Samadi et al. 2021). Landfills have been employed since ancient times and are a prevalent means of disposing of municipal solid waste (MSW), representing the final destination for waste removal. It is widely acknowledged that landfills have adverse impacts on the environment, causing pollution in different environmental aspects. Nevertheless, by implementing protective technologies and adopting appropriate solid waste management approaches, it is possible to alleviate the harmful environmental consequences associated with landfills. (Vaverková et al. 2018). Looking ahead, the future of MSW management shows promise with the adoption of advanced sorting technologies like robotics and artificial intelligence, enhancing recycling efficiency. Smart waste management systems, utilizing sensors, data analytics, and IoT technologies, aim to optimize waste collection routes and overall operational efficiency. Novel waste-to-energy technologies, such as plasma gasification and anaerobic digestion, present viable solutions for efficient energy recovery with reduced environmental impacts (Andeobu et al. 2022). The utilization of non-recyclable municipal solid waste (MSW) as a potential feedstock for liquid fuel production can be achieved through a two-step process involving gasification followed by Fischer–Tropsch (FT) processes. However, due to the diverse composition and properties of MSW materials, the convertibility to liquid hydrocarbon fuels may exhibit significant variability, thereby impacting the overall sustainability of this approach to utilizing non-recyclable MSW for fuel production (Kumar and Samadder 2017; Lee et al. 2023). The utilization of municipal solid waste (MSW) would not only diminish the amount of waste sent to landfills but also address various social and environmental concerns, including soil and water contamination, as well as the emission of greenhouse gases from landfills (Lee et al. 2023). MSW also exerts a notable economic burden. Municipalities and governments must allocate substantial resources to manage the collection, transportation, and disposal of MSW. This entails investments in waste collection vehicles, workforce, and landfill infrastructure. Moreover, improper disposal practices can lead to the squandering of valuable resources that could otherwise have been recycled or repurposed, compounding the economic costs associated with MSW management (Mohsenizadeh et al. 2020). The utilization of MSW compost in salt-affected soils can serve as a sustainable soil conditioner and organic fertilizer, contributing to enhanced crop productivity (Sauve and Van Acker 2020; Liu et al. 2015). In recent times, there has been an increasing acknowledgment of the imperative to manage municipal solid waste (MSW) in a sustainable and conscientious manner. Governments, organizations, and individuals alike are actively exploring various solutions to address this issue. Strategies such as waste reduction and prevention, promoting recycling and composting, and incorporating waste-to-energy technologies are being pursued. By embracing these approaches, it becomes feasible to mitigate the adverse effects of MSW on the environment and society, fostering a more sustainable and healthier future for all (Emara 2023; Misganaw 2023).

In this chapter book, we will explore the characteristics and impacts of MSW. We will discuss the composition of MSW, the sources of MSW, and the challenges associated with its management. We will also examine the environmental, social, and economic impacts of MSW, and explore some of the strategies that are being used to address this critical issue. Ultimately, the chapter aims to foster a comprehensive understanding of MSW and its implications, guiding readers towards the path of sustainable and responsible waste management for a brighter future.

4.2 General Information About MSW

Municipal Solid Waste, denoted as the refuse produced by residences, businesses, educational institutions, and industries within an urban or municipal locality, comprises ordinary objects like paper, cardboard, plastics, glass, metals, organic waste, textiles, and other non-hazardous materials that are discarded by individuals or entities (Aziz et al. 2023; Kiran et al. 2023).

The origins of municipal solid waste can be broadly classified into four main sectors: residential, commercial, institutional, and industrial. Residential waste primarily emanates from households and apartments. Commercial waste originates from various businesses, offices, and retail establishments. Institutional waste is generated by educational institutions, hospitals, and government buildings. Lastly, industrial waste is produced as a result of manufacturing processes and construction activities (Zargar et al. 2023; Kiran et al. 2023). The generation rates of MSW vary depending on factors such as population density, economic activities, lifestyle patterns, and waste management practices. Developed countries typically generate higher amounts of MSW per capita compared to developing countries (Zargar et al. 2023).

The composition of MSW varies from one region to another, as it is influenced by cultural, social, and economic factors. The classification of MSW is illustrated in Table 4.1 However, some common components can be found in most MSW compositions (Zamri et al. 2021; Abdel-Shafy and Mansour 2018). These include:

- Organic waste: Food scraps, yard trimmings, and other biodegradable materials.
- Paper and cardboard: Newspapers, magazines, cardboard boxes, packaging materials, etc.
- Plastics: Bottles, containers, packaging, bags, and various plastic products.
- Glass: Bottles, jars, and other glass containers.
- Metals: Ferrous and non-ferrous metals, such as steel, aluminum, and tin cans.
- Textiles: Clothing, fabrics, and other textile products.
- Miscellaneous: Miscellaneous items like rubber, leather, wood, ceramics, etc.

MSW components can be broadly categorized into three groups based on their source and characteristics:

- Recyclables: These are materials that can be collected, processed, and transformed into new products. Examples include paper, cardboard, glass, plastics, and metals (Cimpan et al. 2015).
- Organics: This category consists of biodegradable waste that can be composted or used for energy generation through processes like anaerobic digestion. Food waste, yard trimmings, and other organic materials fall into this category (Zamri et al. 2021).
- Residuals: Residuals are non-recyclable and non-compostable waste that remains after recycling and organic waste removal. They typically end up in landfills or undergo waste-to-energy processes (El-Saadony et al. 2023).

Table 4.1 MSW classification (Osra et al. 2021)

Waste fractions	Waste components
Organic matter	Food, vegetables, animal excrements
Wood	Wood, garden trimmings
Paper	Newspapers, office paper, bills, magazines, sales notes and receipts
Cardboard	Corrugated cardboard, boxboard
Plastic	HDPE, PVC, Film PE, polyethylene bag, hair, food containers, PS
Metal	Ferrous and non-ferrous material, aluminum cans and foils
Glass	Soda, beer, wine container, window glass, car glass
Textile	Clothes, ropes, sacks, sanitary products, cotton

4.3 Characteristics of Municipal Solid Waste

4.3.1 Physical Characteristics

4.3.1.1 Density and Bulkiness

The density of Municipal Solid Waste (MSW) pertains to the mass of waste per unit volume, and it is subject to variation based on the waste's composition and the extent of compaction. Typically, MSW exhibits low density because of the inclusion of voluminous and lightweight materials like paper, plastics, and textiles. Nevertheless, compaction can elevate the density of MSW, as observed in landfill cells or waste storage containers. The term "bulkiness" is used to describe the amount of space occupied by the waste and is closely associated with its density (Zhu et al. 2021).

4.3.1.2 Moisture Content

Moisture content in Municipal Solid Waste (MSW) refers to the quantity of water present within the waste. This level of moisture can vary due to factors such as environmental conditions, waste composition, and waste management practices. MSW generally contains a specific percentage of moisture, which can have significant implications on its handling, decomposition, and the potential for odors and microbial activity. When MSW has a high moisture content, it can lead to increased weight of the waste and affect its compaction behavior (Tupsakhare et al. 2020).

4.3.1.3 Particle Size Distribution

The particle size distribution of MSW refers to the range of particle sizes present in the waste. MSW can contain a wide range of particle sizes, including large and bulky items, as well as smaller particles and fine materials. The particle size distribution can affect waste management processes such as sorting, shredding, and composting. It can also influence the flow characteristics and compaction behavior of waste (Zhu et al. 2021).

4.3.1.4 Thermal Conductivity

Thermal conductivity is the property of a material that denotes its capacity to conduct heat. In the context of Municipal Solid Waste (MSW), the thermal conductivity can fluctuate based on the waste's composition and moisture content. Materials with higher thermal conductivity, like metals, can significantly influence the overall thermal behavior of the waste mass. This understanding of the thermal conductivity of MSW holds crucial importance for waste-to-energy processes such as incineration or gasification, as well as for the design and operation of landfill systems. Proper knowledge of thermal conductivity aids in optimizing these waste management processes and maximizing their efficiency (Zhu et al. 2021).

4.3.1.5 Compaction and Consolidation Behavior

Municipal Solid Waste (MSW) undergoes compaction and consolidation behavior when exposed to external forces or weight. Compaction involves applying pressure to the waste, leading to a reduction in volume and an increase in waste density. On the other hand, consolidation refers to the gradual settlement of waste over time due to its self-weight. Various factors influence the compaction and consolidation behavior of MSW, including waste composition, moisture content, particle size distribution, and the type of compaction equipment used. Understanding these behaviors is crucial for landfill design, waste transportation, and storage considerations. Proper management of compaction and consolidation in MSW is essential to optimize landfill space usage, minimize environmental impacts, and ensure safe and efficient waste disposal and storage processes (Tupsakhare et al. 2020).

4.3.2 Chemical Characteristics

4.3.2.1 Organic Content

Municipal solid waste (MSW) contains a substantial proportion of organic matter, comprising food waste, yard waste, paper, and other biodegradable materials. These organic components undergo decomposition by microorganisms, either under anaerobic conditions (without oxygen) or aerobic conditions (with oxygen). During this decomposition process, organic waste generates gases like methane (CH_4) and carbon dioxide (CO_2), thereby contributing to greenhouse gas emissions (Chen et al. 2017). However, the organic content of MSW can be a valuable resource if managed effectively. Techniques such as composting or anaerobic digestion can be employed to harness the potential of this organic waste. Composting involves the controlled breakdown of organic matter into nutrient-rich compost, which can be utilized as a natural fertilizer for agricultural purposes. Anaerobic digestion, on the other hand, facilitates the production of biogas, primarily methane, which can be harnessed for energy generation. By implementing these sustainable waste management practices, the organic content of MSW can be effectively utilized to produce renewable energy and environmentally beneficial by-products, thereby reducing greenhouse gas emissions and promoting a circular economy approach to waste management (Zamri et al. 2021; Chen et al. 2017).

4.3.2.2 Inorganic Constituents

The inorganic components found in municipal solid waste comprise substances like glass, metals, plastics, ceramics, and incineration-derived ashes. The quantity and composition of these materials differ based on consumer habits, recycling approaches, and waste management methodologies. Dealing with inorganic waste presents difficulties due to its non-biodegradable nature and the risk of environmental contamination. Nevertheless, it also offers prospects for recycling and recovering resources, which can lessen the need for new materials and mitigate environmental consequences (El-Saadony et al. 2023).

4.3.2.3 Toxic and Hazardous Substances

Municipal solid waste has the potential to contain a wide array of toxic and hazardous elements that originate from household products, industrial waste, or illicit disposal activities. These elements encompass heavy metals like lead, mercury, and cadmium, persistent organic pollutants such as polychlorinated biphenyls (PCBs) and dioxins, as well as pesticides, pharmaceuticals, and household chemicals. Inadequate waste management or improper disposal can result in the release of these substances into the environment, thereby endangering human health and ecosystems. To address this issue, effective waste management practices, such as proper segregation, recycling, and establishment of treatment facilities, are of utmost importance to mitigate the potential hazards associated with these toxic and hazardous substances (Zamri et al. 2021).

4.3.2.4 Leachate Composition

Leachate is a liquid generated when water permeates through waste materials within landfills or waste storage facilities. It constitutes a complex mixture comprising dissolved and suspended organic and inorganic substances. The specific composition of leachate is influenced by factors such as the composition of the waste, the age of the landfill, precipitation levels, and waste management practices (El-Saadony et al. 2023). Within leachate, there may exist elevated levels of organic matter, nutrients, heavy metals, and other contaminants. If not adequately collected, treated, and disposed of, leachate can contaminate soil, groundwater, and surface water, leading to environmental degradation and posing potential risks to human health (Aziz et al. 2023; El-Saadony et al. 2023).

4.3.2.5 Odor Generation and Control

Municipal solid waste is notorious for emitting unpleasant odors, primarily resulting from the decomposition of organic matter and the presence of volatile compounds.

These odors can originate from different sources within waste management systems, including waste storage areas, landfills, and waste treatment facilities. The specific odorous compounds produced can vary, but they commonly include sulfur compounds, amines, organic acids, and volatile organic compounds (VOCs). To effectively manage and minimize odor generation, waste management facilities implement various measures such as covering waste to contain odors, employing gas collection systems in landfills, adopting odor control technologies like biofilters and activated carbon filters, and promoting proper waste handling and storage practices. These efforts aim to alleviate the negative impacts of odors on surrounding environments and communities (Andraskar et al. 2021).

4.3.3 Biological Characteristics

4.3.3.1 Microbial Communities

Municipal solid waste harbors diverse microbial communities that play a significant role in waste decomposition. These microbial communities include bacteria, fungi, and archaea. Different microbial groups thrive under varying environmental conditions and contribute to the breakdown of organic matter present in MSW (Sadeghi et al. 2022). Anaerobic microorganisms, particularly methanogenic archaea, play a crucial role in breaking down organic waste in environments with limited oxygen, such as landfills. In contrast, aerobic microorganisms, including bacteria and fungi, are responsible for decomposing organic matter in the presence of oxygen, often during composting processes. The presence of these distinct microbial communities in municipal solid waste (MSW) is influenced by various factors, such as the composition of the waste, moisture content, pH levels, and temperature conditions. These factors directly impact the composition and activity of the microbial populations involved in the waste degradation processes (Sharma et al. 2023a, b).

4.3.3.2 Biodegradation Processes

Biodegradation is a natural process orchestrated by microorganisms to break down complex organic compounds into simpler substances (Gueboudji 2022). In the context of municipal solid waste (MSW), biodegradation refers to the microbial decomposition of organic materials within the waste. This intricate process involves diverse microbial communities performing enzymatic reactions, leading to the transformation of organic matter into simpler compounds like carbon dioxide (CO_2), water (H_2O), and other organic byproducts. The efficiency and speed of biodegradation are influenced by factors such as waste composition, moisture content, temperature, oxygen availability, and microbial activity. By adopting proper waste management practices, it is possible to optimize biodegradation processes, reducing waste volume and minimizing potential environmental impacts associated with MSW (Sharma et al. 2023a, b).

4.3.3.3 Methane Generation and Emissions

Landfills, particularly municipal solid waste (MSW) landfills, possess a significant biological characteristic in their ability to produce methane (CH_4), a potent greenhouse gas. Methane is generated through the anaerobic decomposition of organic matter by methanogenic microorganisms, primarily in landfill environments. These microorganisms thrive in conditions where oxygen is absent, and favorable factors such as high moisture content and the presence of suitable organic substrates are available (Chen et al. 2017). As a consequence of this methanogenesis process, the generated methane can escape into the atmosphere through various pathways, including landfill vents and surface emissions. Methane emissions from landfills contribute to climate change and global warming since methane is a much more potent greenhouse gas compared to carbon dioxide (CO_2) (Eghbali et al. 2022). To address and mitigate the impact of methane emissions from landfills, landfill gas collection systems can be employed. These systems aim to capture and utilize the methane as an energy source. By collecting methane before it is released into the atmosphere, the greenhouse gas emissions can be significantly reduced. Moreover, the captured methane can be used as an alternative energy resource, providing an eco-friendly and sustainable approach to waste management (Andraskar et al. 2021). Implementing landfill gas collection systems is an essential strategy for combating climate change and reducing the environmental impact of landfills. By adopting such measures, society can take a step towards achieving more sustainable waste management practices and curbing the effects of greenhouse gas emissions on the planet (Saravanan et al. 2022; Andraskar et al. 2021; Chen et al. 2017).

4.3.3.4 Pathogens and Public Health Risks

Municipal solid waste has the potential to harbor various disease-causing agents, such as bacteria, viruses, and parasites, which can pose significant risks to public health. These pathogens may infiltrate MSW from various origins, including contaminated food waste, fecal matter, or medical waste. Inadequate handling and disposal of waste can result in the dissemination of pathogens through direct contact, ingestion, or inhalation of contaminated aerosols (Sadeghi et al. 2022). Among the common pathogens detected in MSW are *Salmonella, Escherichia coli (E. coli),* norovirus, and hepatitis A virus. To mitigate public health hazards, waste management practices should encompass appropriate segregation, treatment, and disposal methods for different waste streams. It is crucial to implement well-designed and operated sanitary landfills and enforce proper personal protective measures to prevent the transmission of pathogens from MSW to humans (Sharma et al. 2023a, b; Sadeghi et al. 2022).

4.4 Environmental Impacts of Municipal Solid Waste

4.4.1 Land Pollution

Land pollution stands out as a significant environmental consequence of municipal solid waste (MSW). Improper waste disposal methods, such as open dumping or inadequately managed landfills, can lead to soil and groundwater contamination. Harmful substances present in MSW, such as heavy metals, organic pollutants, and leachate, have the potential to permeate the soil, reaching underlying aquifers and causing water sources to become polluted. Furthermore, improper waste disposal can degrade land quality, impede agricultural productivity, and disrupt wildlife habitats. To combat land pollution effectively, it is essential to adopt efficient waste management practices that encompass proper landfill design, reliable lining systems, and effective leachate collection (Sauve and Van Acker 2020).

4.4.2 Air Pollution

Municipal solid waste (MSW) significantly contributes to air pollution through various means. One common practice in some areas is open burning of waste, which releases harmful gases and particulate matter into the atmosphere. Additionally, the combustion of waste in incineration facilities can lead to the emission of pollutants like nitrogen oxides (NOx), sulfur dioxide (SO_2), dioxins, and heavy metals. These emissions have adverse effects on air quality and human health, causing respiratory issues, cardiovascular problems, and the formation of smog (Hou et al. 2023; Sharma et al. 2023a, b). To address the air pollution associated with MSW, it is crucial to implement advanced air pollution control technologies in waste incineration processes. These technologies can help reduce the release of harmful substances into the air. Furthermore, promoting waste reduction and recycling initiatives can significantly mitigate air pollution related to MSW, as these practices reduce the amount of waste that needs to be burned or incinerated in the first place (Hou et al. 2023).

4.4.3 Water Pollution

Improper management of municipal solid waste (MSW) can have severe consequences for water bodies, posing risks to aquatic ecosystems and human health. When landfills are not adequately constructed or maintained, they can permit leachate, a liquid formed by water percolating through waste, to enter surface water or infiltrate groundwater. Leachate often contains hazardous substances, including heavy metals, organic compounds, and nutrients, which can contaminate water sources and disrupt aquatic ecosystems. Moreover, inadequate wastewater treatment from waste treatment facilities can also contribute to water pollution. If not properly treated, wastewater from these facilities can introduce pollutants into water bodies, further compromising water quality (Sauve and Van Acker 2020). To safeguard water bodies and mitigate water pollution linked to MSW, it is essential to implement effective leachate collection and treatment systems. These measures can prevent leachate from reaching water sources and minimize its harmful impact on the environment. Additionally, proper wastewater management practices should be employed to ensure that wastewater from waste treatment facilities is adequately treated before discharge into water bodies (Gueboudji et al. 2021; Sauve and Van Acker 2020).

4.4.4 Climate Change and Greenhouse Gas Emissions

Municipal solid waste (MSW) plays a significant role in contributing to climate change due to the release of greenhouse gases (GHGs), particularly carbon dioxide (CO_2) and methane (CH_4). When organic waste decomposes in landfills, it generates methane, which is a potent GHG with a much higher warming potential than CO_2. If not properly collected and treated, landfill gas, which contains methane and other volatile organic compounds, is released into the atmosphere (Eghbali et al. 2022). Furthermore, the energy-intensive processes involved in waste transportation, treatment, and incineration can also lead to CO_2 emissions (Gueboudji and Kadi 2023), adding to the overall carbon footprint associated with MSW management. To effectively mitigate the impacts of climate change, waste management strategies should prioritize waste reduction, recycling, and composting. By reducing the amount of waste sent to landfills, the generation of methane can be curtailed. Additionally, capturing and utilizing methane from landfills as an energy source can help reduce its release into the atmosphere (Mohsenizadeh et al. 2020; Eghbali et al. 2022).

4.4.5 Ecological Consequences

Municipal solid waste (MSW) can indeed have significant negative ecological consequences, affecting biodiversity, ecosystems, and natural habitats. Improper disposal practices, such as open dumping or poorly managed landfills, can lead to the fragmentation and destruction of habitats, displacing wildlife and disrupting ecological balances in the affected areas (Sauve and Van Acker 2020). Landfills, in particular, can have far-reaching ecological impacts. They can alter soil characteristics, nutrient cycling processes, and microbial communities, which can adversely affect the overall ecological health of the surrounding environment. The pollution generated by MSW, such as leachate or hazardous substances, can directly harm organisms and disrupt ecological interactions, leading to detrimental effects on the ecosystem's functioning (Eghbali et al. 2022). To mitigate the ecological consequences of MSW, it is crucial to implement sustainable waste management practices. Waste reduction and recycling efforts can significantly decrease the amount of waste sent to landfills, thus reducing the disturbance of natural habitats. Additionally, conserving and protecting natural habitats and promoting responsible waste disposal methods can help preserve biodiversity and ecological balance in affected regions. By adopting these sustainable waste management approaches, we can work towards minimizing the ecological impacts of MSW and contribute to the preservation and restoration of our natural ecosystems (Sharma et al. 2023a, b; Eghbali et al. 2022; Sauve and Van Acker 2020).

4.5 Health and Social Impacts of Municipal Solid Waste

Workers engaged in various stages of waste management, including collection, sorting, recycling, and disposal, often encounter occupational health risks. These risks stem from exposure to hazardous substances, physical injuries, ergonomic challenges, and respiratory issues, making it crucial to prioritize worker safety in this industry (Dastjerdi et al. 2021). Waste collectors, for instance, face hazards like sharp objects, chemical spills, and heavy lifting, which increase the risk of accidents and injuries during waste collection processes. Workers at recycling facilities may be exposed to harmful substances, such as chemicals or biohazardous materials, while sorting and processing waste materials (Andeobu et al. 2022). To safeguard the health and well-being of waste management workers, it is essential to implement occupational health and safety regulations and provide appropriate training. Equipping workers with personal protective equipment can also significantly reduce their exposure to potential risks. Additionally, the design and layout of waste management facilities should be optimized to minimize hazards and promote safe working conditions (Andeobu et al. 2022; Dastjerdi et al. 2021).

The proximity of communities to poorly managed waste disposal sites, such as landfills or incinerators, can have serious adverse health effects on the residents living nearby. Exposure to air pollution resulting from waste incineration or open burning can lead to respiratory issues, cardiovascular problems, and an increased risk of developing cancer. Additionally, landfills that are poorly designed or maintained may release odors, pathogens, and toxic substances into the surrounding environment, negatively affecting the quality of life and health of the nearby residents. Certain populations, such as children, the elderly, and individuals with pre-existing health conditions, are particularly vulnerable to the health effects of municipal solid waste (MSW) pollution. To safeguard community health, comprehensive waste management planning is crucial. Proper site selection for waste disposal facilities is essential to minimize the potential health impacts on nearby communities. Additionally, effective pollution control measures must be implemented to reduce air and water pollution from waste incineration and landfill sites (Malav et al. 2020; Hoang et al. 2022).

The repercussions of Municipal Solid Waste (MSW) are not equitably spread, giving rise to environmental justice concerns when disadvantaged and marginalized communities shoulder a disproportionate burden of waste management facilities and the related risks. Frequently, waste disposal sites are situated in or close to underprivileged neighborhoods, resulting in unequal exposure to pollution, health hazards, and societal marginalization. Rectifying environmental justice necessitates the fair allocation of waste management facilities, fostering community involvement and empowerment, and integrating principles of social equity into waste management policies and decision-making procedures (Hoang et al. 2022). The effects of MSW on human health and environment are presented in Fig. 4.1.

The success of waste management initiatives can be significantly impacted by public perception and attitudes. Negative views towards waste facilities, apprehensions about health risks, and a lack of trust in waste management practices can impede the effective implementation of waste management strategies. To overcome these challenges, it is essential to undertake public education and awareness programs, ensure transparent communication, and initiate community engagement efforts. By addressing public concerns, building trust, and encouraging cooperation among stakeholders, these measures can foster a more favorable environment for waste management efforts. Additionally, integrating public input into waste management planning processes can lead to solutions that are not only socially acceptable but also sustainable in the long run (Gaur et al. 2022).

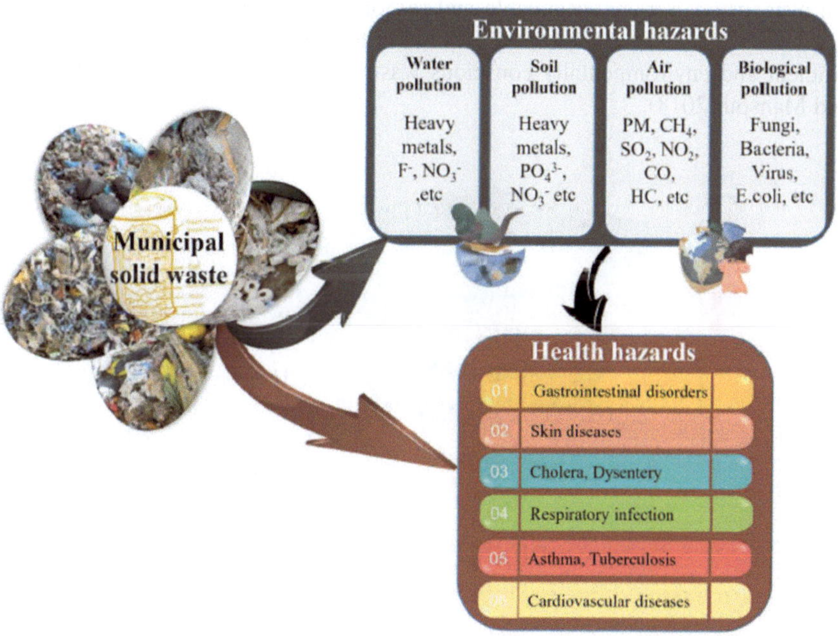

Fig. 4.1 Effects of MSW on human health and environment (Malav et al. 2020; Hoang et al. 2022)

4.6 Waste Management Strategies and Technologies

Figure 4.2 depicts the hierarchical approach to waste management, wherein prevention is regarded as the most favorable option, while disposal represents the least preferred choice.

4.6.1 *Waste Minimization and Source Reduction*

Waste minimization and source reduction play pivotal roles in effective MSW management. This approach focuses on reducing waste generation at its origin by advocating sustainable consumption patterns, enhancing product designs, and utilizing environmentally friendly materials. Waste minimization strategies encompass the implementation of policies and programs that encourage product reuse, repair, and sharing, along with promoting responsible consumer behavior. By adopting waste minimization practices, we can conserve precious natural resources, curtail energy consumption, and lessen the overall environmental impact of waste. This proactive approach not only reduces the amount of waste that requires disposal but also fosters a more sustainable and responsible approach to consumption and production. It addresses the root causes of waste generation, contributing to a more efficient and environmentally conscious waste management system (Abdel-Shafy and Mansour 2018).

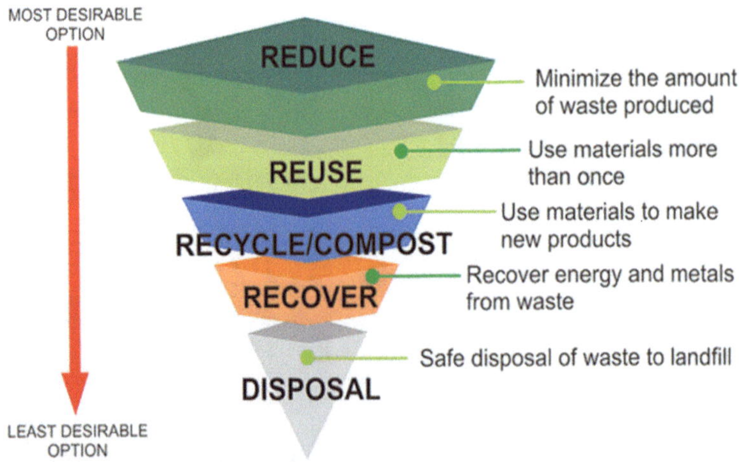

Fig. 4.2 Hierarchy of waste management

4.6.2 Recycling and Material Recovery

Recycling and material recovery constitute fundamental elements of an environmentally sustainable waste management system. This process entails the gathering, categorization, treatment, and transformation of discarded materials into fresh commodities. By doing so, it facilitates the preservation of valuable resources, curtails energy consumption, and diminishes the amount of waste directed to landfills or incineration facilities. Typical recyclable materials encompass paper, plastics, glass, metal, and specific types of electronics. The implementation of advanced sorting technologies, such as optical sorting and automated systems, has notably enhanced the efficacy of recycling endeavors. To foster the adoption of recycling practices, it is imperative to conduct public education and awareness campaigns and establish easily accessible recycling infrastructure (Abdel-Shafy and Mansour 2018).

4.6.3 Composting and Organic Waste Management

Composting and strategies for managing organic waste are centered on diverting organic materials away from landfills and utilizing them to create nutrient-rich compost. Composting is a natural process that involves the breakdown of organic matter, like food scraps, yard waste, and agricultural residues, into a stable substance. This resulting compost can be applied to soil, enhancing its fertility and structure. By composting, the generation of methane in landfills is reduced, landfill space is conserved, and the circular economy is promoted by closing the nutrient loop. It is crucial to employ proper composting techniques, including careful control of moisture and temperature, to ensure the effective management of organic waste (Saravanan et al. 2022).

4.6.4 Waste-to-Energy Conversion

Waste-to-energy (WTE) conversion technologies offer a viable solution for both energy recovery from municipal solid waste (MSW) and the reduction of waste volumes sent to landfills. These technologies encompass incineration, gasification, and pyrolysis. Incineration entails the controlled combustion of waste to produce heat, which can then be utilized to generate steam and electricity. Advanced incineration facilities incorporate pollution control measures to minimize air emissions. On the other hand, gasification and pyrolysis processes involve the thermal decomposition of waste at high temperatures, resulting in the production of syngas or bio-oil. These energy-rich outputs can be harnessed for electricity generation or serve as feedstock for industrial applications. By adopting WTE technologies, greenhouse gas emissions are reduced, and a valuable contribution is made to renewable energy production (Vasileiadou et al. 2023).

4.6.5 *Landfilling and Final Disposal*

Although landfilling is generally regarded as the least favorable waste management approach, it remains indispensable for certain waste types that cannot be efficiently handled through alternative means. Contemporary landfill design and management strategies are focused on mitigating environmental repercussions. Landfills must be engineered with liners and leachate collection systems to safeguard against soil and water contamination. Additionally, methane gas generated within landfills can be captured and utilized as a valuable energy resource, thereby curbing greenhouse gas emissions. To minimize the long-term environmental effects of landfills, it is imperative to exercise appropriate landfill operations, select suitable sites, and provide post-closure care (Saravanan et al. 2022; Aziz et al. 2023; Emara 2023).

4.7 Advancements in Municipal Solid Waste Management: Innovative Approaches and Future Directions

In recent years, there has been a rise in inventive waste management strategies aimed at tackling the challenges associated with municipal solid waste (MSW). One noteworthy example is the integration of waste management systems, which involve combining waste minimization, recycling, composting, and energy recovery to optimize resource usage and reduce environmental impacts. Another innovative approach is the adoption of decentralized waste management systems, wherein waste is handled at the local level through community-driven initiatives, small-scale composting facilities, and recycling centers. Moreover, advancements in waste-to-energy technologies, such as enhanced incineration methods and gasification systems, have facilitated more efficient energy extraction from waste materials (Vaverková et al. 2018).

The future of municipal solid waste (MSW) management appears promising, with the emergence of cutting-edge technologies and innovative approaches. One notable trend is the increasing adoption of advanced sorting technologies, like robotics and artificial intelligence, aimed at enhancing recycling efficiency and maximizing the retrieval of valuable materials from waste streams (Andeobu et al., 2022). Additionally, the development of smart waste management systems utilizing sensors, data analytics, and Internet of Things (IoT) technologies is underway to optimize waste collection routes, monitor container fill levels, and improve operational efficiency. Moreover, the exploration of novel waste-to-energy technologies, such as plasma gasification and anaerobic digestion, shows considerable potential for optimizing energy recovery while minimizing adverse environmental impacts (Andeobu et al., 2022).

4.8 Conclusions

Municipal solid waste (MSW) presents a complex and diverse challenge as it is generated by various entities such as individuals, businesses, and organizations. The characteristics of MSW can differ significantly based on factors like regional location, population density, income levels, and cultural practices. Common components of MSW include food waste, paper, plastics, metals, glass, and textiles. The consequences of improper MSW disposal can have severe impacts on the environment, public health, and the economy. Hazardous substances released during improper disposal can contaminate the air, water, and soil, posing risks to both human health and ecosystems. Landfills, often the primary means of MSW disposal in many countries, release methane, a potent greenhouse gas contributing to climate change. Managing MSW also places significant financial and infrastructural burdens on municipal budgets and resources. Fortunately, there are strategies available to mitigate the negative effects of MSW and even convert waste into a valuable resource. Recycling, composting, and waste-to-energy technologies can reduce the volume of waste going to landfills and extract value from discarded materials. Additionally, community education and participation can raise awareness about the importance of waste reduction, reuse, and recycling, promoting more sustainable behaviors. In summary, managing MSW is a challenging task that requires a comprehensive and integrated approach. By understanding the diverse aspects of MSW, assessing its impacts, and implementing sustainable waste management strategies, communities can progress toward a more circular and sustainable economy that benefits both people and the planet.

Acknowledgements Many thanks to the leader editors Doctor Anouzla Abdelkader and Professor Salah Souabi, Hassan II University Casablanca, Faculty of Sciences and Techniques of Mohammedia, Morocco, for the suggestion to contribute to this book.

References

Abdel-Shafy HI, Mansour MS (2018) Solid waste issue: Sources, composition, disposal, recycling, and valorization. Egypt J Pet 27(4):1275–1290. https://doi.org/10.1016/j.ejpe.2018.07.003

Andeobu L, Wibowo S, Grandhi S (2022) Artificial intelligence applications for sustainable solid waste management practices in Australia: a systematic review. Sci Total Environ 834:155389. https://doi.org/10.1016/j.scitotenv.2022.155389

Andraskar J, Yadav S, Kapley A (2021) Challenges and control strategies of odor emission from composting operation. Appl Biochem Biotechnol 193:2331–2356. https://doi.org/10.1007/s12010-021-03490-3

Aziz HA, Ramli SF, Hung YT (2023) Physicochemical technique in municipal solid waste (MSW) landfill leachate remediation: a review. Water 15:1249. https://doi.org/10.3390/w15061249

Chen M, Yao XZ, Ma RC, Song QC, Long Y, He R (2017) Methanethiol generation potential from anaerobic degradation of municipal solid waste in landfills. Environ Sci Pollut Res 24:23992–24001. https://doi.org/10.1007/s11356-017-0035-x

Cimpan C, Maul A, Jansen M, Pretz T, Wenzel H (2015) Central sorting and recovery of MSW recyclable materials: a review of technological state-of-the-art, cases, practice and implications for materials recycling. J Environ Manage 156:181–199. https://doi.org/10.1016/j.jenvman.2015.03.025

Dastjerdi B, Strezov V, Kumar R, He J, Behnia M (2021) Comparative life cycle assessment of system solution scenarios for residual municipal solid waste management in NSW, Australia. Sci Total Environ 767:144355. https://doi.org/10.1016/j.scitotenv.2020.144355

Eghbali H, Arkat J, Tavakkoli-Moghaddam R (2022) Sustainable supply chain network design for municipal solid waste management: a case study. J Clean Prod 381:135211. https://doi.org/10.1016/j.jclepro.2022.135211

El-Saadony MT, Saad AM, El-Wafai NA, Abou-Aly HE, Salem HM, Soliman SM, AbuQamar SF (2023) Hazardous wastes and management strategies of landfill leachates: a comprehensive review. Environ Technol Innov 2023:103150. https://doi.org/10.1016/j.eti.2023.103150

Emara K (2023) Sustainable solid waste management in rural areas: a case study of Fayoum governorate. Egypt. Energy Nexus 9:100168. https://doi.org/10.1016/j.nexus.2022.100168

Gaur A, Prakash H, Anand K, Kumar G, Hussain A (2022) Evaluation of municipal solid waste management scenarios using multi-criteria decision making under fuzzy environment. Process Integr Optim Sustain 6(2):307–321. https://doi.org/10.1007/s41660-021-00215-1

Gueboudji Z (2022) Effect of storage time on the quality of olive oil mill wastewaters obtained from the cold extraction of olive oil in the region of Khenchela and their biological properties. Doctoral dissertation. Abbes Laghrour University-Khenchela

Gueboudji Z, Bagues M, Kadi K, Nagaz K, Addad D (2021) Effect of storage time on the biodegradability of olive oil mill wastewater from the cold extraction of olive oil system. EuroBiotech J 5(3):142–154. https://doi.org/10.2478/ebtj-2021-0023

Gueboudji Z, Kadi K (2023) Physicochemical characterization and estimation of the pollution degree of olive oil mill wastewaters from the cold extraction system and the traditional system. In: Wastewater from olive oil production: environmental impacts, treatment and valorization. Springer, Cham, pp 143–152. https://doi.org/10.1007/978-3-031-23449-1_6

Hoang AT, Varbanov PS, Nižetić S, Sirohi R, Pandey A, Luque R, Ng KH (2022) Perspective review on Municipal Solid Waste-to-energy route: characteristics, management strategy, and role in circular economy. J Clean Prod 359:131897. https://doi.org/10.1016/j.jclepro.2022.131897

Hou H, Su L, Guo D, Xu H (2023) Resource utilization of solid waste for the collaborative reduction of pollution and carbon emissions: case study of fly ash. J Clean Prod 383:135449. https://doi.org/10.1016/j.jclepro.2022.135449

Kiran DA, Pushkara SV, Jitvan R, Darshan S (2023) Characterization, quantification and management of municipal solid waste in Shivamogga city, Karnataka, India. Waste Manage Bull https://doi.org/10.1016/j.wmb.2023.06.006

Kumar A, Samadder SR (2017) A review on technological options of waste to energy for effective management of municipal solid waste. Waste Manag 69:407–422. https://doi.org/10.1016/j.wasman.2017.08.046

Lee U, Cai H, Ou L, Benavides PT, Wang Y, Wang M (2023) Life cycle analysis of gasification and Fischer-Tropsch conversion of municipal solid waste for transportation fuel production. J Clean Prod 382:135114. https://doi.org/10.1016/j.jclepro.2022.135114

Liu A, Ren F, Lin WY, Wang JY (2015) A review of municipal solid waste environmental standards with a focus on incinerator residues. Int J Sustain Built Environ 4(2):165–188. https://doi.org/10.1016/j.ijsbe.2015.11.002

Malav LC, Yadav KK, Gupta N, Kumar S, Sharma GK, Krishnan S, Rezania S, Kamyab H, Pham QB, Yadav S, Bhattacharyya S, Yadav VK, Bach QV (2020) A review on municipal solid waste as a renewable source for waste-to-energy project in India: current practices, challenges, and future opportunities. J Clean Prod 277:123227. https://doi.org/10.1016/j.jclepro.2020.123227

Misganaw A (2023) Assessment of potential environmental impacts and sustainable management of municipal solid waste using the DPSIRO framework: a case study of Bahir Dar, Ethiopia. Environ Monit Assessment 195(2):297. https://doi.org/10.1007/s10661-023-10929-z

Mohsenizadeh M, Tural MK, Kentel E (2020) Municipal solid waste management with cost minimization and emission control objectives: a case study of Ankara. Sustain Cities Soc 52:101807. https://doi.org/10.1016/j.scs.2019.101807

Osra FA, Ozcan HK, Alzahrani JS, Alsoufi MS (2021) Municipal solid waste characterization and landfill gas generation in Kakia Landfill, Makkah. Sustain 13:1462. https://doi.org/10.3390/su13031462

Pujara Y, Govani J, Patel HT, Pathak P, Mashru D, Ganesh PS (2023) Quantification of environmental impacts associated with municipal solid waste management in Rajkot city, India using life cycle assessment. Environ Adv 12:100364. https://doi.org/10.1016/j.envadv.2023.100364

Sadeghi S, Nikaeen M, Mohammadi F, Nafez AH, Gholipour S, Shamsizadeh Z, Hadi M (2022) Microbial characteristics of municipal solid waste compost: occupational and public health risks from surface applied compost. Waste Manage 144:98–105. https://doi.org/10.1016/j.wasman.2022.03.012

Samadi MT, Mahvi AH, Leili M, Bahrami A, Poorolajal J, Zafari D, Mazaheri Tehrani A (2021) Characteristics and health effects of potentially pathogenic bacterial aerosols from a municipal solid waste landfill site in Hamadan, Iran. J Environ Health Sci Eng 19:1057–1067. https://doi.org/10.1007/s40201-021-00672-3

Saravanan A, Kumar PS, Nhung TC, Ramesh B, Srinivasan S, Rangasamy G (2022) A review on biological methodologies in municipal solid waste management and landfilling: resource and energy recovery. Chemosphere 136630. https://doi.org/10.1016/j.chemosphere.2022.136630

Sauve G, Van Acker K (2020) The environmental impacts of municipal solid waste landfills in Europe: a life cycle assessment of proper reference cases to support decision making. J Environ Manage 261:110216. https://doi.org/10.1016/j.jenvman.2020.110216

Sharma A, Saini H, Thakur B, Soni R, Soni SK (2023a) Consolidated bioprocessing of biodegradable municipal solid waste for transformation into biofertilizer formulations. Biomass Convers Biorefinery 1–15. https://doi.org/10.1007/s13399-023-04110-2

Sharma P, Nadda AK, Kumar S (2023b) Microbial community profiling in bio-stimulated municipal solid waste for effective removal of organic pollutants containing endocrine disrupting chemicals. Microbiol Res 267:127273. https://doi.org/10.1016/j.micres.2022.127273

Tawfik A, Mohsen M, Ismail S, Alhajeri NS, Osman AI, Rooney DW (2022) Methods to alleviate the inhibition of sludge anaerobic digestion by emerging contaminants: a review. Environ Chem Lett 20(6):3811–3836. https://doi.org/10.1007/s10311-022-01465-2

Tupsakhare S, Moutushi T, Castaldi MJ, Barlaz MA, Luettich S, Benson CH (2020) The impact of pressure, moisture and temperature on pyrolysis of municipal solid waste under simulated landfill conditions and relevance to the field data from elevated temperature landfill. Sci Total Environ 723:138031. https://doi.org/10.1016/j.scitotenv.2020.138031

Vasileiadou A, Zoras S, Dimoudi A (2023) A comprehensive experimental study of municipal solid waste (MSW) as solid biofuel and as composite solid fuel in blends with lignite: quality characteristics, environmental impact, modeling, and energy cover. Energy Ecol Environ 8(3):211–240. https://doi.org/10.1007/s40974-023-00271-y

Vaverková MD, Adamcová D, Zloch J, Radziemska M, Boas Berg A, Voběrková S, Maxianová A (2018) Impact of municipal solid waste landfill on environment—A case study. J Ecol Eng 19(4): 55–68. https://doi.org/10.12911/22998993/89664

Zamri MFMA, Hasmady S, Akhiar A, Ideris F, Shamsuddin AH, Mofijur M, Mahlia TMI (2021) A comprehensive review on anaerobic digestion of organic fraction of municipal solid waste. Renew Sustain Energy Rev 137:110637. https://doi.org/10.1016/j.rser.2020.110637

Zargar TI, Alam P, Khan AH, Alam SS, Abutaleb A, Hasan MA, Khan NA (2023) Characterization of municipal solid waste: measures towards management strategies using statistical analysis. J Environ Manage 342:118331. https://doi.org/10.1016/j.jenvman.2023.118331

Zhu Y, Zhang Y, Luo D, Chong Z, Li E, Kong X (2021) A review of municipal solid waste in China: characteristics, compositions, influential factors and treatment technologies. Environ Dev Sustain 23:6603–6622. https://doi.org/10.1007/s10668-020-00959-9

Chapter 5
An Overview on Municipal Solid Waste Characteristics and Its Impacts on Environment and Human Health

Sadia Sikder, Mohammad Toha, and Md. Mostafizur Rahman

Abstract Municipal solid wastes threaten the environment, resulting from landfills, open burning, careless dumping, etc. Waste is being discarded in increasing amounts worldwide and has a diverse composition. Planning, designing, operating, or updating solid waste management systems requires careful consideration of the properties of municipal solid waste (MSW). Understanding local waste characteristics is crucial in developing efficient waste management strategies, as these characteristics are influenced by cultural, climatic, socioeconomic, and institutional capabilities. Inadequate municipal solid waste management has detrimental effects on the environment and poses risks to various forms of life, including human health. Untreated and inadequately managed municipal solid waste constitutes the principal causative factor behind numerous illnesses and diseases prevalent within a specific locality or even at a national level. Landfills produce a considerable amount of the potent greenhouse gas methane. The decomposition of municipal solid waste is the primary cause of this. Therefore, the overabundance of municipal solid waste affects the environment and human health and should be adequately addressed. This chapter covered the characteristics of municipal solid waste, its composition, factors that affect or influence municipal stable waste characteristics, and its impacts on the environment and human health. It aims to address the effects of municipal solid waste along with its complex features.

Keywords Characteristics · Environmental impacts · Human health · Municipal solid waste

S. Sikder · M. Toha · Md. Mostafizur Rahman (✉)
Department of Environmental Science, Bangladesh University of Professionals, Mirpur, Cantonment, Dhaka 1216, Bangladesh
e-mail: rahmanmm@juniv.edu

Md. Mostafizur Rahman
Laboratory of Environmental Health and Ecotoxicology, Department of Environmental Sciences, Jahangirnagar University, Savar, Dhaka 1342, Bangladesh

A. Anouzla and S. Souabi (eds.), *Technical Landfills and Waste Management*, Springer Water, https://doi.org/10.1007/978-3-031-52633-6_5

5.1 Introduction

The rapid rate of industrialization and urbanization, along with the ever-increasing human population, has resulted in an unprecedented amount of waste (Das et al. 2019). Municipal solid waste (MSW) production has increased as a result of population growth and technological development, posing a threat to ecosystems and human health on a global scale (Alfaia et al. 2017).

According to a study funded by the World Bank, cities produce over 1.3 million tons of municipal solid waste yearly. A fourfold increase by 2025 was predicted by the World Bank (Das et al. 2019). Although urbanization is desirable, unregulated growth can lead to environmental problems such as air and water pollution, the destruction of open space and riverbanks, and the accumulation of waste (Troschinetz and Mihelcic 2009). People in developing countries are moving from rural areas to urban ones due to rapid industrialization and population growth, resulting in the daily production of hundreds of tons of municipal solid waste (Sharholy et al. 2008).

Human culture is reflected in MSW's adverse effects on human health and the natural world. People around the globe are discarding more complicated rubbish thanks to the proliferation of plastic and electronic consumer devices. Global urbanization is entirely unprecedented. Urban waste management is complex from an ethical and environmental perspective. Waste management is more effective when cultural, climatic, economic, and institutional factors are considered (Vergara and Tchobanoglous 2012).

Worldwide, a large amount of garbage is generated from cities. Most of it is burned or dumped in unsanitary landfills in developing nations. Waste is nevertheless occasionally dumped or burned in open areas. These actions harm human health, degrade natural resources, including land and water, and increase waste production. Using discounted costs, we find that adjusting how we handle hazardous waste now would be more cost-effective than fixing the damage we're doing to the environment and people's health now (Stafford 2020).

Because of scientific and technological progress, there is now much more garbage, which is harmful. Mistakes in solid waste management are detrimental to both people and the environment. These kinds of acts cause illness and damage the ecosystem. The setting is under stress, leading to accidents, floods, and declining health. Gas explosions, unpleasant scents, bug infestations, and surface and groundwater contamination also adversely affect the environment (Turan et al. 2009).

Cultural norms regarding waste, cleanliness, and filthiness influence the MSW generation. Batagarawa et al. (2015) argues that MSW is not only a product of society but also depends on that society's degree of development, the rate of social and economic change, and its approach to modernity.

Both wealthy and developing countries suffer from the effects of solid waste pollution. Municipal solid waste management has dramatically advanced in the majority of developing countries. Garbage generated by urbanization is frequently mismanaged in these countries. The lack of adequate technical knowledge and financial resources, which often only cover collection and transit costs, is a significant problem for developing country solid waste management systems. Poor solid waste management has been noted by multiple researchers (Sharholy et al. 2008; Imam et al. 2008; Chung and Carlos 2008; Berkun et al. 2005) in urban areas of developing countries. According to two studies (Sharholy et al. 2008; Alavi Moghadam et al. 2009), poor solid waste management costs developing-world cities a lot of money, pollutes their environments, and kills off countless animals.

Asia's garbage management and disposal practices and the region's tropical climate provide significant environmental challenges. The amount of solid waste produced has risen dramatically due to modern consumption patterns and waste materials (Tarmudi et al. 2009).

Garbage has accumulated over the past few decades due to human actions, consumption patterns, and lifestyle choices. All natural resources necessary for human and animal survival are in danger from waste. As a result, in fewer than two decades, people have become increasingly concerned about waste management and the damage it causes. Hossain et al. (2011) cite a plethora of research examining how different waste treatment methods affect pollution levels and resource recovery.

Due to insufficient conveyance and pickup, MSW is now present everywhere. Due to a lack of infrastructure to process and dispose of the massive amounts of MSW produced daily in developing nations, MSW management is at a crucial juncture. Humans and ecosystems alike are put at risk by careless waste management. Unsafe and unregulated MSW dumps are located in low-lying locations. According to Sharholy et al. (2008), MSWM is a significant problem for the planet. Concerns about pollution and improper MSW disposal must be factored into waste management plans (Turan et al. 2009).

The escalating health and environmental dangers posed by MSW have been the primary focus of most research. However, we need to look at the characteristics and consequences of MSW. Characteristics of MSW were outlined in this article. Environmental and health impacts of municipal solid waste were investigated.

5.2 Municipal Solid Waste

The term "municipal solid waste" (MSW) refers to a highly heterogeneous mixture of a wide range of items, including food, paper, wood, textiles, plastics, metal, glass, dirt, ash, and others (Durmusoglu et al. 2006). In another definition, municipal solid waste (MSW) is any solid or semi-solid waste disposed of by businesses and residents. Still, it does not include wastewater or hazardous waste (Vergara and Tchobanoglous 2012).

When compared to waste streams from industrial or agricultural activities, which are more uniform, municipal solid waste (MSW) is the most complex (Troschinetz and Mihelcic 2009). The garbage in a municipality is typically called "municipal solid waste". Most of this solid waste is produced without sorting, which could be dangerous or innocuous. Municipal solid waste is any solid waste created within the boundaries of a municipality, regardless of its physical and chemical makeup or the source of its production. Private residences, lodging facilities, workplaces, retail establishments, and other institutions produce municipal solid waste (Saleh and Koller 2019).

Most processes inevitably result in the generation of waste or excessive or undesired byproducts. Worldwide, 7–9 billion tons of trash are produced annually (Wilson and Velis 2015). Depending on the reporting standard, municipal solid waste (MSW), a specific category originating from households, may also include commercial and industrial wastes (Wilson and Velis 2015). 2 billion tons of the total waste generated in 2016 were MSW. Given its effects on the local, regional, and global environment, its proximity to people and potential health effects, and its potentiality for recovery through circular economy supply chains, it merits special attention (Chen et al. 2020).

Except for industrial and agricultural waste, all municipally produced solid waste includes construction and demolition debris, sludge and ashes from sewage treatment plants, ashes and ashes from municipal solid waste incinerators, food scraps, yard trimmings, and various inorganic wastes like household hazardous wastes (Stafford 2020). Municipal solid waste is the assortment of solid wastes regularly abandoned as rubbish, waste, and refuse by people in urban and rural areas. Municipal solid waste is created when a municipality collects solid waste from people's homes, workplaces, small institutions, and commercial companies (Nanda and Berruti 2020).

5.3 Municipal Solid Waste Generation Rate

Waste generation and composition are affected by many variables, such as geography, socioeconomic level, and collection frequency. (Alavi Moghadam et al. 2009) A collection and disposal system can be better planned with knowledge of quantity change and generation. Solid wastes in developing nations are characterized by unfavorable economic, institutional, legal, technological, and operational constraints. Due to these problems, individuals are irresponsibly dumping piles of rotting debris, plastic bottles, paper and plastic containers, packaging materials, and paper and plastic bags on the streets and drains (Abba et al. 2013).

Only around 33% of the world's 2 billion tons of municipal solid waste produced annually are collected by municipalities. (Waste Atlas 2018). Every day, each person produces 0.74 kg of garbage on average. (Waste Atlas 2018). The World Bank estimates that by 2050, 3.4 billion tons of urban solid waste will have been generated. (The World Bank 2020). The towns collect 70% municipal solid waste, of which 19% is recycled, and 11% is used for energy recovery. The remaining 30% is disposed of in landfills and other disposal sites. Of the 7.6 billion inhabitants today (US Census Bureau 2020), around 3.5 billion do not have access to essential waste management services. (Waste Atlas 2018). In addition, 5.6 billion people will likely need more access to crucial waste management services by 2050. The generation of municipal solid waste is projected to increase globally between 2016 and 2050, as shown in Fig. 5.1. As the day wears on, it is obviously in an escalating phase.

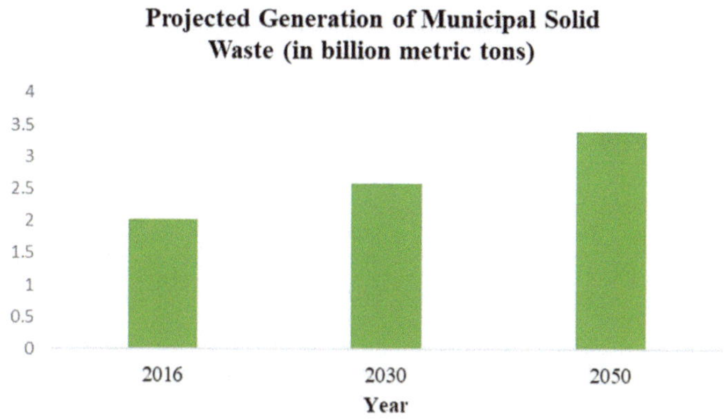

Fig. 5.1 Projected generation of municipal solid waste worldwide from 2016 to 2050 (Anon 2023)

Table 5.1 shows the municipal solid waste generation according to the income level of countries. From this table, it can be clearly understood that the higher-income countries produce more municipal solid waste than any other country. Table 5.2 shows the municipal solid waste generation according to region of countries.

In 1998, 0.76 million tons of municipal solid waste (MSW) were generated daily across Asia, with annual growth rates of 2–3% in developing countries and 3.2–4.5% in industrialized countries, according to Damghani et al. (2008).

India is among the leading producers of MSW in the developing world. High MSW generation is correlated with high GDP per capita in industrialized countries (Das et al. 2019). Figure 5.2 displays the OECD countries' average yearly per capita municipal garbage generation through 2022 (in kilograms). Denmark and the United States produce more municipal solid waste than other OECD nations.

Global waste production has grown from 635 Mt in 1965 to 1999 in 2015 and is projected to reach 3539 Mt by 2050 (Chen et al. 2020). Between 2015 and 2050, the global share of organic waste declined from 47 to 39%, while the shares of

Table 5.1 Municipal solid waste generation according to the income level of countries

Income groups of countries	Average MSW generation rate, 2016 (%)	Average MSW generation rate, 2016 (Kg/capita/day)	Projected MSW generation rate, 2025 (Kg/capita/day)
Lower income countries	5	0.40	0.43
Lower middle-income countries	29	0.53	0.63
Upper middle-income countries	32	0.69	0.83
Higher income countries	34	1.58	1.71

Source World Bank Group (2018), UN-DESA (2018)

Table 5.2 Municipal solid waste generation according to the region of countries

Regions of countries	Average MSW generation rate, 2016 (%)	Average MSW generation rate, 2016 (Kg/capita/day)	Projected MSW generation rate, 2025 (Kg/capita/day)
Middle East and North Africa	6	0.81	0.90
Sub-Saharan Africa (SSA)	9	0.46	0.50
Latin America and Caribbean (LAC)	11	0.99	1.11
North Asia	14	2.21	2.37
South Asia	17	0.52	0.62
Europe and Central Asia	20	1.18	1.30
East Asia and Pacific	23	0.56	0.68

Source Data from WEC (2016) and World Bank Group (2018)

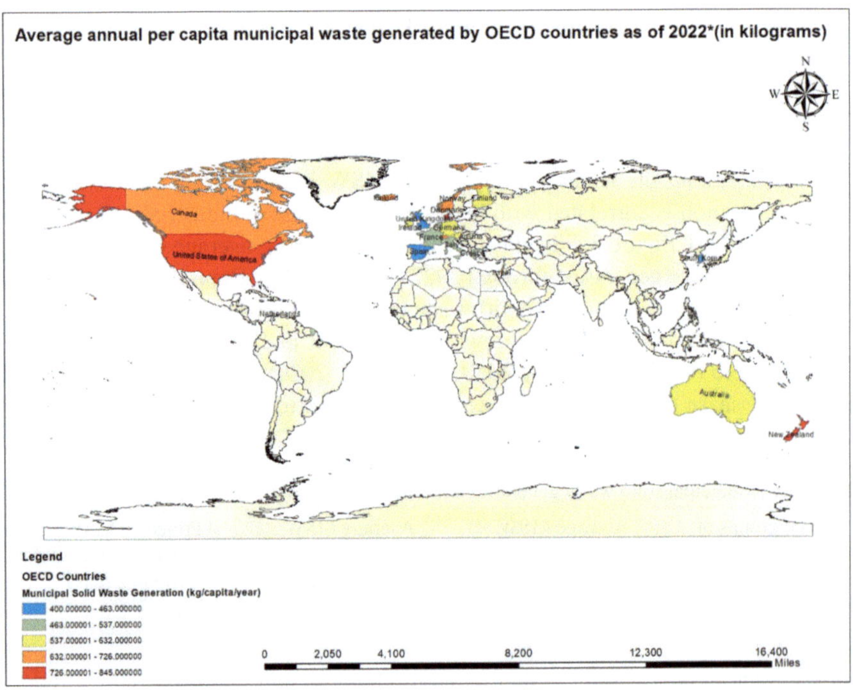

Fig. 5.2 Average annual per capita municipal solid waste generated by OECD countries as of 2022 (Anon 2023)

all other waste, especially paper, increased (Chen et al. 2020). The proportion of garbage processed in dumps declines from 28 to 18% as more regularly employed, environmentally friendly recycling, composting, and energy recovery technologies. We forecast that environmental loads will keep increasing even after the yearly influx of plastic waste into the oceans reaches its peak (Chen et al. 2020).

Two-thirds of the world's population is expected to live in cities by 2025, primarily due to the developing world's rapidly expanding urban people (UN-Habitat and UNESCAP 2015). Globally, 2010 million tons (MT) of MSW were produced in 2016, and by 2050, it is predicted that 3400 MT will have been produced. Eventually, it might lead to an increase in MSW production of almost 70% in just 34 years (Khandelwal et al. 2019). A total of 2.01 BT (0.74 kg/person/day) of MSW was produced worldwide in 2016, of which 33% were managed unsustainable and cautiously. The volume of MSW generation is also expected to increase by 2030, reaching 2.59 BT (7.10 MT/day), and by 2050, reaching 3.40 BT (9.32 MT/day). The average yearly introduction to the solid waste market is 15 MT (Ecoprog 2015).

5.4 Sources and Classification of Municipal Solid Waste

Municipal solid waste (MSW) is divided into three categories based on its origin: urban, industrial, and rural. No matter how many people live in a given location, the SW sources fall within the urban category. Urban areas can be divided into two categories: residential (houses) and non-residential (businesses, institutions, services, building and demolition, and other uses). The second category includes businesses, organizations, and individuals whose operations or products may pose risks to public safety or environmental quality owing to the usage of potentially harmful substances. The industrial sector encompasses all production facilities, whether big or small. The rural sector's agricultural-animal husbandry category contains everything farms and ranches produce (Saleh and Koller 2019).

Municipal solid waste is categorized following where it was produced. Each class produces distinctive solid waste that is divided into the following categories according to the source of generation (Buenrostro et al. 2001):

- **Residential waste** is the garbage produced by homes, such as houses or apartments.
- **Commercial waste**: Waste produced in department shops, supermarkets, eateries, open-air markets, and restaurants.
- **Institutional and service waste**: This includes wastes from public and private workplaces, educational facilities, museums, libraries, archaeological sites, and entertainment venues, including stadiums and movie theatres.
- **Waste from building and demolition projects**: Waste produced at these locations.
- **Specific waste requires** unique management methods due to its relative hazard, condition or state. Sectors including scientific research, healthcare, industrial

and auto repair shops, human and veterinary drugstores, airports and land transportation hubs, among others, produce this garbage.

- **Industrial waste**: Industrial waste is any waste that results from the extraction, utilization, transformation, or creation of products.
- **Agricultural Waste**: Waste produced by agriculture and animal husbandry activities.

There are various types of municipal solid waste, and they are (Sharma and Jain 2020; Joshi and Ahmed 2016):

- **Biodegradable waste**: Food and kitchen garbage, green waste (vegetables, flowers, leaves, and fruits), and paper are all examples of biodegradable waste.
- **Recyclable Materials**: Paper, glass, bottles, cans, metals, some plastics, etc. are recyclable materials.
- **Inert Waste Matter**: Such as dirt and debris.
- **Composite waste**: Composite garbage, including used garments, Tetra packs, and toys made of plastic.
- **Domestic hazardous waste**: Household hazardous waste, also known as domestic hazardous waste, includes blood-stained cotton, disposable syringes, sanitary napkins, etc.
- **Toxic waste**: paints, pesticides, medications, used batteries, prescription drugs, e-waste, chemicals, lightbulbs, fluorescent tubes, spray cans, fertilizer, etc.

Organic materials (mainly from the food and gardening industries), cardboard, paper, plastics and other resins, textile rags, metal, and glass make up the vast majority of trash. Batteries, electric light bulbs and fluorescent tubes, vehicle parts, expired pharmaceuticals and other pharmaceutical products, and other chemicals, such as chlorine, are only some hazardous garbage found. Waste from construction and demolition sites is often collected as well. Therefore, homes and the agricultural, industrial, construction, commercial, and institutional sectors are the principal generators of municipal solid garbage (Saleh and Koller 2019).

Specific categories of primarily generated municipal solid waste are (Zhou et al. 2014):

- **Food residue:** Most municipal solid waste comprises food remnants, which can be further classified into the following five groups: vegetables, fruit peel, bones, starchy foods, and nutshells.
- **Wood waste:** There are four subcategories of wood waste in MSW: wood, bamboo, leaves, and weeds.
- **Paper:** The paper found in MSW can be broken down into three subgroups: cardboard, toilet paper, and printing paper (which includes newspapers, books, and magazines).
- **Textiles:** Cotton, synthetic fibres, and wool are the three subgroups that make up the textiles in MSW.

- **Plastics:** In contrast to other MSW components, plastics are frequently pure. PE (includes high-density polyethene and low-density polyethene), PP (polypropylene), PS, PVC, and PET (polyethene terephthalate) are five types of frequently used plastics.
- **Rubber**: Waste tires were the primary source of rubber in MSW.

5.5 Characteristics and Composition of Municipal Solid Waste

Although it varies greatly in content and classification between various towns worldwide, municipal solid waste comprises both biodegradable and non-biodegradable fractions from organic and inorganic components. Municipal solid waste, however, frequently includes rubbish such as home waste, yard waste, paper and cardboard, plastic and rubber, metal, glass, and electronic waste, as well as inert materials and other garbage. Kitchen and garden waste comprise most of the municipal solid waste considered organic. The most varied type of municipal solid waste is miscellaneous garbage, which includes textiles, fabrics, biological wastes (including sharps and glasses), personal hygiene products, healthcare products, cosmetics, pharmaceuticals, pet litter, leather, rubber, and polymeric residues (Nanda and Berruti 2020).

Globally, organic waste (food and green waste) accounts for 44% of MSW, with paper and cardboard coming in at 17%, plastic at 12%, glass at 5%, metal at 4%, wood at 2%, and other materials at 2% (Sharma and Jain 2020). Figure 5.3. displays the usual breakdown of municipal solid waste generation (in 2016). MSW contains a more significant proportion of organic material than any other component.

Organic matter, paper, cardboard, textiles, polymeric materials (plastics and rubbers), glass, wood, and ferrous and non-ferrous metals are the main components of MSW's physical composition. This mixture contains some biodegradable materials and others that will contaminate the environment by lingering in nature for an extended period. However, the enormous quantities of non-biodegradable materials created present a hazard if they are not correctly handled and wind up in landfills. This issue is even more problematic given that some of these materials, such as food packaging constructed of synthetic polymeric polymers, might take hundreds of years to disintegrate (Alfaia et al. 2017).

Due to poor energy and resource utilization, human activities produce municipal solid waste (MSW). Certain MSWs cannot be directly reused for public benefit because they may harm human health. Paper waste (paper scraps, cardboard, newspapers, magazines, bags, boxes, wrapping paper, phone books, shredded paper, paper drinking cups, etc.) and organic rubbish (food scraps, yard waste, grass, brush, wood, paper from processing residues, etc.) are both included in MSW. Paper (which is technically organic but is not regarded as such unless it has food residue on it), plastic waste (PW) (bottles, packaging, containers, bags, lids, and cups), glass waste (bottles,

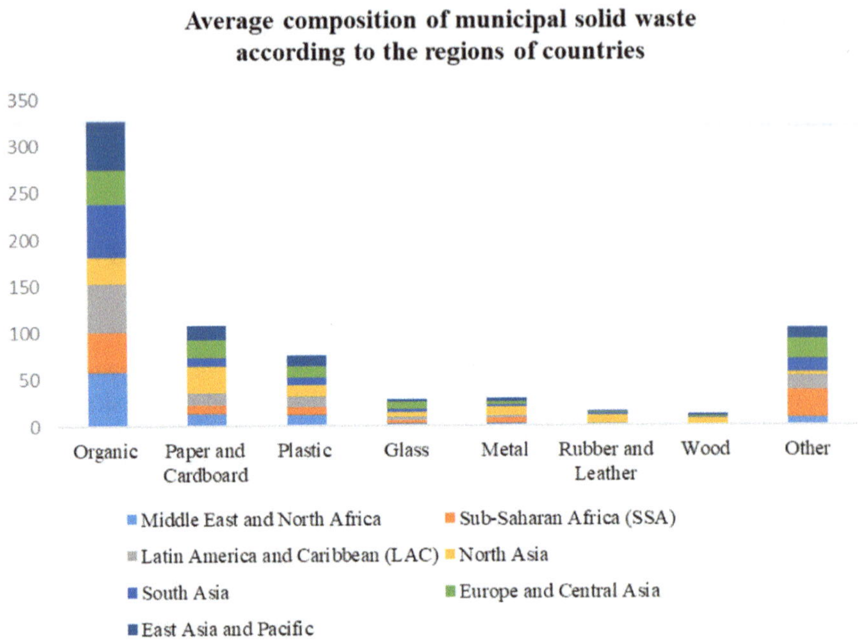

Fig. 5.3 Composition of municipal solid waste according to the regions of countries. *Source* Data from WEC (2016) and World Bank Group (2018)

broken glassware, lightbulbs, coloured glass, etc.), metal waste (cans, foil, tins, non-hazardous aerosol cans, railings, bicycles, etc.), and other debris are all examples of waste (Sharma and Jain 2020).

5.6 Impacts of Municipal Solid Waste on Environment

Creating municipal waste on a global scale has several detrimental effects on the environment, including nitrogen pollution, the buildup of plastic litter in the oceans, and greenhouse gas emissions (Chen et al. 2020). Solid waste production is inextricably linked to air, land, and water pollution (Brunner 2015).

Municipal solid waste has far-reaching effects on the environment and numerous forms of life, including polluting air, water, and soil. The impact of municipal solid waste on land use, odours, and aesthetic qualities have been considered throughout the design phase of waste treatment systems (Saleh and Koller 2019).

Poor solid waste management pollutes the environment by contaminating the air, the surface and ground waters, and the soil. The outcome is determined by the kind of waste produced and how it is disposed of. The traditional landfilling approach generates various landfill gases, including methane, carbon dioxide, carbon monoxide, nitrogen, hydrogen sulfide, and ammonia. Of all greenhouse gases, methane and carbon dioxide make up 40–60% of the total. Among the gases that have disagreeable odours and are dangerous are ammonia and hydrogen sulfide. These unpleasant scents can harm the health of persons close to the dump if an effective collection or venting system is not built (Pokhrel and Viraraghavan 2005).

The creation of municipal solid waste significantly pollutes the environment, mainly due to the release of greenhouse gases like methane and carbon dioxide, high organic content, and frequent incorrect disposal of waste (Saleh and Koller 2019).

Environmental concerns associated with solid waste generation are often overlooked in the context of artificial and economic development. The open dumping of solid waste, commonly observed in low-income and subsistence farming, has detrimental effects on various aspects of terrestrial ecosystems. An "open dump" refers to a substantial area of land designated for storing or disposing of solid waste without covering it. Open dumps are frequently utilized in underdeveloped and emerging nations owing to their straightforward functionality and cost-effectiveness. The global distribution of available dumps is widespread, while their monitoring poses significant difficulties (Das et al. 2019).

The final percentage of solid waste treatment and its global disposal outcome in 2016 are shown in Fig. 5.4. It states that 25% of solid waste is disposed of in landfills, increasing environmental damage, while 33% of solid garbage is thrown in the open. The constraints of processing different solid waste products have been highlighted in Fig. 5.5. These restrictions result in an ever-increasing burden on our environment.

According to the analysis conducted by the International Solid Waste Association (ISWA 2016), it was discovered that there are approximately 50 expansive open dumps scattered throughout various continents, occupying a total area of 2175 acres. The study revealed that the practice of open dumping of solid waste in Islamabad, Pakistan, resulted in significant soil deterioration, leading to a mortality rate

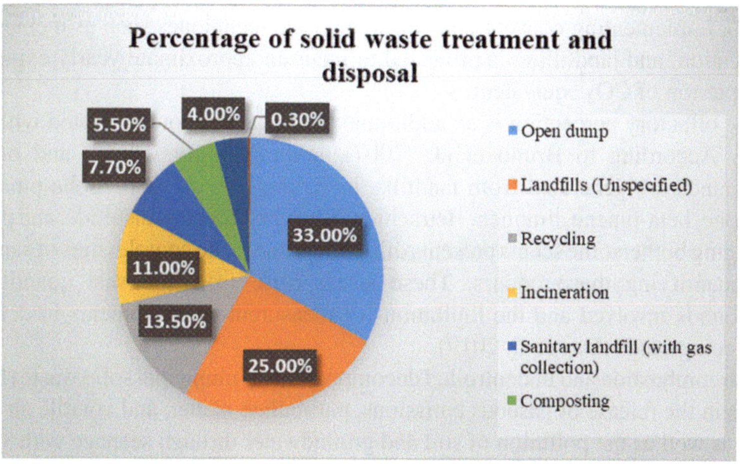

Fig. 5.4 Percentage of solid waste treatment and disposal around the world in 2016 (Sharma and Jain 2020)

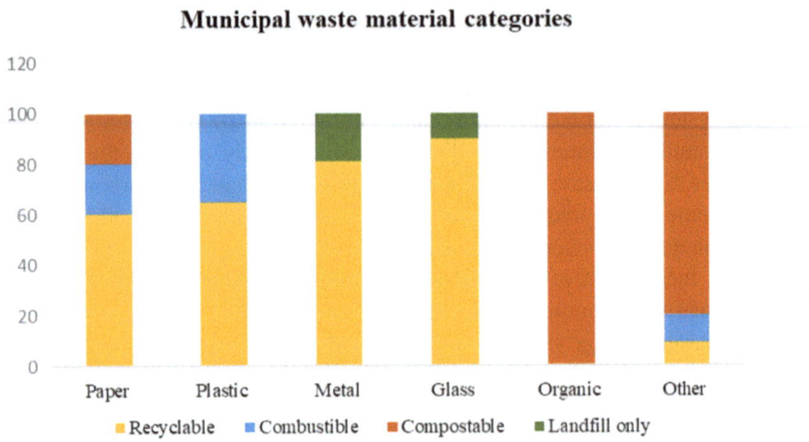

Fig. 5.5 Municipal waste material categories (Daskalopoulos et al. 1998)

of 72% among the vegetation in the vicinity. The improper disposal of solid wastes has the potential to generate harmful pollutants that can negatively impact the soil system and impede the physiological processes of plant roots (Ali et al. 2014). The presence of metal species, including Cd, Cr, Pb, and Ni, in solid waste can be categorized as intrinsic pollutants due to their ability to be assimilated by plants via their root systems. This absorption process obstructs the plant's vascular system, finally resulting in its demise (Jha et al. 2016).

Solid waste landfills contribute around 14% of global methane emissions. Based on the findings of the 17th Conference of Parties (COP) conference, it was observed that the global waste management industry achieved a reduction of around 2.7% in greenhouse gas (GHG) emissions. Based on estimations provided by Das et al. (2019), implementing conventional waste disposal techniques such as incineration, gasification, and landfilling is projected to incur an approximate yearly expense of \$100 per ton of CO_2 equivalent.

The olfactory perception is an additional significant factor associated with solid waste. According to Bruno et al. (2007), several highly volatile and odorous compounds are generated from landfills, including benzoic acid, alpha-pinene, 2-butanone, beta-pinene, limonene, tetrachloroethylene, dimethyldisulfide, and phenol. Managing bothersome scents presents difficulties due to the complexities of sampling and quantifying these odours. These issues arise from the vast quantities of compounds involved and the limitations of measurement instruments in detecting these compounds (Das et al. 2019).

The combustion and uncontrolled decomposition of municipal solid waste (MSW) results in the release of gaseous emissions, particulate matter, and volatile air pollutants, as well as the pollution of soil and groundwater through seepage with various chemical and biological substances. Nevertheless, the process of organic waste decomposition results in the production of a significant quantity of methane, a potent greenhouse gas. This methane emission contributes to the elevation of the global average temperature and subsequently alters the Earth's climate (Pujara et al. 2019; Cetinkaya et al. 2018).

The problem has worsened because of a lack of acceptable fundamental amenities, such as sanitary water supplies, transportation infrastructure, and suitable sanitation facilities. Finding the money to manage garbage adequately is still one of the most significant issues in most countries. As a result, poisonous solids, liquid, air emissions, leachates, and disagreeable odours are present in the neighbourhood. These problems have been endangering the environment and human health locally and globally. As a result, solid waste generation will continue to be an issue and significantly impact climate change and global warming (Abba et al. 2013).

Unmanaged waste harms wildlife and contaminates our oceans, clogs sewers, spreads illnesses, worsens respiratory conditions, and impedes the growth of the world's economy. In 2016, the treatment and disposal of MSW resulted in approximately 1.6 billion tons (BT) of GHG emissions (CO_2 eq.), primarily due to open dumping and landfilling. If no adjustments are made to the SWM business, this amount is anticipated to rise by almost 2.6 BT by the year 2050. (Sharma and Jain 2020).

Most airborne contaminants from municipal waste discharged for a period are hydrocarbons. A far wider variety of chemicals are often present in the air and recently buried waste. It has been demonstrated that landfill emissions of higher molecular weight alkanes, alkenes, and aromatic compounds increase with time when anaerobic conditions become well-established. The early stages of waste disintegration are when most of these compounds are lost to the air. Others may evaporate straight from the waste, while some of them, like chlorinated aliphatic hydrocarbon solvents, may be produced by microbial decomposition. Hydrogen sulfide, carbonyl sulfide, and carbon disulfide emissions are most likely linked to garbage less than six months old or places where anaerobic degradation is not occurring effectively (Ari Setyan et al. 2016).

Because waste is heterogeneous and complicated, it includes a variety of toxins and produces leachate that seeps into groundwater. These dumpsites get many fluxes of garbage from numerous sources. Leachate migration may cause heavy metal pollution of soils, including lead, cadmium, zinc, iron, nickel, manganese, and chromium. The migration of heavy metal contaminants from municipal solid waste dumpsites to neighbouring soil results from a geochemical process. This is controlled by the properties of the solid waste phases in or on which metals can be bound, the pH, the redox potential, and complexing agents such as chlorides, sulfates, carbonates, and organic acids. Due to these metals' longevity in the environment, hazardous concentrations could develop. These metals could build up in locally grown plants, which could then be eaten by grazing animals and finally make their way to people through the food chain (Ihedioha et al. 2016).

The environmental effects of the various waste management techniques continuously degrading our environment are shown in Table 5.3. The landfill is where most waste ends as it releases a tremendous amount of greenhouse gases (GHGs) into the atmosphere.

Table 5.3 Environmental impacts of various waste technologies

Environmental sink	Pollutants emitted from various waste technologies					
	Landfill	Incineration	Composting	Transport	Recycling	Land application
Air	CO_2, CH_4, odor, noise, VOCs, GHGs (CO_2, CH_4, N_2O)	SO_2, N_2O, HCl, CO, CO_2, dioxins, furans, PAHs, VOCs, GHGs, Hg	Odor, GHGs (minor)	CO_2, SO_2, NO_x, odor	GHGs (minor)	Bioaerosols, odor, GHGs (minor)
Soil	Heavy metals, organic compounds	Fly ash, slag	Minor impact	Heavy metals, Hydrocarbons	Landfilling of residues	Bacteria, viruses, heavy metals, PAH, PCBs
Water	Leachate, heavy metals, organic compounds	Fallout of atmospheric pollutants	Leachate	Fallout of atmospheric pollutants (e.g., nitrate)	Wastewater from processing	Bacteria, viruses, heavy metals

Reference Vergara and Tchobanoglous (2012)

5.7 Impacts of Municipal Solid Waste on Human Health

Without adequate research, the disposal of municipal solid waste (MSW) poses significant risks to the environment and public health. Developing countries need help with a complex, multifaceted solid waste management system due to a lack of administrative commitment, financial resources, poor management, and scientific input. Health risk assessment is an essential technique for identifying the detrimental effects of municipal solid waste on human health. There are both cancer-causing and non-cancerous effects of contaminated soil. The three potential exposure routes, ingestion, inhalation, and dermal, are considered when assessing risk (Gujre et al. 2021).

High organic content in MSW promotes the growth of microbial infections, increasing the risk of infectious and chronic diseases in garbage workers, rag collectors, and nearby homeowners. Burning MSW or dangerous microbes has been connected to several health problems, including allergies, gastrointestinal problems, lung problems, skin irritation, eyes and nose problems, and psychological problems. In Mumbai, 25% of waste collectors and 71% of Kolkata had respiratory conditions (Di Maria et al. 2021) on the occupational health and safety of workers employed by the municipal corporations in Kolkata and Mumbai. Additional problems like chemical poisoning, low birth rates, cancer, congenital deformities, nausea and vomiting, and neurological disorders are also brought on due to the careless mixing of hazardous and technological wastes with solid garbage. The risk of getting Hepatitis B and HIV increases when household biomedical waste and medical supplies are mixed with domestic waste and thrown in untreated landfills. Among the most often noted health impacts in populations exposed to incinerator emissions include congenital disabilities, congenital deformities, generalized and site-specific malignancies, non-cancer disorders, death, and hospital admission (Di Maria et al. 2021).

A few studies revealed favourable connections between congenital and total malformations, cardiac malformations, urinary malformations, facial clefts and breast cancers due to MSW (Di Maria et al. 2021). In reality, enterprises and incinerators for industrial and medical waste were present in addition to the ten municipal waste incinerators in one study on sarcoma (Zambon et al. 2007). Leukemia (Minichilli et al. 2016), Hodgkin disease, pleural and bladder cancer (Ranzi et al. 2011), or any other illness linked with MSW (Di Maria et al. 2021).

The weather at the dump will last for a limited period, and the pollution will only harm the health of the community's most vulnerable residents, who could feel uneasy. Residents who reside near garbage incinerators have also reported various detrimental effects, including cancer-causing and damaging reproductive effects. In many of these instances, the public's worries have been sparked by information about the toxicological properties of emitted pollutants that have been made public, generally in the daily news (Njoku et al. 2019).

These metals could build up in locally grown plants, which could then be eaten by grazing animals and finally make their way to people through the food chain. These heavy metals pose serious health hazards since they are xenobiotics. Lead and cadmium can damage the liver and kidneys, whereas zinc, copper, and nickel are essential minerals but can be hazardous in far higher doses. Thirty per cent of the kids tested showed red blood cell abnormalities that indicated significant exposure to heavy metal poisoning. About fifty per cent of the kids had blood levels of lead that were equivalent to or higher than globally recognized dangerous levels. According to a medical study of the population, children and teenagers living and attending school close to a landfill in Kenya had a high frequency of disorders connected to excessive exposure to heavy metal pollution (Ihedioha et al. 2016).

5.8 Municipal Solid Waste Management

Municipal solid waste is managed in various ways by municipalities, towns, states, and countries. However, the main steps in managing municipal solid waste are (1) producing waste, (2) collecting, handling, and transporting waste, and (3) disposing of, processing, and treating waste (Nanda and Berruti 2020).

Landfilling is still considered the most popular method of managing MSW, even if recycling efforts have increased and new treatment techniques have recently been developed. It is the most straightforward and practical choice currently available. Effective landfill design criteria require knowledge of the geotechnical properties of the deposited MSW. Quantifying waste material characteristics, mainly when working with heterogeneous materials like MSW, can be difficult. (Durmusoglu et al. 2006). Large cities around the world need help managing their solid waste. (SW). This is especially true in developing nations due to the rapid growth in SW generation brought on by rapid urbanization, industrialization, and economic development. In many urban areas of developing countries, government and municipal authorities administer the SW system from collection to processing. However, most enterprises fail to provide high-quality service for various reasons. Important MSW components cannot be easily recycled or composted, which could lead to pollution, a loss of aesthetic value, and financial losses. Inadequate SW management may also result in severe, sanitary, and environmental problems, such as groundwater contamination from leachate percolation, an unpleasant odour, and explosion risk in landfill zones. The resources and energy that could be recycled and produced from a sizable amount of solid waste are also wasted by inefficient SW collection and management techniques (Damghani et al. 2008).

Municipal solid waste (MSW) management strategies in most developed countries are based on the "Waste Management Hierarchy" theory. This thinking strongly emphasizes prevention/minimization, materials recovery, landfilling, and incineration. How much each alternative is used in a particular country depends on various factors, including terrain, population density, transportation infrastructures, socioeconomics, and environmental constraints. Anaerobic digestion, composting, and other biological treatment techniques are reemerging as economically viable strategies to permanently reduce the amount of organic material in the waste stream. Because the success of these technologies depends on keeping a steady market for the treated product, nations are putting regulatory procedures in place to ensure that compost quality is appropriate for the product's intended application. Typically, this has resulted in the decision to exclusively handle the putrescible fraction of the waste stream rather than processing mixed solid waste (Cucchiella et al. 2014).

Since composting biodegradable waste could replenish nutrients in the soil in the form of fertilizer for farming, its proper use is essential for sustainable growth (Alfaia et al. 2017). Over the past few decades, the problems with municipal solid waste (MSW) management have gotten dangerously worse in developing countries. The high population growth rate has increased economic activity in rising country metropolises, and a lack of training in modern solid waste management methods impedes efforts to improve solid waste management services. Although developing nations produce less solid waste per person than developed countries in urban residential areas, they are less able to collect, process, dispose of, or recycle that garbage more cost-effectively than developed nations. The waste generated by human settlements and the problems they create are comparable in developing countries. However, there are distinctions between regions and locations due to geographic, social, industrial, infrastructural, legal, and environmental factors (Ahsan et al. 2014).

Human activities like manufacturing, agriculture, household, and industrial activities all produce waste. Environmental and public health threats can result from how the waste is handled, stored, collected, and disposed of. Most industrialized countries, including Britain, the United States, and Canada, have much economic activity, leading to increased waste production. These countries do, however, have efficient waste management systems. This situation cannot be compared to developing nations like Nigeria, Ghana, and Cameroon, which have inefficient waste management systems despite continually expanding populations and rising rates of rubbish creation from human and industrial activity. Any current management system faces difficulties due to a lack of an appropriate management plan, an institutional structure, and financial resources. It is also challenging to implement an efficient waste management system due to the high rate of urbanization and the rise in the number of people relocating to urban regions. City residents, particularly those close to the disposal sites, face many health risks because of the potential for rubbish to contaminate water, food sources, soil, air, and vegetables (Ihedioha et al. 2016).

Municipal solid trash is typically disposed of in dumps and landfills since it is the most straightforward, cost-effective, and low-tech option. Greenhouse gases are produced during the decomposition of organic matter, which makes up the vast majority of municipal solid waste in anaerobic landfills. According to the available data, integrated solid waste management is the best solution. However, this approach must be adjusted to account for the growing proportion of organic waste in municipal landfills. When executed correctly, this can have a positive economic impact in addition to helping reduce emissions of greenhouse gases. Systems for managing municipal solid waste sustainably are excellent and intriguing study subjects to examine to assess current consumption trends worldwide and safeguard the environment (Saleh and Koller 2019).

Improving solid waste disposal methods is far more complex in developing countries and emerging regions with mixed economies than in wealthy countries. Most municipal administrations in developing countries need more resources and expertise to provide their rapidly growing populations with the solid waste management infrastructure and services required to ensure an adequate level of life. Emerging nations now seek to establish a balance between economic growth and environmental advances in the context of sustainable development. These countries have given different institutional and legislative components of environmental transformation priority as part of their ecological planning. This will require increased environmental efforts by the federal government, local governments, and the business sector and collaboration to build the necessary environmental infrastructure in urban and industrial areas (Turan et al. 2009).

5.9 Conclusion

Municipal solid waste is a serious environmental issue that needs to be addressed immediately. The world population is growing, and rising consumption patterns are the leading causes of the alarming rate at which solid waste is produced. The inappropriate handling of this waste has several detrimental effects on the environment, such as the contamination of the air and water, the emission of greenhouse gases, and the depletion of natural resources (Hoornweg and Bhada-Tata 2012).

Although waste will always be produced by the people, making intelligent decisions about what constitutes rubbish (i.e., minimizing wastage), lowering production, and implementing efficient management techniques by the government are essential steps in averting irreversible ecological problems (Alfaia et al. 2017).

Reduction, reuse, recycling, and appropriate disposal techniques must all be used in conjunction for municipal solid waste management to be effective. Governments and other stakeholders should collaborate to create policies and programs that support sustainable waste management practices, such as recycling and waste reduction, to reduce the quantity of garbage produced and lessen its environmental impact.

By adopting environmentally friendly behaviours, including lowering waste generation, recycling, and properly disposing of waste, people can play a significant part in controlling municipal solid waste. Together, we can lessen the quantity of garbage created and ensure that it is handled in a way that promotes sustainability and the preservation of the environment.

References

Abba AH, Noor ZZ, Yusuf RO, Din MFMD, Hassan MAA (2013) Assessing environmental impacts of municipal solid waste of Johor by analytical hierarchy process. Resour Conserv Recycl 73:188–196. https://doi.org/10.1016/j.resconrec.2013.01.003

Ahsan A, Alamgir M, El-Sergany MM, Shams S, Rowshon MK, Daud NNN (2014) Assessment of Municipal Solid Waste Management System in a Developing Country. Chin J Eng 2014:1–11. https://doi.org/10.1155/2014/561935

Alavi Moghadam MR, Mokhtarani N, Mokhtarani B (2009) Municipal solid waste management in Rasht City, Iran. Waste Manage 29(1):485–489. https://doi.org/10.1016/j.wasman.2008.02.029

Alavi Moghaddam MR, Sadeghcheh T (1998) A review on solid waste management in Iran (In Persian)

Ali SM, Pervaiz A, Afzal B, Hamid N, Yasmin A (2014) Open dumping of municipal solid waste and its hazardous impacts on soil and vegetation diversity at waste dumping sites of Islamabad city. J King Saud Univ Sci 26:59–65. https://doi.org/10.1016/j.jksus.2013.08.003

Anon (2023) Per capita MSW generation by OECD country. Statista. https://www.statista.com/statistics/478928/leading-countries-by-per-capita-generated-municipal-waste/

Batagarawa R, Williams JB, Potts JS, Brown JC (2015) Use of analytic hierarchy process (AHP) as an instrument to develop a solid waste management assessment tool. Global J Adv Eng Technol 4(2):70

Berkun M, Aras E, Nemlioglu S (2005) Disposal of solid waste in Istanbul and along the Black Sea coast of Turkey. Waste Manage 25:847–855

Bruno P, Caselli M, de Gennaro G, Solito M, Tutino M (2007) Monitoring of odor compounds produced by solid waste treatment plants with diffusive samplers. Waste Manage 27:539–544. https://doi.org/10.1016/j.wasman.2006.03.006

Brunner PH, Rrechberger H (2015) Waste to energy–Key element for sustainable waste management. Waste Manag 37:3–12

Buenrostro O, Bocco G, Cram S (2001) Classification of sources of municipal solid wastes in developing countries. Resour Conserv Recycl 32(1):29–41. https://doi.org/10.1016/s0921-3449(00)00094-x

Cetinkaya AY, Bilgili L, Levent Kuzu S (2018) Life cycle assessment and greenhouse gas emission evaluation from Aksaray solid waste disposal facility. Air Qual Atmos Health 11(5):549–558

Chen DM-C, Bodirsky BL, Krueger T, Mishra A, Popp A (2020) The world's growing municipal solid waste: trends and impacts. Environ Res Lett 15(7):074021. https://doi.org/10.1088/1748-9326/ab8659

Chung SS, Carlos WHL (2008) Local waste management constraints and waste administrators in China. Waste Manage 28(2):272–281

Cucchiella F, D'Adamo I, Gastaldi M (2014) Strategic municipal solid waste management: a quantitative model for Italian regions. Energy Convers Manage 77:709–720. https://doi.org/10.1016/j.enconman.2013.10.024

Damghani AM, Savarypour G, Zand E, Deihimfard R (2008) Municipal solid waste management in Tehran: current practices, opportunities and challenges. Waste Manage 28(5):929–934. https://doi.org/10.1016/j.wasman.2007.06.010

Das S, Lee S-H, Kumar P, Kim K-H, Lee SS, Bhattacharya SS (2019) Solid waste management: scope and the challenge of sustainability. J Clean Prod https://agris.fao.org/agris-search/search.do?recordID=US201900360488

Daskalopoulos E, Badr O, Probert SD (1998) An integrated approach to municipal solid waste management. Resour Conserv Recycl 24(1):33–50. https://doi.org/10.1016/s0921-3449(98)00031-7

Di Maria F, Mastrantonio M, Uccelli R (2021) The life cycle approach for assessing the impact of municipal solid waste incineration on the environment and on human health. Sci Total Environ 776:145785. https://doi.org/10.1016/j.scitotenv.2021.145785

de Alfaia RGSM, Costa AM, Campos JC (2017) Municipal solid waste in Brazil: a review. Waste Manage Res 35(12):1195–1209. https://doi.org/10.1177/0734242x17735375

Durmusoglu E, Sanchez IM, Corapcioglu MY (2006) Permeability and compression characteristics of municipal solid waste samples. Environ Geol 50(6):773–786. https://doi.org/10.1007/s00254-006-0249-6

Ecoprog (2015) Waste to Energy 2015/2016—Technologies, plants, projects, players and backgrounds of the global thermal waste treatment business, Extract 8th edition, https://ecoprog.com/pdf/Waste-to-Energy_2022-2023_-_extract_-_ecoprog.pdf

Gujre N, Mitra S, Soni A, Agnihotri R, Rangan L, Rene ER, Sharma MP (2021) Speciation, contamination, ecological and human health risks assessment of heavy metals in soils dumped with municipal solid wastes. Chemosphere 262:128013. https://doi.org/10.1016/j.chemosphere.2020.128013

Hoornweg D, Bhada-Tata P (2012) What a waste: a global review of solid waste

Hossain MdS, Santhanam A, Nik Norulaini NA, Omar AKM (2011) Clinical solid waste management practices and its impact on human health and environment—A review. Waste Manage 31(4):754–766. https://doi.org/10.1016/j.wasman.2010.11.008

Ihedioha JN, Ukoha PO, Ekere NR (2016) Ecological and human health risk assessment of heavy metal contamination in soil of a municipal solid waste dump in Uyo, Nigeria. Environ Geochem Health 39(3):497–515. https://doi.org/10.1007/s10653-016-9830-4

Imam A, Mohammed B, Wilson DC, Cheeseman CR (2008) Solid waste management in Abuja, Nigeria. Waste Manage 28(2):468–472

ISWA (2016) A roadmap for closing waste dumpsites the world's most polluted places. International Solid Waste Association, Vienna

Jha P, Sasmal A, Santra S, Dewanji A (2016) Heavy metal accumulation potential of some wetland plants growing naturally in the city of Kolkata, India. Am J Plant Sci 7:2112–2137

Joshi R, Ahmed S (2016) Status and challenges of municipal solid waste management in India: a review. Cogent Environ Sci 2(1). https://doi.org/10.1080/23311843.2016.1139434

Khandelwal H, Thalla AK, Kumar S, Kumar R (2019) Life cycle assessment of municipal solid waste management options for India. Biores Technol 288:121515. https://doi.org/10.1016/j.biortech.2019.121515

Kirkeby JT, Birgisdottir H, Hansen TL, Christensen TH, Bhander GS, Hauschild M (2006) Evaluation of environmental impacts from municipal solid waste management in the municipality

of Aarhus, Denmark (EASEWASTE). Waste Manage Res 24(1):16–26. https://doi.org/10.1177/0734242x06062598
Management. Urban development series knowledge papers. 15:1–98
Minichilli F, Santoro M, Linzalone N, Maurello MT, Sallese D, Bianchi F (2016) Studio epidemiologico di coorte residenziale su mortalità e ricoveri ospedalieri nell'area intorno all'inceneritore di San Zeno. Arezzo. Epidemiol. Prev. 40:33–43
Nanda S, Berruti F (2020) Municipal solid waste management and landfilling technologies: a review. Environ Chem Lett. https://doi.org/10.1007/s10311-020-01100-y
Njoku PO, Edokpayi JN, Odiyo JO (2019) Health and environmental risks of residents living close to a landfill: a case study of Thohoyandou Landfill, Limpopo Province, South Africa. Int J Environ Res Public Health 16(12):2125. https://doi.org/10.3390/ijerph16122125
Pokhrel D, Viraraghavan T (2005) Municipal solid waste management in Nepal: practices and challenges. Waste Manage 25(5):555–562. https://doi.org/10.1016/j.wasman.2005.01.020
Pujara Y, Pathak P, Sharma A, Govani J (2019) Review on Indian Municipal Solid Waste Management practices for reduction of environmental impacts to achieve sustainable development goals. J Environ Manage 248:109238. https://doi.org/10.1016/j.jenvman.2019.07.009
Ranzi A, Fano V, Erspamer L, Lauriola P, Perucci CA, Forastiere F (2011) Mortality and morbidity among people living close to incinerators: a cohort study based on dispersion modeling for exposure assessment. Environ Health 10:2
Saleh H, Koller M (2019) Introductory chapter: municipal solid waste. Municipal Solid Waste Manage. https://doi.org/10.5772/intechopen.84757
Sharholy M, Ahmad K, Mahmood G, Trivedi RC (2008) Municipal solid waste management in Indian cities—A review. Waste Manage 28(2):459–467. https://doi.org/10.1016/j.wasman.2007.02.008
Sharma KD, Jain S (2020) Municipal solid waste generation, composition, and management: the global scenario. Soc Responsibil J 16(6):917–948. https://doi.org/10.1108/srj-06-2019-0210
Setyan A, Patrick M, Wang J (2016) Emissions of airborne pollutants from the municipal solid waste incineration plants of Giubiasco (TI) and Hinwil (ZH). Semantic Scholar. https://www.semanticscholar.org/paper/Emissions-of-airborne-pollutants-from-the-municipal-Setyan-Patrick/e4d92014e87418ed0e4645d55b2ae16c21bdeef8
Stafford WHL (2020) WtE best practices and perspectives in Africa. In: Municipal solid waste energy conversion in developing countries, pp 185–217. https://doi.org/10.1016/b978-0-12-813419-1.00006-1
Tarmudi Z, Abdullah L, Osman A (2009) An overview of municipal solid wastes generation in Malaysia an overview of municipal solid wastes generation in Malaysia. Jurnal Teknologi 51:1–15. https://core.ac.uk/download/pdf/300360192.pdf
The World Bank (2020a) Trends in solid waste management. https://datatopics.worldbank.org/what-a-waste/trends_in_solid_waste_management.html Accessed 21 Apr 2023
Troschinetz AM, Mihelcic JR (2009) Sustainable recycling of municipal solid waste in developing countries. Waste Manage 29(2):915–923. https://doi.org/10.1016/j.wasman.2008.04.016
Turan NG, Çoruh S, Akdemir A, Ergun ON (2009) Municipal solid waste management strategies in Turkey. Waste Manage 29(1):465–469. https://doi.org/10.1016/j.wasman.2008.06.004
U.S. Census Bureau (2020) U.S. and world population clock. https://www.census.gov/popclock/. Accessed 22 Apr 2023
UN Habitat and UNESCAP (2015) UN-Habitat and UNESCAP to strengthen Asian and Pacific partnerships. UN-Habitat
UN-DESA (2018) Population division (2019) World urbanization prospects: the 2018 revision (ST/ESA/SER.A/420) [WWW document]. United Nations, Department of Economic and Social Affairs (UN DESA). Available at https://doi.org/10.4054/demres.2005.12.9
Vergara SE, Tchobanoglous G (2012) Municipal solid waste and the environment: a global perspective. Annu Rev Environ Resour 37(1):277–309. https://doi.org/10.1146/annurev-environ-050511-122532

Waste Atlas (2018) What a waste: an updated look into the future of solid waste management. https://www.worldbank.org/en/news/immersive-story/2018/09/20/what-a-waste-an-updated-look-into-the-future-of-solid-waste-management. Accessed 15 Apr 2023

WEC (2016) World energy resources-2016 [WWW document]. World Energy Council. Available at https://www.worldenergy.org/wp-content/uploads/2016/10/World-Energy-Resources-Full-report-2016.10.03.pdf

Wilson DC, Velis CA (2015) Waste management—still a global challenge in the 21st century: an evidence-based call for action Waste Manage. Res 33:1049–1051

World Bank Group (2018) "What a waste 2.0: a global snapshot of solid waste management to 2050-The urban development series [WWW document]. International Bank for Reconstruction Devloment/World Bank 1818 H Str. NW, Washington, DC. Available at https://openknowledge.worldbank.org/handle/10986/30317

Zambon P, Ricci P, Bovo E, Casula A, Gattolin M, Fiore AR, Chiosi F, Guzzinati S (2007) Sarcoma risk and dioxin emissions from incinerators and industrial plants: a population-based case-control study (Italy). Environ Health 6:19

Zhou H, Meng A, Long Y, Li Q, Zhang Y (2014) Classification and comparison of municipal solid waste based on thermochemical characteristics. J Air Waste Manag Assoc 64(5):597–616. https://doi.org/10.1080/10962247.2013.873094

Chapter 6
Landfills in Developing Economies: Drivers, Challenges, and Sustainable Solutions

Pablo Emilio Escamilla-García

Abstract This book chapter provides a comprehensive analysis of the utilization and functioning of sanitary landfills within developing economies, focusing on their intricate social and environmental repercussions. The chapter delves into the complex dynamics that underlie the operation of these waste disposal sites and examines the multifaceted challenges they engender, as well as the potential avenues for positive change they offer. Through an exploration of the social dimensions, the chapter highlights how sanitary landfills often exacerbate existing inequalities, impact livelihoods, and pose health risks to marginalized communities. It emphasizes the significance of integrating local perspectives, community engagement, and socioeconomic strategies into waste management policies to ensure equitable outcomes. By amalgamating insights from both social and environmental spheres, the chapter concludes with a call for comprehensive waste management strategies that holistically address the complexities of sanitary landfill use in developing economies. It advocates for collaborative efforts between policymakers, practitioners, and researchers to reshape waste management policies, incorporating innovative engineering solutions while considering cultural contexts. Ultimately, the chapter aims to inspire informed dialogue, innovative policy formulation, and transformative action for sustainable waste management practices that safeguard the well-being of communities and the environment.

Keywords Landfill · Developing countries · Emerging economies · Waste management · MSW

The original version of the chapter has been revised: The author name Pablo Emilio Escamilla-García has been updated. A correction to this chapter can be found at https://doi.org/10.1007/978-3-031-52633-6_11

P. E. Escamilla-García (✉)
Instituto Politécnico Nacional, Taxqueña 1620, Paseos de Taxqueña, Coyoacán, 04250 Mexico City, Mexico
e-mail: peescamilla@ipn.mx

A. Anouzla and S. Souabi (eds.), *Technical Landfills and Waste Management*, Springer Water, https://doi.org/10.1007/978-3-031-52633-6_6

6.1 Some Statistics About MSW Worldwide

According to data from the World Bank (Kaza et al. 2018), globally, the annual production of MSW stands at around 2.01 billion tonnes, with at least 33% of this waste being managed in environmentally unsafe ways. On a daily basis, the average waste generated per person is about 0.74 kg, though this figure varies widely, ranging from 0.11 to 4.54 kg. Despite accounting for just 16% of the world's population, high-income countries contribute about 34%, equivalent to 683 million tonnes, of the total global waste. Looking ahead, it is projected that worldwide waste will escalate to approximately 3.40 billion tonnes by 2050. A noteworthy correlation between waste generation and income level is evident, with daily per capita waste in high-income countries predicted to rise by 19% by 2050 (see Fig. 6.1). In contrast, low- and middle-income countries are anticipated to experience an increase of around 40% or more.

The pace of waste generation tends to accelerate more rapidly for incremental income changes in lower income brackets compared to higher income levels. Low-income countries are expected to observe a more than threefold increase in total waste generated by 2050. Regionally, the East Asia and Pacific area leads in waste generation, accounting for 23% of the global total, while the Middle East and North Africa contribute the least at 6%. Interestingly, the regions with the fastest-growing waste production are Sub-Saharan Africa, South Asia, and the Middle East and North Africa, where waste generation is projected to nearly triple, double, and double respectively by 2050. Within these regions, over half of the waste is presently openly dumped, underscoring the significance of waste growth trajectories for environmental, health, and economic considerations, necessitating immediate action.

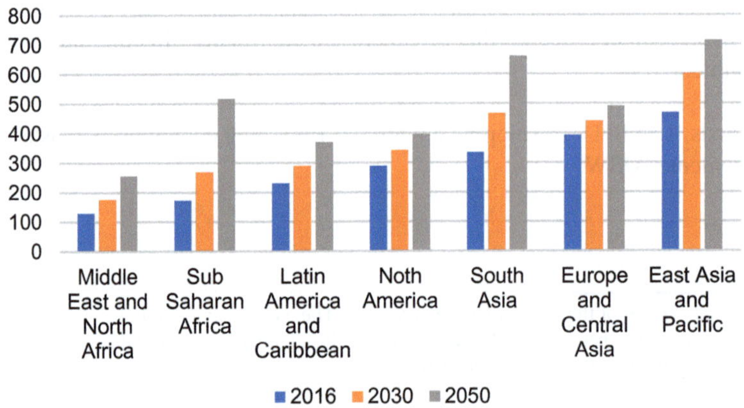

Fig. 6.1 Total projected waste generation by region (Mllions of tonnes per year) (Kaza et al. 2018)

Fig. 6.2 Collection rates by income level (Kaza et al. 2018)

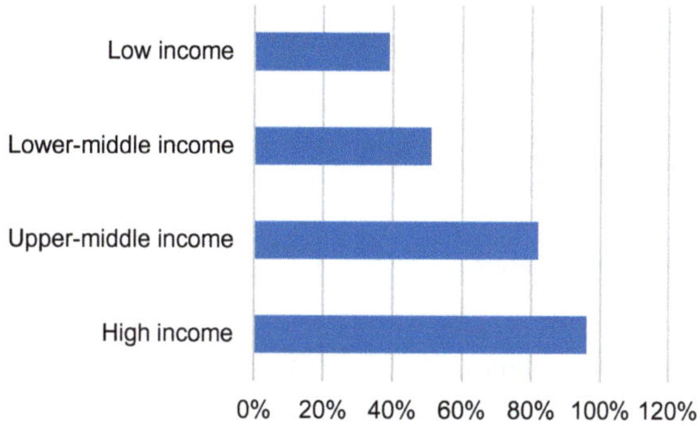

An essential element in waste management is waste collection, although rates vary significantly based on income levels. Upper-middle and high-income countries offer nearly universal waste collection, whereas low-income countries collect approximately 48% of waste in urban areas, dropping to a mere 26% outside urban regions. Across different regions, Sub-Saharan Africa collects roughly 44% of waste, whereas Europe and Central Asia, along with North America, manage to collect at least 90% of waste (see Fig. 6.2).

Waste composition diverges across different income levels, mirroring distinct consumption patterns. High-income countries produce proportionally less food and green waste, constituting 32% of the total waste, while generating a higher share of recyclable dry waste such as plastics, paper, cardboard, metals, and glass, accounting for 51%. In contrast, middle- and low-income countries yield 53 and 56% of food and green waste, respectively, with the proportion of organic waste increasing as economic development diminishes. Notably, low-income countries only allocate 16% of their waste stream to materials that could be recycled (see Fig. 6.3).

Across various regions, waste streams exhibit limited variation beyond those associated with income levels. Almost all regions generate approximately 50% or more organic waste on average, except for Europe and Central Asia as well as North America, where a higher proportion of dry waste is generated. A common misconception is that technology singularly solves the challenge of growing and unmanaged waste. However, technology is not a universal solution but merely one factor among many in effective waste management. Successful progress is more probable when countries transitioning from open dumping and rudimentary methods opt for locally suitable solutions. Globally, a substantial portion of waste is currently dumped or disposed of in landfills. Roughly 37% of waste is landfilled, with 8% managed in sanitary landfills featuring landfill gas collection systems. Open dumping comprises about 33%, while 19% is recuperated through recycling and composting, and 11% is incinerated for final disposal. Adequate waste treatment or disposal, including controlled landfills or more rigorous facilities, is primarily the domain of high- and upper-middle-income nations. In contrast, lower-income countries often resort to open dumping, with 93% of waste in the former and merely 2% in the latter being dumped. Upper-middle-income countries exhibit the highest landfill percentage at

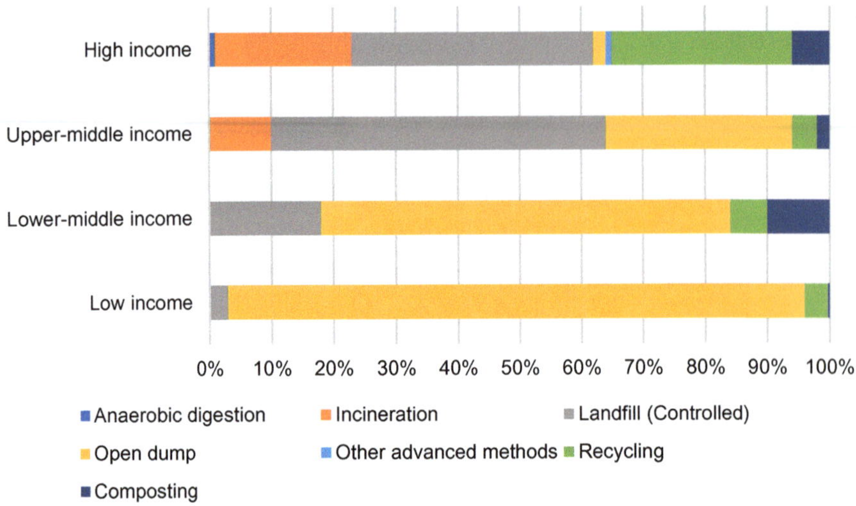

Fig. 6.3 Disposal methods by income (Kaza et al. 2018)

54%, which declines to 39% in high-income countries. These high-income countries divert 35% of waste to recycling and composting and 22% to incineration, a method primarily used in countries with substantial capacity, income, and land constraints.

Based on waste volume, composition, and management, an estimated 1.6 billion tonnes of carbon dioxide (CO_2) equivalent greenhouse gas emissions resulted from solid waste treatment and disposal in 2016. This was largely due to open dumping and landfill disposal without gas capture systems, constituting around 5% of global emissions. Without sector improvements, emissions related to solid waste are projected to rise to 2.6 billion tonnes of CO_2-equivalent annually by 2050. In most nations, solid waste management operations are generally a local responsibility, with almost 70% of countries establishing institutions responsible for policy development and regulatory oversight in the waste sector. About two-thirds of countries have implemented targeted legislation and regulations for waste management, though enforcement varies considerably. Direct involvement of central governments in waste services, apart from oversight or fiscal support, is uncommon. Approximately 70% of waste services are managed directly by local public entities. Around half of these services, spanning from primary waste collection to treatment and disposal, are operated by public entities, and about one-third involve public–private partnerships. However, successful partnerships with the private sector for financing and operations tend to thrive only under specific conditions with appropriate incentive structures and enforcement mechanisms, making them an ideal solution only in certain cases.

6.2 A Brief History of Landfills

Landfills, as integral components of modern waste management systems, have a long history that traces back to the earliest human settlements. Throughout the ages, the practice of waste disposal has evolved from rudimentary methods to complex engineered systems. The origins of landfills are rooted in the intrinsic human need to manage waste while reflecting the cultural, technological, and environmental contexts of different eras (CHEJ 2016). The concept of landfills can be traced to ancient civilizations, where waste disposal was often driven by practicality rather than environmental concern. Ancient Mesopotamians, for instance, disposed of waste in designated areas outside city walls, creating mounds of refuse that served as early prototypes of modern landfills (Hill 2016). These early landfills primarily consisted of organic waste and naturally decomposable materials, minimizing their long-term environmental impact. As societies advanced and urbanized, waste generation increased, leading to more organized waste disposal methods. The ancient Romans developed communal pits for waste disposal, a rudimentary form of landfill (Hawkins and Muecke 2002). This approach, however, still lacked the environmental considerations necessary to address the growing scale of waste accumulation. The shift towards industrialization in the eighteenth and nineteenth centuries brought about significant changes in waste generation and disposal practices (EPA 2014). Rapid urbanization and increased production of non-biodegradable materials necessitated more systematic waste management solutions. The advent of the Industrial Revolution further exacerbated waste-related challenges, as factories and production facilities churned out unprecedented amounts of waste and pollution (Barles 2014). Open dumping and unregulated disposal became increasingly unsustainable, giving rise to the idea of sanitary landfills. The modern concept of sanitary landfills emerged in the mid-twentieth century as a response to the growing concern over environmental degradation caused by waste mismanagement (Ramsy 2022). These engineered landfills were designed to minimize groundwater contamination, air pollution, and other negative impacts on surrounding ecosystems. Proper waste segregation, compaction, and containment measures were introduced to curtail the spread of diseases and limit environmental harm (Kovacs 1993). However, even as sanitary landfills represented a marked improvement over their predecessors, their proliferation still raised environmental and social issues. Land scarcity, improper waste segregation, and inadequate landfill management practices often resulted in soil and water pollution, emission of greenhouse gases, and negative impacts on local communities (Diaz 2011; Mondal et al. 2023). As awareness of these issues grew, efforts to implement more sustainable waste management strategies gained traction. In recent decades, the principles of reduce, reuse, and recycle (3Rs) have gained prominence, with governments, organizations, and individuals striving to minimize waste generation and promote resource efficiency (Jibril et al. 2012). Additionally, the concept of "zero waste" has gained momentum, aiming to send as little waste as possible to landfills or incineration facilities (Sakai et al. 2011). Advances in technology and innovations such as waste-to-energy facilities and composting systems have also contributed to more sustainable waste management practices (Prajapati et al. 2021). As can be seen, the origin of landfills can be traced back to the earliest human settlements, evolving through various historical and technological contexts. From ancient waste mounds to the sophisticated sanitary landfills of today, the journey of waste disposal reflects humanity's changing attitudes towards environmental stewardship. As we continue to grapple with the consequences of waste accumulation, it is imperative that we learn from the past and adopt innovative strategies to minimize the impact of landfills on our environment and communities.

6.3 Key Factors for the Proliferation of Landfills

The extensive use of landfills in developing economies has been a result of a complex interplay of economic, social, and infrastructural factors. While these waste disposal sites have provided a seemingly convenient solution for managing waste in resource-constrained settings, their utilization has come with significant challenges and environmental consequences. Exploring the reasons behind the widespread use of landfills in developing economies sheds light on the need for sustainable waste management solutions.

Socio-Economic Factors:

- Rapid Urbanization and Population Growth: Developing countries have experienced significant urbanization and population growth over the past decades. The resultant surge in waste generation has often outpaced the development of proper waste management infrastructure, leading to the reliance on landfills as a quick and seemingly low-cost solution (Ogbonna et al. 2007).
- Limited Financial Resources: Many developing countries face economic challenges and limited resources, constraining their ability to invest in advanced waste management technologies. Landfills offer a relatively inexpensive way to manage waste compared to more sophisticated alternatives, making them appealing in resource-constrained environments (Voukkali et al. 2023).
- Informal Economy: Informal waste pickers play a crucial role in waste management across developing countries, recovering recyclable materials from waste streams. This informal sector, while providing income for marginalized communities, may contribute to the continuation of landfill-centric waste disposal due to the economic incentives associated with informal recycling (Khajuria et al. 2008).

Infrastructural and Social Factors:

- Lack of Infrastructure: Developing economies may lack the necessary infrastructure for efficient waste collection, separation, and recycling. Inadequate waste collection systems can result in a large portion of waste ending up in landfills due to logistical challenges (Vij 2012).
- Population Density: High population densities in urban areas of developing economies make it challenging to allocate space for waste management facilities.

Landfills, while problematic from an environmental perspective, might seem like the most feasible solution in densely populated regions (Patel and Meka 2013).

- Limited Public Awareness: Public awareness about the environmental impact of landfills and the benefits of sustainable waste management practices might be limited in developing economies. This lack of awareness can impede the adoption of alternative waste management approaches (Dutta and Jinsart 2020).

Inadequate waste management systems:

- Weak Waste Collection Systems: Inadequate waste collection and transportation systems in many Latin American countries result in improper waste disposal practices. The absence of comprehensive collection networks makes it challenging to redirect waste away from landfills (Sharma and Jain 2018).
- Lack of Regulation and Enforcement: Inconsistent waste management regulations and limited enforcement capacity contribute to the persistence of landfills as the primary waste disposal method. The absence of penalties for improper waste disposal and inadequate monitoring mechanisms hinder the adoption of more sustainable practices (Hoornweng et al. Kennedy 2013).
- Land Scarcity: Rapid urbanization and limited available land for waste management facilities exacerbate the landfill issue. As urban areas expand, finding suitable sites for landfills becomes increasingly difficult, forcing authorities to rely on existing facilities (Ramos, Martinho and Pina 2023).

Challenges and Environmental Consequences:

- Environmental Degradation: Improperly managed landfills can lead to soil and water contamination, air pollution, and habitat destruction. The lack of proper containment measures and leachate control systems in many developing countries exacerbates these issues (Khan et al. 2022).
- Health Risks: Landfills are associated with various health risks for nearby communities due to the emission of methane and other harmful gases, as well as the attraction of pests and disease vectors (Hajam et al. 2023).
- Resource Depletion: Landfills represent missed opportunities for resource recovery and recycling. Valuable materials that could be reused or recycled are buried, contributing to resource depletion and energy inefficiency (Uttara, Bhuvandas and Aggarwal 2012).

Towards Sustainable Solutions:

- Integrated Waste Management: Developing economies need to transition from reliance on landfills to integrated waste management systems that incorporate waste reduction, recycling, composting, and waste-to-energy technologies (Zargar et al. 2023).
- Awareness and Education: Raising public awareness about the environmental and health impacts of landfills is crucial. Education campaigns can promote waste reduction, segregation, and responsible disposal practices.

- Technology Transfer: International collaboration and technology transfer can help bring advanced waste management technologies to developing economies, making sustainable alternatives more accessible (Khan et al. 2022).
- Policy and Regulation: Governments should enact and enforce regulations that promote sustainable waste management practices and discourage the excessive use of landfills. Incentives for recycling and penalties for improper disposal can drive change (Khajuria et al. 2013).

As stated in the previous commented aspects, the extensive use of landfills in developing economies is a result of economic, infrastructural, and social challenges. While landfills might offer a short-term solution for waste disposal, the associated environmental and health consequences necessitate a shift towards sustainable waste management practices. By addressing these challenges and embracing integrated waste management approaches, developing economies can mitigate the negative impacts of landfills and pave the way for a more environmentally conscious future. Waste management in these countries has long been a complex and multifaceted challenge, leading to the widespread use of landfills as the predominant waste disposal strategy. The intricate interplay of socio-economic factors, inadequate infrastructure, and governance issues has contributed to the prevalence of landfills in the region. However, this approach comes with substantial environmental, social, and health consequences, highlighting the urgent need for more sustainable waste management solutions.

6.4 Challenges in Eradicating Landfills in Developing Countries

The eradication of landfills remains a complex and daunting challenge in many developing economies, despite growing global awareness about sustainable waste management practices. A convergence of historical, economic, social, and infrastructural factors has contributed to the persistence of landfills as predominant waste disposal solutions. Among the main factors that have contributed to the predominate utilization of landfills in these countries we can discuss the following aspects:

Historical Legacies and Cultural Norms:

- Ingrained Practices: Landfills have historically been a common method of waste disposal in these economies. Long-standing cultural practices and familiarity with landfill operations have made it challenging to shift to more sustainable alternatives (Hill 2016).
- Limited Environmental Awareness: Historically, environmental concerns may not have been prioritized in the same way they are today. The evolution of waste management practices has lagged behind changing environmental attitudes and global best practices (Budihardjo et al. 2023).

Economic Considerations:

- Cost Constraints: Developing and transitioning to advanced waste management systems, such as waste-to-energy facilities or comprehensive recycling programs, requires substantial financial investments. Many developing countries face economic limitations that hinder their ability to make these investments, making landfills a seemingly more affordable option (Nika et al. 2023).
- Resource Allocation: In the face of limited resources, governments may prioritize other pressing issues, leading to a lack of funding and support for the development of sustainable waste management infrastructure (Voukkali et al. 2023).

Infrastructure Challenges:

- Inadequate Collection Systems: Proper waste collection systems are essential for diverting waste away from landfills. However, inadequate waste collection networks in some regions prevent the efficient routing of waste to alternative facilities (Yong et al. 2019).
- Urbanization Pressure: Rapid urbanization exacerbates the challenges of finding suitable locations for waste management facilities. The lack of available land for establishing new waste treatment and disposal sites perpetuates the reliance on existing landfills (Khan et al. 2022).

Governance and Regulatory Issues:

- Lax Enforcement: Weak regulatory frameworks and insufficient enforcement mechanisms can lead to non-compliance with waste management regulations. Without strict adherence to rules, the eradication of landfills becomes even more challenging (Elgarahy et al. 2023).
- Fragmented Governance: The responsibility for waste management might be distributed among multiple local and regional authorities, causing inconsistencies and hindering the implementation of holistic waste management strategies (Patel and Meka 2013).

Social and Cultural Dynamics:

- Informal Economy: Informal waste picking and recycling activities play a significant role in many developing countries, providing livelihoods for marginalized communities. The presence of this informal sector can complicate efforts to transition away from landfill-centric waste management (Patel and Meka 2013).
- Public Perception: Landfills might be perceived as an efficient solution due to their familiarity and historical usage. Public resistance to change, combined with a lack of awareness about the environmental consequences of landfills, can hinder efforts to eradicate them (Zhang et al. 2022).

The eradication of landfills in developing countries is a multi-faceted challenge that requires addressing historical legacies, economic constraints, infrastructure limitations, governance issues, and social dynamics. While transitioning to sustainable waste management practices is a complex endeavor, it is not insurmountable. By adopting a holistic approach that combines regulatory reforms, targeted investments, awareness campaigns, and community engagement, these countries can work towards a future where landfills are replaced by more environmentally responsible and resource-efficient waste management solutions.

6.5 Strategies to Reduce Landfill Usage in Developing Countries: Navigating Economic and Social Realities

Developing countries, facing unique economic and social contexts, can employ a range of strategies to effectively reduce the reliance on landfills for waste disposal. Recognizing the economic constraints and social dynamics prevalent in the region, these strategies should be designed to be feasible, inclusive, and sustainable. In this section we explore potential approaches that can contribute to the reduction of landfill usage while considering the economic and social complexities.

1. Invest in Integrated Waste Management: Developing comprehensive waste management systems that integrate waste reduction, recycling, composting, and waste-to-energy technologies can significantly decrease landfill dependency (Mondal et al. 2023). While initial investments are required, the long-term economic and environmental benefits are substantial. Governments can collaborate with private sector entities to develop infrastructure and ensure efficient waste collection and processing (Nika et al. 2023).
2. Support Informal Recycling Sectors: Acknowledge and formalize the role of informal waste pickers and recyclers. Provide training, resources, and protective measures to enhance their contribution to recycling efforts (Mahinroosta and Allahverdi 2018). By integrating informal workers into the formal waste management system, countries can boost recycling rates and create economic opportunities for marginalized communities (Yong et al. 2019).
3. Public Awareness and Education: Launch public awareness campaigns to inform citizens about the environmental and health impacts of landfills. Promote waste reduction, recycling, and responsible consumption habits. These campaigns can be tailored to address cultural sensitivities and encourage communities to actively participate in waste management efforts (Tahir, Hussain and Behaylu 2015).
4. Incentivize Recycling and Circular Economy Practices: Implement policies that reward recycling initiatives and promote the development of a circular economy. Introduce deposit-refund systems for beverage containers and provide tax incentives to businesses that prioritize eco-friendly packaging and products. Such incentives can stimulate innovation and create a culture of sustainability (Elgarahy et al. 2023).
5. Encourage Community Engagement: Engage local communities in waste management decisions and solutions. Establish community-based waste collection and recycling programs that empower residents to take ownership of

waste reduction efforts. By fostering a sense of responsibility and ownership, communities become active participants in waste management (Zhang et al. 2022).

6. Promote Extended Producer Responsibility (EPR): Shift the burden of waste management onto producers by implementing EPR programs. Manufacturers are responsible for the lifecycle of their products, including their eventual disposal. This encourages the design of products that are easier to recycle and reduces the burden on landfills (Ogbonna et al. 2007).
7. Establish Public–Private Partnerships: Collaborate with private sector entities to leverage their expertise and resources in waste management initiatives. Public–private partnerships can help drive innovation, optimize operational efficiency, and ensure the sustainable management of waste streams (Hill 2016).
8. Implement Gradual Landfill Bans: Introduce gradual bans on certain types of waste that are suitable for recycling or alternative disposal methods. This approach allows the waste management infrastructure to adapt to changing practices while still addressing the region's economic realities (Tahir et al. 2015).
9. Promote Eco-Industrial Parks: Develop eco-industrial parks that encourage industrial symbiosis, where one industry's waste becomes another's raw material. This minimizes waste generation and fosters resource efficiency among industries, reducing the need for landfilling (Vij 2012).
10. Prioritize Research and Innovation: Invest in research and development to find innovative waste management solutions that are tailored to the region's economic and social context. Encourage local universities and research institutions to collaborate on waste-related projects (Nika et al. 2023).

As noted, developing countries can effectively reduce their reliance on landfills by adopting a combination of strategies that consider the economic and social conditions unique to the region. By investing in integrated waste management, supporting informal recycling sectors, raising public awareness, incentivizing recycling, engaging communities, and collaborating with the private sector, these nations can pave the way for a more sustainable and responsible waste management future. These strategies not only address environmental concerns but also promote economic growth, job creation, and community well-being.

6.6 Final Thoughts

In conclusion, the analysis of the use and operation of sanitary landfills in developing economies reveals a complex interplay of social and environmental impacts that necessitates a comprehensive and multi-faceted approach. Throughout this chapter, we have delved into the intricate dynamics that shape the functioning of these waste disposal systems, shedding light on the multifarious challenges they pose and the potential opportunities they offer. Undoubtedly, the social consequences of sanitary landfills in developing economies are profound and far-reaching. From exacerbating existing disparities in marginalized communities to influencing livelihood patterns, the human dimensions of landfill operation demand immediate attention. Our analysis underscores the importance of incorporating local perspectives, engaging communities in decision-making processes, and implementing robust socio-economic mitigation strategies. By fostering partnerships among stakeholders and adopting participatory planning, we can cultivate a more inclusive waste management framework that prioritizes the well-being of all residents.

Simultaneously, the environmental ramifications of sanitary landfills underscore the pressing need for sustainable waste management strategies. As we have explored, inadequate waste containment practices and the release of hazardous substances into the environment can lead to long-term ecological damage. To counteract this, embracing advanced landfill technologies, optimizing waste segregation and recycling efforts, and implementing stringent monitoring mechanisms emerge as vital steps towards reducing the ecological footprint of these disposal sites. In charting a way forward, it is imperative that policymakers, practitioners, and researchers collaborate to refine waste management policies and practices in developing economies. By integrating innovative engineering solutions with sociocultural considerations, we can strike a balance between effective waste disposal and social equity. Recognizing the intricate web of relationships between socioeconomic factors, environmental sustainability, and public health will undoubtedly guide the development of holistic waste management strategies that promote a healthier and more equitable future.

In essence, this chapter highlights the intricate tapestry of challenges and opportunities that define the use and operation of sanitary landfills in developing economies. By acknowledging the intricate interplay between social and environmental dimensions, we can embark on a transformative journey towards sustainable waste management practices that safeguard the well-being of both communities and the planet. As we move forward, it is our hope that the insights presented here will serve as a foundation for fostering dialogue, inspiring policy innovation, and driving positive change on a global scale.

References

Barles S (2014) History of waste management and the social and cultural representations of waste. In Agnoletti M, Neri Serneri S (eds) The basic environmental history. Springer, Cham. https://doi.org/10.1007/978-3-319-09180-8_7

Budihardjo M, Humaira N, Ramadan B, Wahyuningrum I, Huboyo H (2023) Strategies to reduce greenhouse gas emissions from municipal solid waste management in Indonesia: the case of Semarang City. Alex Eng J 69:71–783. https://doi.org/10.1016/j.aej.2023.02.029

CHEJ (2016) Landfills, trashing the earth. center for health, environment & justice. Retrieved from https://chej.org/wp-content/uploads/PUB-063-Landfills.pdf. Accessed 5 Aug 2023

Diaz L (2011) The 3Rs as the basis for sustainable waste management: moving towards zero waste. Retrieved From https://www.uncrd.or.jp/content/documents/Text%20for%20Keynote%20Address%20for%203rd%20Regional%203R%20Forum_Luiz%20Diaz.pdf. Accessed 12 July 2023

Dutta A, Jinsart W (2020) Waste generation and management status in the fast-expanding Indian cities: a review. J Air Waste Manag Assoc 70(5):491–503. https://doi.org/10.1080/10962247.2020.1738285

Elgarahy A, Eloffy M, Alengebawwy A, El-Sherif D, Gaballah M, Elwakeel K, El-Qelish M (2023) Sustainable management of food waste; pre-treatment strategies, techno-economic assessment, bibliometric analysis, and potential utilizations: a systematic review. Environ Res 225:115558. https://doi.org/10.1016/j.envres.2023.115558

EPA (2014) Municipal solid waste landfills, economic impact analysis for the proposed new subpart to the new source performance standards. U.S. Environmental Protection Agency. Retrieved from https://www3.epa.gov/ttnecas1/docs/eia_ip/solid-waste_eia_nsps_proposal_07-2014.pdf. Accessed 09 July 2023

Hajam Y, Kumar R, Kumar A (2023) Environmental waste management strategies and Vermi transformation for sustainable development. Environ Challenges 13:100747. https://doi.org/10.1016/j.envc.2023.100747

Hawkins G, Muecke S (2002) Culture and Waste, the creation and destruction of value. Rowman & Littlefield

Hill S (2016) Making garbage, making land, making cities: a global history of waste in and out of place. Glob Environ 9(1):166–195

Hoornweng D, Bhada-Tata P, Kennedy C (2013) Environment: waste production must peak this century. Nature 502:615–617. https://doi.org/10.1038/502615a

Jibril J, Sipan I, Sapri M, Shika S, Isa M, Adbullah S (2012) 3Rs critical success factor in solid waste management system for higher educational institutions. Procedia—Soc Behav Sci 65:626–631. https://doi.org/10.1016/j.sbspro.2012.11.175

Kaza S, Yao L, Bhada-Tata P, Van Woerden F (2018) What a waste 2.0, a global snapshot of solid waste management to 2050. The World Bank. Retrieved from https://openknowledge.worldbank.org/server/api/core/bitstreams/92a50475-3878-5984-829e-0a09a6a9badc/content. Accessed:12 June 2023

Khajuria A, Yamamoto Y, Morioka T (2008) Solid Waste management in Asian countries: problems and issues. WIT Trans Ecol Environ 109:643–653. https://doi.org/10.2495/WM080661

Khajuria A, Yamamoto Y, Morioka T (2013) Estimation of municipal solid waste generation and landfill area in Asian developing countries. J Environ Biol 31(5):649–654

Khan A, López E, Khan N, Gómez L, Munshi F, Alsabhan A, Perveen K (2022) Current solid waste management strategies and energy recovery in developing countries—State of art review. Chemosphere 291:133088. https://doi.org/10.1016/j.chemosphere.2021.133088

Kovacs W (1993) Solid waste management: Historical and future perspectives. Resour Conserv Recycl 8(1–2):113–130. https://doi.org/10.1016/0921-3449(93)90023-9

Mahinroosta M, Allahverdi A (2018) Hazardous aluminum dross characterization and recycling strategies: a critical review. J Environ Manage 223:452–468. https://doi.org/10.1016/j.jenvman.2018.06.068

Mondal T, Choudhury M, Kundu D, Dutta D, Samanta S (2023) Landfill: an eclectic review on structure, reactions and remediation approach. Waste Manage 164:127–142. https://doi.org/10.1016/j.wasman.2023.03.034

Nika M, Alygi N, Arvaniti O, Thomaidis N (2023) Non-target screening of emerging contaminants in landfills: A review Author links open overlay panel. Curr Opin Environ Sci Health 32:100430. https://doi.org/10.1016/j.coesh.2022.100430

Ogbonna D, Amangabara G, Ekere T (2007) Urban solid waste generation in Port Harcourt metropolis and its implications for waste management. Manag Environ Qual 18(1):71–88. https://doi.org/10.1108/14777830710717730

Patel V, Meka S (2013) Forecasting of municipal solid waste generation for medium scale towns located in the state of Gujarat, India. Int J Innov Res Sci Eng Technol 2(9):4707–4716

Prajapati P, Varjani S, Singhania R, Patel A, Awasthi M, Sindhu R, Zhang Z, Binod P, Awasthi S, Chaturvedi P (2021) Critical review on technological advancements for effective waste management of municipal solid waste—Updates and way forward. Environ Technol Innov 23:101749. https://doi.org/10.1016/j.eti.2021.101749

Ramos M, Martinho G, Pina J (2023) Strategies to promote construction and demolition waste management in the context of local dynamics. Waste Manage 162:102–112. https://doi.org/10.1016/j.wasman.2023.02.028

Sakai S, Yoshida H, Hirai Y, Asari M, Takigame H, Takahashi S, Tomoda K, Peeler M, Wejchert J, Unterseh T, Ravazzi A, Hathaway R, Hylander L, Fischer C, Oh G, Jinhui L, Chi N (2011) International comparative study of 3R and waste management policy developments. J Mater Cycles Waste Manage 13:86–113. https://doi.org/10.1007/s10163-011-0009-x

Sharma K, Jain S (2018) Overview of municipal solid waste generation, composition, and management in India. J Environ Eng 145(3):1–15. https://doi.org/10.1061/(ASCE)EE.1943-7870.0001490

Ramsy D (2022) Principles of Waste Management and its History. Res Rev: J Ecol Environ Sci 10:3–4. https://doi.org/10.4172/23477830.10.S4.002

Tahir M, Hussain T, Behaylu A (2015) Scenario of Present and future of solid waste generation in India: a case study of Delhi mega city. J Environ Earth Sci 5(8):83–91

Uttara S, Bhuvandas N, Aggarwal V (2012) Impacts of urbanization on environment. Int J Res Eng Appl Sci 2(2):1637–1645

Vij D (2012) Urbanization and solid waste management in India: present practices and future challenges. Procedia Soc Behav Sci 37:437–447. https://doi.org/10.1016/j.sbspro.2012.03.309

Voukkali I, Papamichael I, Loizia P, Zorpas A (2023) Urbanization and solid waste production: prospects and challenges. Environ Sci Pollut Res (in press). https://doi.org/10.1007/s11356-023-27670-2

Yong Y, Lim Y, Ilankoon I (2019) An analysis of electronic waste management strategies and recycling operations in Malaysia: challenges and future prospects. J Clean Prod 224:151–166. https://doi.org/10.1016/j.jclepro.2019.03.205

Zargar T, Alam P, Khan A, Alam S, Abutaeb A, Hasan M, Khan N (2023) Characterization of municipal solid waste: Measures towards management strategies using statistical análisis. J Environ Manage 342:118331. https://doi.org/10.1016/j.jenvman.2023.118331

Zhang Z, Malik M, Khan A, Ali N, Malik S, Bilal M (2022) Environmental impacts of hazardous waste, and management strategies to reconcile circular economy and eco-sustainability. Sci Total Environ 807:150856. https://doi.org/10.1016/j.scitotenv.2021.150856

Chapter 7
The Environmental Pressure by Open Dumpsites and Way Forward

Pabasari A. Koliyabandara, D. D. P. Preethika, Asitha T. Cooray, Sudantha S. Liyanage, Chamika Siriwardana, and Meththika Vithanage

Abstract Environmental hazard from open dumping is a severe problem occurring around the world. The composition of waste from most of developing countries accounts for more than 50% of organic matter. Prolonged incomplete handling of solid waste or lack of management at disposal sites has caused significant environmental and social challenges across the world in these regions such as pollution of nearby water supplies, pollution of groundwater, enhanced mosquito breeding sites, unintentional landfill gas fire, microbial pollution from atmospheric dust, atmospheric contaminantscreating unpleasant and highly toxic gases and Volatile Organic Carbons (VOCs), emission of green house gases (GHGs) damage to vegetation, air pollution leading to global warming etc. Among the sources of water pollution, open dumpsites have been identified as one of the significant threats. A significant amount of metal release namely Cd, As, Cr, Fe can be occurred by the disposal methods such as open dumps. Disastrous events like "dumpslides", fires have been recorded worldwide causing casualties due to the mismanagement of open dumpsites. Composting, resource recovery, waste mining, use of bio covers are useful techniques which can be applied to existing open dumpsites. This review will address the environmental pressure from open dump sites as and the futuristic approaches in waste management aligning with Sustainable Development Goals (SDG) and circular economy.

P. A. Koliyabandara (✉) · D. D. P. Preethika · S. S. Liyanage
Department of Civil and Environmental Technology, Faculty of Technology, University of Sri Jayewardenepura, Pitipana, Homagama, Sri Lanka
e-mail: arundathi@sjp.ac.lk

A. T. Cooray
Department of Chemistry, Faculty of Applied Sciences, University of Sri Jayewardenepura, Gangodawila, Nugegoda, Sri Lanka

A. T. Cooray · C. Siriwardana
Faculty of Applied Sciences, Instrument Centre, University of Sri Jayewardenepura, Gangodawila, Nugegoda, Sri Lanka

M. Vithanage
Ecosphere Resilience Research Center, Faculty of Applied Sciences, University of Sri Jayewardenepura, Gangodawila, Nugegoda, Sri Lanka

A. Anouzla and S. Souabi (eds.), *Technical Landfills and Waste Management*, Springer Water, https://doi.org/10.1007/978-3-031-52633-6_7

Keywords Municipal solid waste · Landfill · Sustainable development goals · Circular economy

7.1 Municipal Solid Waste (MSW) Management

Increase in population, industrialization, urbanization along with inadequate waste management practices cause the waste disposal a burning issue in the modern world. A close relationship can be observed between national gross domestic product (GDP) and waste generation per capita in developed economies and the amount of waste production keeps on growing for decades (Laner et al. 2012). Addressing the MSW problem in a sustainable way is a must for a healthy world. MSW or the everyday waste and industrial waste should be managed following the hierarchy of waste management (Banar et al. 2009). The United States Environmental Protection Agency (USEPA) defines MSW as the waste generated by everyday activities which do not include construction and demolition waste, hazardous waste or industrial waste (USEPA 2020). Impacts created by MSW both on society and environment has to lead the path to create more awareness regarding the issue (Calvo et al. 2005). Different waste management practices such as recycling, reducing waste, waste recovery for reuse, incineration, use of sanitary landfilling other than landfilling or open dumping are carried out around the world (Narayana 2009). The use of landfills for MSW are popular among developing economies due to reduced treatment costs and cost related to alternative methods for the management of waste (Brunner and Fellner 2007). Controlled waste dumping is a major need for the waste management in developing economies. Controlled landfills are an improved method for waste management compared to open dumpsites as it reduces the access of scavenging animals, presence of staff for management of site can be observed for activities like compaction. Sanitary landfills are modern engineering landfill waste can be decomposed to biologically and chemically inert material in an isolated environmental setting (Chen et al. 2003).

However, waste management has become a serious challenge to developing nations compared to developed countries since there are limited facilities, infrastructure, lack of regulations to control pollutants, lack of technical knowledge, improper waste collection system, absence of rules and regulations, etc. Further, researchers argued that unplanned urbanization growth creating severe challenges especially for meeting the increasing demand. For example, 45% of the world's population are still lacking in meeting the basic sanitation and hygiene requirements (Khatib 2010). Hence, all these issues mainly result to make a complexity on controlling, disposing of wastes, meeting healthy municipal services in many nations.

According to Laner et al., USA has been disposing 54% of the 250 $\times$ 10^6 metric tons of MSW generated in 2008 in landfills (Laner et al. 2012). Similarly, in the year of 2008, 55% waste from UK, 51% from Finland, 77% from Greece has been landfilled. It is recorded that in Australia, about 70% of untreated MSW has been directed to landfills in 2002. The recorded data in countries like Taiwan, Korea and Poland states

95, 52 and 90% of MSW has been landfilled (Renou et al. 2008; Mukherjee et al. 2015). Countries like Finland Australia, USA, United Kingdom depends heavily on landfilling. USA in 2008, have generated 54% of MSW and 33% have been recycled and composted.

According to Organization for Economic Co-operation and Development (OECD) data reports, countries like Australia has moved forward in reducing the amount of waste landfilled as shown in Germany, Netherlands, Sweden, Denmark, and Austria records values below 5% for landfilling in the year 2008 (Eurostat 2010). Considering the data recorded from Brazil, their waste disposal is divided into three major sections. Sanitary landfills account for 57% of waste disposal on average, controlled landfills account for 24% and the open dumping is calculated as 18% average (Wilson et al. 2015).

Moreover, Menikpura et al. (2012) has identified that 85% of MSW has been disposed of in Sri Lankan open dumpsites, and only 27% of waste has been collected while generating approximately 7210 tons/ day. Researchers have identified that 10% and 5% of waste have been composted and recycled, respectively (Saja et al. 2021). In Cambodia, nearly 361,000 tonnes of waste was estimated in open dumpsites, 2008 and 635,000 tonnes in 2015 (Seng et al. 2018). Similarly, households' residents are burned and buried their waste because of the absence of a proper MSW regulation system. Comparably, in the year 2005, Palestinians have generated 2728 tons/day of MSW, and it was recorded that people were aware of the environmental consequences associated with open dumpsites and more than 41.6% of the population have been already suffered from these sites (Al-Khatib et al. 2015). Moreover, in Maputo, inhabitants have been transported 0.5 kg/day of generated waste into dumpsites during the last 40 years, and scholars debate that open fires and auto ignition of waste creating a common challenge for their residents (Ferronato and Torretta 2019). In Thailand, around 60% of SW largely depends on open dumping sites and most of them received 25 tons/day (Ferronato and Torretta 2019). With that context, the situation has become unmanageable at alarming rates in recent decades. The lack of an overall perspective of the current situation especially in developing nations poses a significant challenge in identifying and exploring potential economically and technically viable solutions. As a result, this study aims to bridge this knowledge gap and propose paths forward towards achieving sustainability.

7.1.1 Disposal of Municipal Solid Waste Using Open Dumpsites

Conventional final disposal methods include open burning, open dumping while at present sanitary landfills are known to the best option as it helps in reducing the adverse environmental impacts which will be discussed later (Nagendran et al. 2006; Erses et al. 2005). Table 7.1 discusses about the differences between commonly used disposal methods.

Table 7.1 A brief comparison of MSW management strategies (Joseph 2002; Idris et al. 2004)

Management method	Engineering measures	Landfill gas management	Leachate management
Open dumpsites	None	None	None
Semi controlled dumps	Compaction of waste Surface water monitoring No engineering measures	None	None
Controlled landfills	Waste registration and proper placement Compaction of waste Daily coverage of waste Surface and ground water monitoring Infrastructure and liner in place	Passive ventilation or flaring	Containment and some level of leachate treatment Reduced leachate volume through waste cover
Sanitary landfills	Waste registration and proper placement Compaction of waste Daily cover of waste Measures for final top cover and closure Proper siting, infrastructure; liners and post-closure plan	Proper gas management by flaring with or without energy recovery	Proper leachate treatment mechanism

Most of the developing countries use non-engineered open dumpsites for MSW management because its low operational and management costs (Aboyeji and Eigbokhan 2016; Nagendran et al. 2006). South and South-East Asia countries dispose more than 90% of MSW in open dumpsites (Visvanathan et al. 2002). The waste generation per day is around 20,000 tons for cities like Delhi, Kolkata, Mumbai and Chennai in India (Chattopadhyay et al. 2009). The Estrutural landfill in Brasilia, Brazil (136 ha) has been using over 50 years which is even known as one of the largest landfills in the world. It receives nearly 2 million metric tons per day and the estimated waste amount present in the landfill is 30 million metric tons (Wang 2020). Among the largest landfills in the world, Apex regional landfill in USA is of 2200 acres, Bordo Ponient landfill, Mexico 927 acres, Laogan landfill in China 830 acres and Delhi landfill is 560 acres in size (Wang 2020). All these can be categorized into sanitary landfills. Borg El-Arab sanitary landfill, Egypt is located parallel to the Mediterranean sea shoreline (El-Salam and Abu-Zuid 2015), Olusosun open dumpsite, Nigeria lies in low-lying, flat towards and gently slopes seaward (Aboyeji and Eigbokhan 2016), Karsara landfill situated close to agricultural lands and it is nearly 1 km north to river Ganga (Mishra et al. 2018). Location of open dumps can be observed in areas such as wetlands, marshes, beaches and areas adjacent to water which helps for the dilution of waste output such as leachate (Gunawardana et al. 2009; Joseph et al. 2004; Esakku et al. 2007). Examples of some open dumpsites in developing countries are listed in Table 7.2.

Table 7.2 Examples for some of the open dumping in selected developing countries

Location of dumpsites	Salient features	References
Banju, Gambia	Located in a populated area	Sanneh et al. (2011)
Thailand	330 dump sites 4500 tons per day to Bankok city landfills	Chiemchaisri et al. (2007)
Phnom Penh, Cambodia	Major disposal method—open dumping buried, or dumped about 361,000 tons of MSW—2008 635,000 tons—2015	Seng et al. (2018)
Nablus, Salfit, Ramallah, and Al-Bireh areas, Palestine	133 MSW dumpsites—2001 MSW generation about 2,723 T per day in 2005	Al-Khatib et al. (2015)
Maputo, Mozambique	Main dumpsite in action more than 40 years	dos Muchangos et al. (2015)
Tiruchirappalli, India	400–600 tons per day	Ferronato and Torretta (2019)
Dar es Salaam, Tanzania	Industrial waste is also dumped with MSW within open dumpsite	Mbuligwe and Kaseva (2006)
Mediterranean sea shoreline, Northern Coast Road: Alexandria-Matrouh, Egypt	~~2700 tons /day~~ Borg El-Arab (0.75 km^2 area, 7 cells) and El-Hammam (1.19 km^2 area, 13 cells) landfill sites Cell capacity—1.5 million tons Waste generation is nearly 1 million tons/year Leachate production—6000m^3/month	Abd El-Salamab and Abu-Zuid (2015)
Delhi, Hyderabad, and Kadapa, India	Major disposal methods—open dumping (Okhla, Ghazipu, Gurgaon landfills), engineered landfill (Hyderabad landfill), integrated waste management facility (Narela-Bawana landfill) Total waste quantity—0.2 million tons-15 million tons Less amount of biodegradable waste and most of them gets degrades in the initial 3–5 years Most of the landfills have high pollution index	Mohit Somania et al. (2019)

(continued)

Table 7.2 (continued)

Location of dumpsites	Salient features	References
Chennai, India	Major dumpsites–Kodungaiyur and Perungudi (disposal rate of approximately 2200–2400 tons/day and height of dumpsite is around 10–15 m) City waste generation approximately 6404 metric tons of MSW/day MSW consists of 36% by weight bio-waste, around 30% recyclable material, 62% combustible materials 25–33% of plastics and 28–55% of mixed residue having a high energy content Major issues—subjects to air intrusion and rain events which enhances the biological and chemical oxidation of organic waste materials, increasing the temperature of waste piles, and fire events	Peter (2019a, b)
Meethotamulla, Sri Lanka	Open dumpsite—290,000 metric tons of MSW/year and collapsed in 2017	Jayaweera et al. (2019a, b)
Sub Saharan, Africa	MSW types—hazardous waste, medical waste Major issues—groundwater contamination, uncontrolled burning, and visibility impediments	Idowuac et al. (2019)
Malaysia	Dumpsite types—active landfills (Ulu Maasop and Kampung Keru) and closed landfill (Pajam) Major issue—leachate have high content of BOD, COD, and potentially toxic heavy metals (As and Cr)	Hussein et al. (2019)
Gohagoda, Sri Lanka	2.5 ha, 130 tons/day of MSW Approximately 59% food waste from daily input Located adjacent to Mahaweli river	Wijesekara et al. (2014)
Karadiyana, Sri Lanka	10 ha, 575 tons/day of MSW Located highly urbanized and environmentally sensitive area closer to Bolgoda river Major issue—Leachate have high content of TDS, ammonia, COD, nitrate and heavy metals	Koliyabandara et al. (2022)

Bold information states the presence of main open dump sites in particular locations

Dumpsites receive different types of waste such as MSW, heath care waste, waste electric and electronic equipment (WEEE), hazardous waste, used batteries, construction and demolished waste (CD), used tires (Esakku et al. 2003; Oyelami et al. 2013; Joseph et al. 2004; Sawyerr et al. 2017; Ferronato and Torretta 2019). The composition of waste from most of the developing countries accounts for more than 50% of organic matter where Afghanistan (Kabul) is about 70%, Bangladesh (Dhaka), 60–75%, China 60%, Kenya 75%, Zimbabwe 47% and Gaza strip 52% (Harir et al. 2015).

7.2 Environmental Impacts of Open Dumpsites

Global Waste Management Outlook report in 2015 suggested that around 2 billion people in world is deprived of proper waste collection and they are served by dumpsites (ISWA 2016). Prolonged incomplete handling of solid waste or lack of management at disposal sites has caused significant environmental and social challenges across the world in these regions, pollution of nearby water supplies, degradation of waste dumps by scavengers, enhanced mosquito breeding sites, unintentional landfill gas fire, airborne contaminants creating unpleasant odors, damage to vegetation, pollution of groundwater, air pollution leading to global warming (Calvo et al. 2005; Umar et al. 2010) The presence of hazardous substances in waste has worsen the impacts by open dumps increasing the health risk from leachate and gaseous emissions (Erses et al. 2005). Leachates, gaseous emissions are common issues occur despite the waste management practice either open dump site or in a sanitary landfill. The process of mitigating such issues in a proper manner in sanitary landfills help to reduce the burden on environment.

7.2.1 Environment Pollution by Landfill Leachate

The production of leachate is associated with precipitation, as rainwater percolating through waste mass generates leachate. The biological and chemical reactions occur inside the waste pile lead to the liquid formation along with the rainfall. The composition of leachate is greatly influenced by the type of waste and its water content (Samuding et al. 2009; Renou et al. 2008; Fatta et al. 1999; Naveen et al. 2017). The quantity and quality of leachate produced throughout the season varies. Biological and physiochemical processes release both organic and inorganic contaminants in waste. Nature of waste and the chemical reactions that take place directly affects the quality of leachate (Fatta et al. 1999). Leachate can contain large quantities of organic and inorganic components. The chemical makeup of leachate comprises of a number of elements as discussed in Table 7.3.

Table 7.3 Composition of landfill leachate

Parameter		Land fill age	Concentration range	References
Organic matter	BOD	–	1000–57,700 mg/L	Kargi and Pamukoglu (2003)
	COD	–	1500–71,100 mg/L	Kargi and Pamukoglu (2003)
	Volatile Organic Acid	> 10 years	100–3000 mg/L	Aziz et al. (2010)
	Total organic carbon	30 years	30–29,000 mg/L	Kjeldsen et al. (2002)
	BOD_5/COD	20–30 years	Ratio—0.69	El-Salam and Abu-Zuid (2015)
Heavy metals	Zinc	30 years	0.03–1000 mg/L	Kjeldsen et al. (2002)
	Iron	13 years	0.4–11.49 mg/L	Han et al. (2014)
	Arsenic	–	0.6–1.7 mg/L	Hossain et al. (2014)
	Chromium	– 7 years	1.81–5.36 mg/L < 0.06 mg/L	Ibezute and Erhunmwunse (2018) Przydatek and Kanownik (2019)
	Cadmium	–	0.17–0.62 mg/L	Ibezute and Erhunmwunse (2018)
	Lead and Copper	7 years	< 0.2 mg/L	Przydatek and Kanownik (2019)
Nitrogen	Nitrate	> 10 years Operated since 1982 > 10 years	55.82 ± 12.31 mg/L 18.6 mg/L, 19.2 mg/L, 20.3 mg/L 0–17 mg/L 100 mg/L	Ibezute and Erhunmwunse (2018), Bhalla et al. (2013) Tatsi and Zouboulis (2002) Han et al. (2014)
Inorganic compounds	Ammonia	30 years-	50–2200 mg/L 500–1500 mg/L 10.1 ± 0.1 mg/L 2675 mg/L	Kjeldsen et al. (2002) Tatsi and Zouboulis, (2002) Sawyerr et al. (2017) Mor et al. (2006)
	Ammonia-nirogen	4 years	6–4095 mg/L	Kulikowska and Klimiuk (2008)

(continued)

Table 7.3 (continued)

Parameter		Land fill age	Concentration range	References
	Chloride	20–30 years	9500 to 16,250 mg/l Gazipur landfill, India—28–737 mg/L open dumpsite, Nigeria—0 to 167 mg/L	El-Salam and Abu-Zuid (2015) Aboyeji and Eigbokhan (2016), Mor et al. (2006)
	Phosphate	Operated since 1960	5–20 mg/L (acetogenic leachate) 1000–3000 mg/L (methanogenic leachate)	Wijesekara et al. (2014)
		> 10 years < 5 years	9 mg/L 170 mg/L	Fatta et al. (1999), Guasch et al. (2004)

Moreover, landfill leachate may contain very high concentrations of particular matter, pesticides, aromatics, phenols like xenobiotic organic matter (XOM) (El-Salam and Abu-Zuid 2015; Hossain et al. 2014; Han et al. 2014; Kjeldsen et al. 2002; Mor et al. 2006; Devare and Bahadir 1994). The technology which is used in a landfill and its age too contribute to the content of leachate (Baker and Curry 2004). Compounds such as aromatic hydrocarbons, pesticides, phenols, heavy metals, ammonium, nitrates, halogenated compounds, xenobiotics such as benzene and naphthalene sulfonates in leachate are considered as potential hazardous compounds (Naveen et al. 2017; Riediker et al. 2000). These compounds are mainly conditioned by precipitation penetrating into waste because of some meteorological conditions. Hence, as the landfill age increases, most pollutants can percolate through soil and subsoil, causing high contamination of groundwater table.

Surface and ground waters near open dumpsites are highly susceptible to pollution because of the potential of mixing with landfill leachate rich in harmful organic and inorganic constituents. Areas around sanitary landfills have lower potential of water pollution due to the proper management of leachate either by biological, physical, chemical methods or integrated methods. Ground and surface water resources polluted with landfill leachate poses a substantial risk to the natural environment (Mor et al. 2006). Landfills which have traditionally been constructed without engineered liners and collection systems for leachate pose a major threat to groundwater pollution. In many countries, the most recent regulations required the installation of the liner system, the collection of leachate and the treatment system (Kjeldsen et al. 2002). However, toxic substances may be present in variable concentrations and are associated with the nature of the waste deposited. Pollution intensity is determined by the concentration of contaminants, groundwater flow direction and depth to the water table (Aderemi et al. 2011).

7.2.1.1 Pollution by Inorganic Macro Components

Both young and old leachate having a significant amount of ammonia nitrogen (NH_4^+-N) because of the hydrolysis and fermentation of the nitrogenous fraction of biodegradable waste (Kulikowska and Klimiuk 2008; Tatsi and Zouboulis 2002). Deamination during the breakdown of organic compounds and amino acids cause the young and older leachate to have NH_4^+-N respectively (Han et al. 2014). Unionized ammonia is present in equilibrium with ammonium ions (NH_3 (aq)) in an aqueous solution. According to the dissociation equation:

$$NH_4^+ + H_2O \overset{Ka}{\longleftrightarrow} NH_{3(ap)} + H_3O \quad (7.1)$$

The cumulative value of un-ionized and ionized ammonia is termed as total ammonia. With rising pH, the toxic effect of total ammonia rises, suggesting that unionized ammonia is the main toxic source (Osada et al. 2011). More significantly, landfill leachate with a high concentration of ammonia proved to be too difficult and too costly for municipalities to handle, particularly during biological treatment at wastewater treatment plants. (Heidenreich and Kleeberg 2003). As proof, leachate from a municipal solid waste landfill in a Mediterranean climate žThessaloniki, Greece have recorded principal pollutants in the leachate samples were organic and ammonia loads.(Tatsi and Zouboulis 2002) Moreover, during the methanogenic fermentation nitrogen gets reduced to ammonia. The presence of ammonia in the water source is often associated with contaminants due to wastewater from like use of nitrogen fertilizers or waste from livestock.

Excretion of toxicants into water by aquatic organisms gets problematic when ammonia levels increase in water bodies. It makes the toxicants to accumulate in internal tissues and in blood stream which leads to the death of such aquatic species. Factors like pH, temperature affect the ammonia toxicity of aquatic species (Chaudhary et al. 2010; Manassaram et al. 2005). Ammonia is also used as a tracer for leachate in surface and groundwater. Natural ammonia concentration levels in ground and surface water are typically less than 0.2 mg/L (Canter 2019; Taha et al. 2004).. Hence, the higher concentrations of ammonia are likely to affect reducing conditions in groundwater (Wang et al. 2007). Ammonia can cause a diverse range of diseases, including metabolic disorders, neurological malfunctions, coma, and even death (Thrane et al. 2013).

Natural nitrate levels in groundwater are generally very low, but caused by human activities such as agriculture, manufacturing, domestic effluents, nitrate levels keep on rising. In soil and wastewater, nitrates tend to move relatively slowly and there is a period of about 20 years between chemical activity and contaminant detection in groundwater (Manassaram et al. 2005; Canter 2019; Kross 2002; Chaudhary et al. 2010). Dissolved nitrogen, which is present in the form of nitrates, is known to be one of the most common contaminants present in groundwater (Chaudhary et al. 2010). Nitrate becomes hazardous when it is reduced to nitrite in the intestine. Hemoglobin is converted to methemoglobin by nitrite which blocks oxygen transferand results in methemoglobinemia, /blue baby syndrome (Kross 2002). Nitrate is lost through denitrification to nitrogen gas after the degradation of organic matter consumes dissolved oxygen and disappears. It occurs slowly by reducing insoluble Mn (IV) species to soluble Mn (II), reducing insoluble Fe (III) to soluble Fe (II), reducing sulfate to sulphide and eventually reducing carbon dioxide to methane. The oxidation of ammonium to nitrite by the nitrification cycle allows nitrates to form in leachate samples. Erudition on nitrate and phosphate is helpful in interpreting nutrient status of waters. These appear as a result of mineralization and decomposition of organic matter.

In addition, phosphorous can also be identified as a pollutant contains in landfill leachate and the natural occurrence of phosphorus in surface waters is minimal. Small amounts of phosphorous that become soluble in natural waters can induce a significant amount of algal blooms. The presence of nitrogen, carbon is also required for the development of algal blooms. These blooms lower light penetration and dissolved oxygen levels. In the meanwhile, the aesthetic value of waters too degrades with it. A marginal amount of phosphorus reaching a water body will lead to significant algal boom, while nitrogen (N) and carbon (C) are also required for algal growth, minimizing light infiltration and dissolved oxygen levels, as well as causing the surface water bodies to degrading esthetically (Mylavarapu 2008). Leaching and drainage are the primary mechanisms through which phosphate is transported into rivers, streams, canals, wetlands, or groundwater. Surface runoff is the portion that the soil does not hold or accumulate but runs down the slope. Internal sources of nutrients are derived from the lake/reservoir sediments. Sediments release phosphate into the water column when the level of dissolved oxygen is low (anoxic) in the water. This process stimulates algae growth.

A number of studies have also found leachates with high concentrations of chloride (Chen et al. 2017). Total dissolved Solids in leachate has a major input from chlorides (Komárck et al. 2007). Sources such as leachate, domestic effluents, fertilizers affect the high amount of chlorides in groundwater (Mor et al. 2006). Chlorides are typically not attenuated by the soil and are highly mobile under all circumstances, as the tracer portion of the leachate plume connecting the groundwater has a special significance (Umar et al. 2010; Kumar and Alappat 2005). When considering chloride concentrations in drinking water, no health-based guideline value is proposed, although it is stated that chloride concentrations over about 250 mg/L can give rise to a detectable taste in water (WHO 2003). In its natural form, chloride ions are incredibly stable in water. Leaching of sedimentary soils and rocks, weathering, leaching of dumpsite effluents can be stated as reasons for the presence of chlorides in groundwater (Prasanth et al. 2012; Logeshkumaran et al. 2015). Bromides have a conservative nature. They can infiltrate into groundwater easily (Soltermann et al. 2017, 2016). A very low acute toxicity characterizes bromide. The Acceptable Daily Intake of 0.4 mg/kg body weight marks an allowable total daily intake of 24 mg per person for a 60 kg person (WHO 2009). A derivation from the acceptable daily limit has concluded that maximum bromide level in drinking water for children who is of 10 kg consuming 1L/day, the value would be up to 2 mg L^{-1}. Numerous tests to reach the toxicity of bromide in freshwater living organisms present a guideline value of 1 mg L^{-1} for sodium bromide in both surface and groundwater (WHO 2006).

7.2.1.2 Potentially Toxic Elemental Pollution

Typically, waste is deposited in non-engineered dumpsites from which leachates receive heavy metals as As, Hg, Cd, Pb, Zn, Ni, Cu, Cr and Mn (Al Raisi et al. 2014; Qingjie et al. 2008). High metal concentrations occur in leachate due to the slow release of these metals in an acidic environment. It is not possible to completely prevent the production of leachate in a landfill but leachate generation could be decreased or could even be treated to mitigate the environmental hazard (Ifeanyichukwu 2008). The appearance of a high iron level in leachate demonstrates the leachate's dark brown color. It is a result of ferric oxidation and the structure of ferric hydroxide colloids and humic acid complexes (Al Raisi et al. 2014). Metals also precipitate phosphates and hydroxides (Meunier et al. 2006). Hydroxide precipitates form when pH is at or above neutral, which is typically the case in methanogenic leachate (Bilgili et al. 2007; He et al. 2006; Kjeldsen et al. 2002). This might explain the low levels of metals observed in both surface and groundwater, rather than in leachate. Trace metals, such as Zn, Cu, also have been documented with the potential for bioaccumulation in the food web for various health problems (Pohland et al. 1993; Mohan et al. 1996). As most of the developed nations have engineered dumpsites, pollution of ground water is mitigated.

7.2.1.3 Goundwater and Soil Pollution

Open dumpsites in Nigeria have also recorded groundwater quality data which lies below WHO and EPA standards (Oyelami et al. 2013). Disposal of waste electric and electronic equipment (WEEE) too have caused significant effect on environment due to leaching of metals. Generation of 18,000 tons of WEEE from Banglore city, India has been recorded and most of them are illegally dumped into sites (Garlapati 2016). Well water closer to Lagos state Nigeria where WEEE and used batteries are disposed, records Pb 2.77 mg/L, Cr 0.520 mg/L, Cd 0.035 mg/L and Ni concentration in soil in the range of 35.45 mg kg^{-1} to 85.43 mg kg^{-1}(Olafisoye et al. 2013). A study in Mexico has analyzed soil near a WEEE disposed open dumpsite. The results show the presence of 304 mg kg^{-1} Cu, 4.7 mg kg^{-1} Cr, 74 mg kg^{-1} Pb. The values of Geoaccumulation index showed the site falls into the category of very polluted with Cu and Pb (Nava-Martínez et al. 2012).

Several studies have shown that a lower level of metals is due to the fact that the leachate was produced and entered in low quantities primarily due to processes favorable to metal immobilization, such as sorption, chemical precipitation of metals and higher pH values (Song et al. 2016; Banat et al. 2005; Jumbe and Nandini 2009; Bourg and Bertin 1996; Al Raisi et al. 2014). Metals concentrations recorded in Onibu-Eja dumpsite in Nigeria were below standards while on Mn showed values above ground and surface water (Oyelami et al. 2013). Table 7.4 discusses metal concentrations recorded in selected studies.

Many toxic metal ions may bind to the dissolved organic carbon, which allows them to transport easily. Wells used by humans for drinking or sanitary purposes are

Table 7.4 Metal concentrations in groundwater closer to dumpsites

	Landfill	Leachate (mg/L)	Groundwater (mg/L)	WHO (mg/L)	References
Cr	Landfill in Warminsko-Mazurskie Province in the northeastern Poland		0.019	0.05	Talalaj and Biedka (2016)
Fe				0.3	
Cu			0.038	0.1	
Cd			0.005	0.01	
Cr	Bawari Nigeria		0.03 0.01	0.05	WHO (2011), Sawyerr et al. (2017)
Fe			0.2 0.1	0.3	
Cu			0.2 0.01	0.1	
Cd			0.2 0.07	0.01	
Cr	Gazipur landfill India	0.29	Below detection	0.05	Mor et al. (2006)
Fe		70.62	0.04–2.48	0.3	
Cu		0.93	–	0.1	
Cd		0.06	Below detection	0.01	
Cr	Landfill site in Nagpur India	0.0088		0.05	Pujari and Deshpande (2005)
Fe		0.83		0.3	
Cu				0.1	
Cd				0.01	
Cr	Payatas Landfill Philippines	< 0.06		0.05	Chounlamany et al. (2019)
Fe		–		0.3	
Cu		–		0.1	
Cd		< 0.09		0.01	

readily polluted by the mixture of leachate from landfill (El-Salam and Abu-Zuid 2015; Han et al. 2014). Aquifers polluted with landfill leachate are characterized with iron and manganese reduction. These are known to be major redox reactions in such waters (Rügge et al. 1999; Heron et al. 1994; Röling et al. 2001). When iron oxyhydroxides and manganese oxides are reduced, soluble metals are released into water. Certain metals, together with other reduced species such as ammonia, ammonium and hydrogen sulfide, can pose a significant threat to drinking water (Lovley 1997). Serious impacts occur on both human and animals, primarily aquatic as it does create adverse effects on food chains (Hossain et al. 2014). High metal concentrations occur in leachate due to the slow release of these metals in an acidic environment. It is not possible to completely prevent the production of leachate in a landfill but leachate generation could be decreased or could even be treat to mitigate the environmental hazard (Ifeanyichukwu 2008). As the environmental standards imposed on quality of ground and surface water become strict, the treatment of leachate should be considered. As groundwater pollution is one the major issues from leachate, monitoring these metals can help in developing suitable remedial measures (Kanmani and Gandhimathi 2013; Esakku et al. 2003). A significant amount of metal release can be occurred by the disposal methods such as open dumps, landfills, incinerators (Kanmani and Gandhimathi 2013; Iwegbue et al. 2010; Waheed et al. 2010).

7.2.2 *Air Pollution from Dumpsites*

In different phases of a landfill, the generation of gases occur. Biodegradation of organic matter at the anaerobic acid fermentation, generates volatile organic acids and CO_2. VOCs (Volatile Organic Carbons) are converted to CH_4 and CO_2 in the methanogenic phase. It is recorded that the temperature between 32 and 38 °C is preferable for the generation of CH_4 and organic pollutants in a landfill (Koide et al. 2012; Kjeldsen et al. 2002). Landfill gases are characterized with toxic VOCs. The state of California Air resources board have recorded the emission of hazardous chemicals of 35 kg per million kg of refuse from landfill waste and these gaseous emissions are known for health related issues (Bennett 1987; Sivertsen 2006).Sanitary landfills have methods such as flaring with or without energy recovery in order to mitigate the air pollution from landfilled waste while open dumpsites do not (Idris et al. 2004; Joseph et al. 2004).

When the degradation process moves from aerobic condition to anaerobic, production of methane rises while the level of carbon dioxide too increases (Aljaradin and Persson 2012). Methane is a highly flammable gas could lead to possible fire and explosion hazard if not appropriately managed (Chun 2001). Both CO_2 and CH_4 contributes to global warming. Open dumpsites are known for the release of CH_4, CO_2, N_2O, bioaerosol emissions and other greenhouse gases (Zhang et al. 2019; Njoku et al. 2019). As the temperature increases in the landfill after placement of waste, gases like H_2S, organic compounds containing sulphur like mercaptans tend to release. The objectionable odour arises from dumpsites are due to the presence of ammonia and hydrogen sulfide (Sivertsen 2006). Burning and fires occur in landfills produces particulate Matter (PM), polyaromatic hydrocarbons (PAHs), dioxins like hazardous pollutants. Climatic conditions such as rainfall and wind direction increase the distribution of odours from landfills (Zhang et al. 2019). Considering all gaseous emissions from landfills nearly 45% accounts for CO_2. It is recorded that the emitted amount of CO_2 from landfills in Germany has been reduced during last fifteen years by nearly two thirds. Still the current annual emission accounts for more than 10 Mtonnes (Ritzkowski and Stegmann 2007). N_2O is considered as greenhouse gas which influences global warming 300 times than CO_2. Low concentrations(4ppm) are toxic by inhalation. Potency of CH_4 is 21 times greater than CO_2 in related to its effect on global warming and nearly 50% from landfill gas is CH_4. It is highly flammable and explodes at minimum concentrations of 50,000 ppm (USEPA 2017; Terraza and Willumsen 2009).

USA records that landfills are the largest source of CH_4 source next to ruminant animals. It is estimated that from the worldwide anthropogenic greenhouse gas emissions, landfill CH_4 generation is responsible for 1–2% (Bogner et al. 2011). The methane generation potential for sanitary landfills (Air Hitam sanitary landfill, the Jeram sanitary landfill) have been calculated as 151.7 m^3 t^{-1} and 75.9 m^3 t^{-1} for open dumping (Sungai Sedu) in Malaysia (Abushammala et al. 2015). Indian landfills produce nearly 16 tons of CO_2 per year and it is estimated to be increased to a value closer to 20 tons by the year 2020 (Singh et al. 2016). SO_2 emissions from landfills is a root cause for acid rains and particulates present in atmosphere (Rim-Rukeh 2014). Landfill gases are estimated to contain 1,000,000 and 10,000,000 μg/L of ammonia (Wallingford 1987).

Open dumps are identified as a significant source of bioaerosols. Recorded values in Olusosun open dumpsite, Nigeria for indicator groups are as approximately 4.8–105 CFU/m^3 total bacteria, 1.4–106 CFU/m^3 of gram negative bacteria in 96 m^3 of air (Akpeimeh et al. 2019). Open dumpsites are estimated to generate 1000 kg CO_2-eq. t^{-1} while 70 kg CO_2-eq. t^{-1} from landfills with low organic carbon waste (Ferronato and Torretta 2019). Recorded values for landfill gas emissions in selected countries are recorded in Table 7.5.

Fires occur in dumpsites which lead to the burning of dumped tires too produce hazardous air emissions. Gases like SO_2, NO_2, CO, PAH, VOCs are emitted from these burnings. (Ferronato and Torretta 2019). Fires occur accidently at the uncontrolled dumping sites, and burning of residential waste regarded as the highest dioxin emitter in USA (Rim-Rukeh 2014). Nepal records CO level of 21–49 g kg^{-1}, SO_2 102–820 g kg^{-1} NO_2 emission was 3–9 g Kg^{-1} from tire burning (Shakya et al. 2008). Due to long term exposure to gaseous emission and chemicals from landfills many people living closer to dumpsites are prone to health issues such as flue, asthma, skin irritations, lung problems etc. Therefore it is a need to establish proper gas recovery systems, daily coverage of dumped waste to minimize the odor related issues (Vrijheid 2000; Njoku et al. 2019; Ferronato and Torretta 2019). In March 2016, smoke blanketed Jamaica's capital due to a fire occurred in an open dumpsite (ISWA 2016). A similar incident has been reported in Ghazipur and Bhalswa landfills of India in April 2016 in releasing toxic gases affecting country's capital Delhi (ISWA 2016; Sunny 2017).

7.2.3 *Landfill Collapses—"Dumpslides"*

Landfill collapses with casualties and fatalities are common in developing countries. Some of the reported collapses include Guatemala City dumpsite, Guatemala in 2016 (Press 2016), Meethotamulla dumpsite, Sri Lanka in 2017 (Jayaweera et al. 2019a, b), Leuwigajah dumpsite, Indonesia in 2005 (Koelsch et al. 2005), and Hongao dumpsite in China in 2015 (Xu et al. 2017), Payatas, Manila, Philippines in 2000 (Merry et al. 2005), Tuban, East Java, Indonesia 2012 (Lavigne et al. 2014) etc. Loosing of stability in a landfill due to prolonged rainfall (Yang et al. 2017), presence of poorly landfilled construction and demolished waste, weakened subsoil (Koelsch et al. 2005), continuous explosions from biogas and its release (Lavigne et al. 2014)

Table 7.5 Gas emissions from dumpsites in selected countries

Dumpsite	CO	CO_2	NO_2	N_2O	NH_3	CH_4	H_2S	SO_2	$PM_{2.5}$	PM_{10}	References
Nigeria	133.7 and 141.6 mg/L	401 and 404.5 mg/L		21.0 mg/ Land 27.3 mg/L	14.7 to 19.5 mg/L	2310–2771 mg/L		27.7 and 37.1 mg/L			Rim-Rukeh, (2014)
Solid waste burning in Thailand 2016	86.4 kt	418.73 kt		0.514 kt		22.29 kt		1.0 kt,	PM—19.6 kt/year		Pansuk et al. (2018)
Faisalabad, Pakistan		9.026×10^7 m^3/year				1.354×10^8 m^3/year					Rafiq et al. (2018)
Bhalswa 2011–2012						91.23 Gg/yr					Singh et al. (2016)
Ghazipur 2011–2012						3845.20 Gg/yr					
Okhla 2011–2012						77.42 Gg/yr					
Permissible levels	10 mg/L	397 mg/L	40–60 mg/L	< 40	< 25	1000 mg/L	10 mg/L	100 mg/L			Rim-Rukeh (2014)
	9 mg/L		53 mg/L					0.5 mg/L	35µg/m3	150 µg/m^3	USEPA (1990)

have been identified as main reasons for the major collapses and avalanches of landfills around the world. It is a must to consider the geological features, groundwater levels, capacity of landfill and the composition of waste, proper drainage systems, gas management systems to minimize the risk of landfill collapse (Chathumani et al. 2019; Sunny 2017; Jayaweera et al. 2019a, b).

7.2.4 Emerging Contaminants

Dumpsites are a common place for the disposal of consumer products such as fire retardants, disinfectants, preservatives, pharmaceuticals etc. The organic compounds present in these materials which can be either synthetic or naturally present chemicals/compounds are known as Emerging Contaminants (ECs). These are present in low concentrations. ECs have potential health risk to human as well as to natural environment (Andrews et al. 2012; Ramakrishnan et al. 2015). A study carried out in the city Elkhart; Indiana reported the presence of ECs in low concentrations (parts per billion range) in groundwater where a nearby dumpsite is closed unlined. The ability of ECs to be adsorbed to solids or to be dissolved in leachate is a factor to consider (Buszka et al. 2009). Andrews et al., recorded that the presence 13 household and industrial compounds, 7 hydrocarbons (mostly polycyclic aromatic hydrocarbons (PAHs), pesticides etc. in Norman Landfill, Oklahoma in 2000 (Andrews et al. 2012). Even after the closure of landfill, the transport of ECs were detected from waterbodies closer to the site where lesser concentrations of ECs in the 2009 sampling were present than in the 2000 sampling indicate natural attenuation of these compounds caused by transport and degradation processes in the aquifer. The further analysis of the presence of ECs in landfill leachate and its presence in close by waterbodies should be researched to decide its toxicity level for environment.

Annual global production of plastic was recorded around 322 million tons for the year 2016 and the value was around 288 million tons per year in 2013 (Bellasi et al. 2020). Significant amounts of plastics are buried in landfills. Fragmentation of plastics can be occur with time due to adverse environmental conditions including pH and temperature gradients, microbial degradation in landfills (He et al. 2019; Kjeldsen et al. 2002). Microplastics (MPs) entering into surface waterbodies can originate from landfills due to the large amount of landfilled amount of plastic and hazardous waste (Imhof et al. 2013; Sundt et al. 2014). In landfills such as Landfill Shanghai, in China which is active since 2013 and landfill Changzhou (2003-now) have recorded the presence of MPs. Different types as lines, flakes, fragments, pellets and foams were reported (He et al. 2019). Concentration was 0.42–24.58 items/L in leachate taken from the southern China, while it was 0–4.51 items/L in the leachates from Nordic countries (Praagh et al. 2018).

Table 7.6 Recorded levels of pathogens in leachate, groundwater and surface water

	Landfill site	Groundwater CFU	Surface water CFU	WHO CFU	References
Escherichia coli	Nigeria Bawari	33		0	Sawyerr et al. (2017)
Shigella		27		0	
Heterotrophic bacteria		410		110	
Fecal Coliform	Landfill site in Nagpur		10.588		Pujari and Deshpande (2005)
TC (CFU/100 mL)			338.82		

7.2.5 *Pathogens*

Several studies identified overall density of coliforms ranging from 15 to 20,000 CFU /100 mL and E. Coli concentration of from 15 to 15,199 CFU /100 mL (Grisey et al. 2010). Microbial activity in landfill leachate depends on the availability of electron acceptors. Due to the difficulty of detecting low concentrations of pathogenic bacteria and viruses, coliform bacteria are used to determine fecal contamination of water supplies (Klinck and Stuart 1999). Pseudomonas recorded in rivers and lakes in concentrations of 10/100 mL to > 1000/100 mL. The bacterial counts are expressed in CFU which stands for Colony Forming Counts. The physio-chemical environment and microbial communities play vital role in transformations of organic and inorganic compounds that helps in leachate decomposition and mineralization. (Röling et al. 2001). Studies have shown the presence of different kinds of bacteria in leachate. Zhang has identified the bacteria present in leachate samples using advanced molecular analysis. (Zhang et al. 2009). The presence of coliform bacteria such as *E.coli, Shigella* sp and *salmonella* sp have been recorded in leachate and leachate contaminated aquifers (Adeyemi et al. 2007). Measuring anthropogenic stresses and its contributors have been shown by the Total coliforms counts (Pujari and Deshpande 2005) (Table 7.6).

7.3 Way Forward

All mentioned investigations have been proven that the final destination of waste gains less attention in terms of its environmental impacts. Hence, It is important to find out an effective way to minimize the impact of solid waste on the environment as well as on humans living on earth despite of being a root cause for increasing global warming, acidification, groundwater pollution, resource exploitation, etc. Considering different environmental issues, it is a major requirement to compare different technologies and different ways to rebuild a sustainable society. In such a context, life cycle assessment (LCA) and circular economy (CE) concepts can be identified as holistic techniques which allow people to compare different systems with associated environmental impacts. These have been, considerably applied in order to evaluate environmental burden associated with the waste management systems.

LCA is an objective framework for evaluating or assessing the environmental impact associate with products, processes, services, materials, or systems across its entire lifespan from cradle to grave (Goulart Coelho and Lange 2018). For example, steel is one of the significant raw material which commonly used for various manufacturing processes and when it becomes no longer to use, discarded steel can be recycled back to the process. However, these actions are no longer proceed and unfortunately, a large portion of waste going to end up in a landfill or dumpsite. So each of these stages has specific inputs and outputs, hence it is a prerequisite requirement to fully understand entire lifecycle impacts and apply the concept to evaluate alternate solid waste management practices (Turner and Kemp 2016). In such a setting, LCA can give sustainable solutions as it can explore more opportunities to improve the conventional waste management systems beyond its current perspective. This consists of energy production from waste material (which is known to be a realistic approach to minimize fossil fuel consumption), improving virgin materials usage from recycled waste materials, organic compost extraction as a byproduct from biological treatment, etc. (Dan Wang 2020).

Since 1995, LCA has been recorded as an evaluation tool for waste management practices (Dan Wang 2020). As a case in point, Habib has assessed the global warming potential of solid waste management practices in Alborg, Denmark through the period from 1970 to 2010 using the LCA approach. This study was mainly conducted by evaluating the historical implications regarding MSW composition and different treatment technologies (incineration, composting, recycling) and results show that greenhouse gas emissions rate from the MSWM system have significantly decreased during the last 40 years from net emission of 618 kg CO_2-eq. tonne to a net saving of 670 kg CO_2-eq. tonne of MSWM. Habib believed that this was possible by changing the waste dumping from non-sanitary landfills into more specialized waste treatment technologies including composting, recycling, incineration over given time period (Komal Habib and Christensen 2013). Similarly, Zhou has evaluated the environmental performance, challenges, and issues associated with the municipal solid waste management system in Hangzhou, China from 2007 to 2016 using the LCA methodology. This was primarily focused on novel treatment technologies and source-separated collection. Moreover, these scholars found that waste incineration has high environmental performance than in landfills in terms of global warming potential (GWP), acidification potential (AP), nutrient enrichment potential (NEP), POP regardless of source-separated collection (Zhaozhi Zhou et al. 2018). In addition, Francesco compared waste disposal alternatives (landfill, landfill with biogas recovery, MSW sorting plant, incineration) from a life cycle perspective, examining landfills in Roma, and results discovered landfills system as the worst management system while sorting plants coupled with electricity and biogas production would be the effective method for MSW since it controls the effect of harmful emissions (NO_x,

PM10, heavy metals, etc.). Simultaneously, these (electricity and biogas production) are able to cover the Italian electricity demand while providing environmental advantages at a wide-scale (Cherubini and Ulgiati 2009).

However, data on the environmental assessments of open waste dumpsites in developing nations are extremely limited. In Sri Lanka, LCA study was conducted to assess the environmental impacts on open waste dumps (Bloemendhal and Gohagoda) mining for resource recovery. The analysis highlights the starting of waste valorization during the early stage of waste degradation of dumpsites is effective in the GWPs' perspective and further emphasizes the necessity of avoiding waste transportation to gain profits. Moreover, the study introduced plasma gasification is more beneficial instead of yield incineration. In addition, energy & material recovery systems through efficient technologies can be recognized to minimize the impact of traditional dumpsite remediation including excavation, cleaning up, landfilling the excavated waste, etc. (Maheshi 2015). Even so, non of the investigated studies is completely able to avoid the environmental impacts, and those can be used in the decision-making process regarding the choice of an effective waste management system.

Besides, CE can also be recognized as a national strategy for sustainable waste management. In contrast to the traditional linear pattern of using natural resources consisting primarily of take, make, use, discard steps; circular economy maintains a high resources efficiency in terms of focusing on concepts such as eco-design, distribution, reuse, repair, recycling of products (IUGA 2016). Similarly, after the make and use stages, CE completely makes sure that materials are recovered and indeed proceed or return into a new/next lifecycle of production. With that remark, solid waste management has an equally prominent role in the circular economy since it makes the other half of the cycle complete (Korhonen 2018). However, along the way and at the end of the one production cycle, materials can also be fully returned to the biological cycle with no harm to humans and the environment. These can be further proceeded by aerobic composting, anaerobic digestion, to recover resources such as organic matter, nutrients, and energy (Logan and Visvanathan 2019). Correspondingly, product user stages can be further extend and discarded assembly parts can be used for refurbishment and remanufacture of the products. Comparably, another aspect is to examine the design stage, whereas producers need to make sure that they made all decisions about materials, assembly methods, and usage (UNEP 2015). In addition, recycling is usually identified as downcycling, whereby materials lose their quality in each cycle. However, it is possible for recycling to upgrade into upcycling, whereby quality improves in each cycle (Korhonen et al. 2018b). Hence, lot of innovations will be needed in various disciplines of fields. Although, well-designed materials are particularly important in order to produce high-quality end products. Similarly, it is important to consider the end of use stage when designing a product which means that determining the way of product to be repaired, remanufactured, refurbished, disassembled (Ness and Xing 2017). Moreover, new relationships are needed along the supply chain in order to share information about new materials and parts (Ezeudu 2019). In Romania, CE approach has been changing the current waste system by enhancing disposable patterns, selling wastes as a product, paying for waste treatment, recycling waste into recourse, and mainly consider waste as a tactic to develop a zero-emission environment (IUGA 2016). Further, Tisserant has discovered new waste treatment and waste footprint scales in order to design and evaluate the rules and regulations for CE (Tisserant et al. 2017). Similarly, the European Union promoted new ambitious goals by performing CE functions to their waste management system in 2018. The concept has been spread through various regional perspectives by enhancing reusable and recycled products usage in order to achieve sustainable MSWM (Tsai et al. 2020). Globally, Japan and China can be identified as major key players who follow CE concepts officially, therefore, Gang has researched CO_2 reduction methodologies for MSWM in Kawasaki, Japan (Geng et al. 2010). All these aspects equally contribute to managing the ultimate destination of waste to reduce its environmental pressure. However, in order to address the issues efficiently, advanced waste management solutions (which can easy to implement and have positive environmental outcomes) are indeed required (Abdel-Shafya et al. 2018).

7.3.1 *Leachate Treatment and Reuse*

Generated leachate from sites should be properly treated to reduce environmental burden from it. Many different methods are currently in use to treat the landfill leachate. Conventional treatment methods have three main steps leachate transfer, biodegradation and physiochemical methods. Physiochemical methods such as adsorption, coagulation/flocculation, air stripping, sedimentation and etc. are used for the process and use of aerobic and anaerobic methods are observed in biodegradation (Renou et al. 2008).

Biological treatment is used widely considering the cost effectiveness and reliability. It shows efficient results in removing organic and nitrogenous (Bove et al. 2015; Salem et al. 2008; Costa et al. 2019). Sequencing Batch Reactors (SBR), activated sludge processes, aerated lagoons have been adopted worldwide for the treatment of leachate using aerobic method (Li and Zhao 2001). Carbon adsorption, ion exchange methods are known as advanced techniques which are also used for leachate treatment. For better treatment, integrated methods of combining chemical, physical and biological steps, are required (Wiszniowski et al. 2006).

7.3.2 *Resource Recovery*

Resource recovery from open dumpsites can reduce the quantities of MSW directed to landfilling while lowering the operating costs. Materials like textile, metals plastic, leather can be used for recycling (Harir et al. 2015). Composting of organic waste is a popular method which triggers the rate of biological decomposition of organic materials (Saheri et al. 2017). Use of manure from compost in urban agriculture helps reducing the pollution from leachate, land remediation, restoring soil nutrients and loss of water and CO_2 reduces the solid waste volume by 25–60% (Babu et al. 2013; Farrell and Jones 2009; Guerrero et al. 2013).

Functioning open dumpsites should be retrofitted to minimize the groundwater pollution by establishing leachate collection systems, applying cover soils for reduction of leachate production and GHG emissions. Evapotranspiration (ET) covers can be used to reduce the amount of waste percolating into waste. Thickness of the material using as ET, ability for retaining water and air are important characters when designing such covers (Harshani et al. 2015; Abichou et al. 2003). Use of soil cover systems is a conventional technology which can be upgraded with soil capping. It helps in landfill gas (LFG) optimization while reducing the leachate generation. Landfill sites which are large can use bio covers. Effective degradation takes place when use of bio covers and it reduces emission of CH_4 (Jayawardhana et al. 2016; Sadasivam and Reddy 2014). Use of bio covers, biofilter systems and bio windows are effective alternatives to soil covers (Huber-Humer et al. 2008). Bio covers like compost, sewage sludge, peat are high in organic matter (Scheutz et al. 2009).

Existing dumpsites can be used for bio transformation with proper segregation of waste. Bio oil, char and gas are such outcomes can be achieved from current dumpsites (Kaur et al. 2014). Pyrolysis and composting are the common practices used to overcome the issue of waste dumps. Biomass is known to be a resource for the generation of fuel and chemical products with low environmental impact (Jayawardhana et al. 2016). Production of bio char by thermal decomposition of carbonaceous material without oxygen have shown promising results for the use against environmental pollution from landfills such as by emission reduction (LFG, leachate), reducing the quantity of waste, and alteration of soil quality (Verma et al. 2014, Oonk and Boom 1995, Chen et al. 2016). Composting too can be benefited from biochar as it triggers the intake of CH_4 and organic compound which are non-methane (Reddy et al. 2014).

The growing problem of MSW is mainly triggered due to lack of source separation, ability of open dumping, lack of financial support and education and lack of stringent law enforcement. Minimizing the amount of waste that is generated is the basic of waste management. (Peter et al. 2019a, b). Use of biodegradable packaging, political decisions such as "pay as you throw system", highlighting and creating proper markets for resource recovery are effective when working towards waste regarded sustainability (Zhang et al. 2010).

Landfill mining or landfill reclamation is a process used by countries to generate income from waste. It is a process by which excavation of dumped waste for reducing the landfill masses and recovery of resources for combustion or recycling. It was previously used for redevelopment of sites, reduction of area of the landfill and for removing potential polluting sources (Raga and Cossu 2014; Van der Zee et al. 2004; Jain et al. 2013). The reclaimed land can be converted to a sanitary landfill by a step wise process while the excavated soil can be used as covers (Mutz et al. 2017; Jain et al. 2013). Landfill mining could use several steps as direct selling of refuse derived fuel (RDF) as an alternative fuel source for coal to be used for cement industry and electricity generation. Both these options have shown benefits both environmental and economical in compared to leaving the dumpsite without any reclamation. It is recommended to carry on more research to look into the processes which can develop open waste dump mining as a clean development (Zhou et al. 2014; Winterstetter et al. 2015; Maheshi 2015).

7.3.3 Conversion of Open Dumpsites into Sanitary Landfills

Use of sanitary landfills in contrast to open dumps/ controlled landfills are more preferred by developed economies. Three steps has been described by ISWA as, closing the dumpsite by improving in to sanitary landfill, closing the dumpsite by covering waste and extracting waste and closing the dump site (ISWA 2016). According to ISWA roadmap for open dumpsite closure, closing the site by improving can be done with topsoil layer cover and with a low permeability cap. It can be used for tree planting later (Brunner 1971). Mechanisms to collect gas generating from waste should be properly managed and proper leachate collection/ treatment should be implemented. The second option is the in-place closure. It is implemented when no addition space is available for further improvements. Covering waste with a soil layer and vegetation is carried out on top of it. Removal of waste for the closure of a dump site is carried out when a suitable off the site is established such as a sanitary landfill. The excavated land can be used as recreational parks or sites for development. After the removal and clean-up, the former land use as a waste dump should be noted in land records and the land can be treated as a brownfield redevelopment site or as a passive recreation park facility (ISWA 2016; Yousuf and Rahman 2009).

7.3.4 Waste to Wealth

As global pressure rise to find solutions, researchers have shown the possibility of utilizing waste as raw material or alternative fuel by substituting primary raw material or fossil fuel sources. It is essential to have long-term solutions for the end-of-life disposal of waste. Ironically, every single part of waste has its unique value and most residue can be transformed into valuable products which can promote towards a sustainable and bio-economy (Oncioiu et al. 2020). In particular, organic wastes from bio-based industries and households can be utilized using biological treatment with anaerobic digestion which converts waste into other useful by-products with the help of microorganisms (Lange Lene et al. 2021). On the one hand, biogas output can be applied as a replacement for fossil fuels in many fields especially these gases can be used as LP gas which can further use for cooking purposes, heat up water, etc. On the other hand, waste output from the process which known to be bio-digest can be applied as organic fertilizer instead of an alternative for chemicals that improve the fertility of the soil and yields of crops (Antje Klitkou and Capasso 2019). Similarly, garbage that contains high percentage of biodegradable fraction, as well as moisture content, can divert to the bio-gas unit and this provides an efficient solution for solid waste management which also has extra benefits including improving the quality of soil, air, water thus ensuring a lavish environment for all organisms (Mikael Lantz et al. 2007). Sweden is one of a major country that processes large scale biogas plants and produced gas (almost 60% (3 PJ/year) of the total biogas production) consists methane (50–80%), carbon dioxide (20–50%), and hydrogen sulphide (0–0.4%) (Berglund 2006). Researchers suggest these products can be chosen for energy services such as combined heat and power (CHP), vehicle fuel, and further decisions can be also made to introduce energy into the national grid as possible. However, biogas production in landfills accounts for approximately 30% (1.5 PJ/year) of the total biogas production and these contain high amount of impurities thus further treatment is essential before usage (Mikael Lantz et al. 2007).

In addition, the biorefinery approach whereas bioresources (eg: food waste residues, straw, and stover from plant production, sludge from wastewater treatment, residues from fishing, aquaculture industry, and residues from forest-based) utilized to produce value-added products which can be identified as a successful route for landfill waste management (Karthikeyan et al. 2017). For instance, the European Commission discovered that the annual food waste generation was 89 million tonnes in 2011 and this could produce 0.22 EJ of energy, representing almost 0.5% of the total EU final energy demand of 46.19 EJ (Antje Klitkou and Capasso 2019). Although, scholars evaluate a large portion of agricultural residues are recorded as straw and stover from grain crops such as wheat, barley, maize, and these have a significant potential energy output range from 0.8 to 2.64 EJ which emphasizes 1.7 and 5.7% of EU final energy consumption. Similarly, authors estimate that energy from forest residues was from 0.51 EJ to 2.7 EJ/year and the majority (over 90%) of energy was supplied by the crop and forest residues. Thus, the study has summarized that food, crops, and forestry residues can offer a range of energy demand from 1.55 to 5.56 EJ per year and proving that adding a considerable amount of value to waste leads to reduce the environmental pressure from landfills/dumpsites (Bettina Kretschmer et al. 2013).

Further, biorefineries have the ability to convert biomass into energy and other beneficial by-products such as protein and phosphorus (Antje Klitkou and Capasso 2019). Moreover, waste management in the context of recycling (upcycling) can also be decided as a sustainable value-added method compared with the traditional waste disposal. This operation mostly encourages to increase the value of reprocessed materials than the value of original products and consolidating waste reduction and money saving altogether (EPA 2020). Similarly, there are a wide range of value-added activities such as creation of consumer products including handbags, jewellery, flower pots/garbage cans from discarded wastes (tires), conversion of wastes to animal feed, recycled paper, etc. (Nick Dale et al. 2012). In such a context, we can believe that people have a significant role to follow eco-friendly environmental practices in order to increase waste management efficiency. In parallel, government can provide supportive policies, legislations, and incentives to support such types of further developments in society.

7.4 Conclusion

The paper discusses about the continuous environmental pressure arising from waste dumping. The pollutant loads released to the environment imparing the water quality, air and land quality directly affect the overall health of livingbeins. Especially developing nations are facing the issue of waste management due the unavailability of proper waste management scenario, lack of funds and awareness. Source segraration, the waste upcyling aligning with the SDGs would reduce the waste collected at dumpsites. The urgent need of resource recovery, promoting waste to value, conversion of open dumps into sanitary landfills is highly emphazied by the current work streamlining the the pollutant loads discharge dto environment by open dumps.

Conflict of Interest The authors declare that there is no conflict of interest regarding the publication of this paper.

References

Abd El-Salam MM, Abu-Zuid GI (2015) Impact of landfill leachate on the groundwater quality: a case study in Egypt. J Adv Res 6:579–586

Abdel-Shafy HI, Mansour MS 2018. Solid waste issue: sources, composition, disposal, recycling, and valorization. Egyptian J Petrol 27: 1275–1290.

Abichou T, Langoni G Tawfiq K (2003) Assessment of alternative earthen final covers for Florida landfills. Florida Center for Solid and Hazardous Waste Management, University of Florida, Report, pp 03–05

Aboyeji OS, Eigbokhan SF (2016) Evaluations of groundwater contamination by leachates around Olusosun open dumpsite in Lagos metropolis, southwest Nigeria. J Environ Manage 183:333–341

Abushammala MF, Basri NEA Younes MK (2015) Methane generation from landfills: Malaysia specific parameters. In: 5th International conference on environment science and engineering, pp 14–19

Adeyemi O, Oloyede O, Oladiji A (2007) Physicochemical and microbial characteristics of leachate-contaminated groundwater. Asian J Biochem 2:343–348

Aderemi AO, Oriaku AV, Adewumi GA, Otitoloju AA (2011) Assessment of groundwater contamination by leachate near a municipal solid waste landfill. Afr J Environ Sci Technol 5:933–940

Akpeimeh G, Fletcher L, Evans B (2019) Exposure to bioaerosols at open dumpsites: a case study of bioaerosols exposure from activities at Olusosun open dumpsite, Lagos Nigeria. Waste Manage 89:37–47

Al Raisi SAH, Sulaiman H, Suliman FE, Abdallah O (2014) Assessment of heavy metals in leachate of an unlined landfill in the Sultanate of Oman. Int J Environ Sci Dev 5:60

Aljaradin M, Persson KM (2012) Environmental impact of municipal solid waste landfills in semi-arid climates-case study–Jordan. Open Waste Manage J 5:28–39

Al-Khatib IA, Hammad AA, Sharkas OA, Sato C (2015) Public concerns about and perceptions of solid waste dump sites and selection of sanitary landfill sites in the West Bank, Palestinian Territory. Environ Monitor Assessment 187:186

Andrews WJ, Masoner JR, Cozzarelli IM (2012) Emerging contaminants at a closed and an operating landfill in Oklahoma. Groundwater Monitor Remediat 32:120–130

Antje Klitkou AMF, Capasso M (2019) From Waste to Value: Valorisation pathways for organic waste streams in bioeconomies. Routeledge studies in waste management and policy

Aziz SO, Aziz HA, Yusoff MS, Bashir MJ, Umar M (2010) Leachate characterization in semi-aerobic and anaerobic sanitary landfills: a comparative study. J Environ Manage 2608–2614

Babu NV, Rao PJ, Prasad I (2013) Impact of municipal solid waste on groundwater in the environs of Greater Visakhapatnam Municipal Corporation Area, Andhrapradesh, India. Int J Eng Sci Invent 2

Baker A, Curry M (2004) Fluorescence of leachates from three contrasting landfills. Water Res 38:2605–2613

Banar M, Cokaygil Z, Ozkan A (2009) Life cycle assessment of solid waste management options for Eskisehir, Turkey. Waste Manage 29:54–62

Banat K, Howari F, Al-Hamad A (2005) Heavy metals in urban soils of central Jordan: should we worry about their environmental risks? Environ Res 97:258–273

Bellasi A, Binda G, Pozzi A, Galafassi S, Volta P, Bettinetti R (2020) Microplastic contamination in freshwater environments: a review, focusing on interactions with sediments and benthic organisms. Environments 7:30

Bennett GF (1987) Air quality aspects of hazardous waste landfills. Hazard Waste Hazard Mater 4:119–138

Berglund M (2006) Biogas production from a systems analytical perspective. Doctoral dissertation. Environmental and Energy System Studies, Lund University, Lund

Bhalla B, Saini M, Jha M (2013) Effect of age and seasonal variations on leachate characteristics of municipal solid waste landfill. Int J Res Eng Technol 2:223–232

Bilgili MS, Demir A, Ince M, Özkaya B (2007) Metal concentrations of simulated aerobic and anaerobic pilot scale landfill reactors. J Hazard Mater 145:186–194

Bogner JE, Spokas KA, Chanton JP (2011) Seasonal greenhouse gas emissions (methane, carbon dioxide, nitrous oxide) from engineered landfills: daily, intermediate, and final California cover soils. J Environ Qual 40:1010–1020

Bourg AC, Bertin C (1996) Diurnal variations in the water chemistry of a river contaminated by heavy metals: natural biological cycling and anthropic influence. Water Air Soil Pollut 86:101–116

Bove D, Merello S, Frumento D, Arni SA, Aliakbarian B, Converti A (2015) A critical review of biological processes and technologies for landfill leachate treatment. Chem Eng Technol 38:2115–2126

Brunner PH, Fellner J (2007) Setting priorities for waste management strategies in developing countries. Waste Manage Res 25:234–240

Brunner DR (1971) Closing open dumps. US Environmental Protection Agency, Solid Waste Management Office

Buszka P, Yeskis D, Kolpin D, Furlong E, Zaugg S, Meyer M (2009) Waste-indicator and pharmaceutical compounds in landfill-leachate-affected ground water near Elkhart, Indiana, 2000–2002. Bull Environ Contam Toxicol 82:653–659

Calvo F, Moreno B, Zamorano M, Szanto M (2005) Environmental diagnosis methodology for municipal waste landfills. Waste Manage 25:768–779

Canter LW (2019) Nitrates in groundwater. Routledge

Chathumani D, Singhe DW, Gunarathna I (2019) Decades to accumulate, seconds to fall: a case study on Meethotamulla Garbage Dump Collapse in Sri Lanka

Chattopadhyay S, Dutta A, Ray S (2009) Municipal solid waste management in Kolkata, India—A review. Waste Manage 29:1449–1458

Chaudhary V, Kumar M, Sharma M, Yadav B (2010) Fluoride, boron and nitrate toxicity in ground water of northwest Rajasthan, India. Environ Monit Assess 161:343–348

Chen Y-C, Chen K-S, Wu C-H (2003) Numerical simulation of gas flow around a passive vent in a sanitary landfill. J Hazard Mater 100:39–52

Chen X, Wong JT, Ng CW, Wong M (2016) Feasibility of biochar application on a landfill final cover—a review on balancing ecology and shallow slope stability. Environ Sci Pollut Res 23:7111–7125

Chen G, Grasel P, Millington G, Hallas J, Ahmad H, Tawfiq K (2017) Chloride removal from landfill leachate by the ultra-high lime with aluminum process. J Urban Environ Eng 11:3–8

Cherubini F, Ulgiati SBS (2009) Life cycle assessment (LCA) of waste management strategies: landfilling, sorting plant and incineration. Energy 2116–2123

Chiemchaisri C, Juanga J, Visvanathan C (2007) Municipal solid waste management in Thailand and disposal emission inventory. Environ Monit Assess 135:13–20

Chounlamany V, Tanchuling MA, Inoue T (2019) Water quality and pollution loading of a river segment affected by landfill leachate and domestic waste. Int J Environ Stud 76:379–395

Chun S (2001) Sanitary landfills: toward sustainable development. Lund University

Costa AM, Alfaia RGDSM, Campos JC (2019) Landfill leachate treatment in Brazil—An overview. J Environ Manage 232:110–116

Dan Wang JH, Tang Y-T, Higgitt D, Robinson D (2020) Life cycle assessment of municipal solid waste management in Nottingham, England: past and future perspectives. J Clean Prod

Devare M, Bahadir M (1994) Biological monitoring of landfill leachate using plants and luminescent bacteria. Chemosphere 28:261–271

Dos Muchangos LS, Tokai A, Hanashima A (2015) Analyzing the structure of barriers to municipal solid waste management policy planning in Maputo city, Mozambique. Environ Dev 16:76–89

El-Salam MMA, Abu-Zuid GI (2015) Impact of landfill leachate on the groundwater quality: a case study in Egypt. J Adv Res 6:579–586

Epa (2020). Best practices for solid waste management: a guide for decision-makers in developing countries. United States Environmental Protection Agency

Erses AS, Fazal MA, Onay TT, Craig WH (2005) Determination of solid waste sorption capacity for selected heavy metals in landfills. J Hazard Mater 121:223–232

Esakku S, Palanivelu K, Joseph K (2003) Assessment of heavy metals in a municipal solid waste dumpsite. In: Workshop on sustainable landfill management, Citeseer, pp 139–145

Esakku S, Karthikeyan OP, Joseph K, Nagendran R, Palanivelu K, Pathirana K, Karunarathna A, Basnayake B (2007) Seasonal variations in leachate characteristics from municipal solid waste dumpsites in India and Srilanka. In: Proceedings of the international conference on sustainable solid waste management, pp 2–8

Ezeudu OB, Ezeudu TS (2019) Implementation of circular economy principles in industrial solid waste management: case studies from a developing economy (Nigeria). Recyling 4

Farrell M, Jones D (2009) Critical evaluation of municipal solid waste composting and potential compost markets. Biores Technol 100:4301–4310

Fatta D, Papadopoulos A, Loizidou M (1999) A study on the landfill leachate and its impact on the groundwater quality of the greater area. Environ Geochem Health 21:175–190

Ferronato N, Torretta V (2019) Waste mismanagement in developing countries: A review of global issues. Int J Environ Res Public Health 16:1060

Garlapati VK (2016) E-waste in India and developed countries: Management, recycling, business and biotechnological initiatives. Renew Sustain Energy Rev 54:874–881

Geng Y, Tsuyoshi F, Chen X (2010) Evaluation of innovative municipal solid waste. J Clean Prod 993–1000

Goulart Coelho LM, Lange LC (2018) Applying life cycle assessment to support environmentally sustainable waste management strategies in Brazil. Resour Conserv Recycl 2016:438–450

Grisey E, Belle E, Dat J, Mudry J, Aleya L (2010) Survival of pathogenic and indicator organisms in groundwater and landfill leachate through coupling bacterial enumeration with tracer tests. Desalination 261:162–168

Guasch H, Navarro E, Serra A, Sabater S (2004) Phosphate limitation influences the sensitivity to copper in periphytic algae. Freshw Biol 49:463–473

Guerrero LA, Maas G, Hogland W (2013) Solid waste management challenges for cities in developing countries. Waste Manage 33:220–232

Gunawardana EG, Basnayake BF, Shimada S, Iwata T (2009) Influence of biological pre-treatment of municipal solid waste on landfill behaviour in Sri Lanka. Waste Manage Res 27:456–462

Han D, Tong X, Currell MJ, Cao G, Jin M, Tong C (2014) Evaluation of the impact of an uncontrolled landfill on surrounding groundwater quality, Zhoukou, China. J Geochem Explor 136:24–39
Harir AI, Kasim R, Ishiyaku B (2015) Exploring the resource recovery potentials of municipal solid waste: a review of solid wastes composting in developing countries. Int J Sci Res Publ 5:2250–3153
Harshani H, Nawagamuwa UP, Senanayake A (2015) Evaluation of cover soil properties of solid waste dumpsites in Colombo District, Sri Lanka. J Natl Sci Found Sri Lanka 43.
He P-J, Xiao Z, Shao L-M, Yu J-Y, Lee D-J (2006) In situ distributions and characteristics of heavy metals in full-scale landfill layers. J Hazard Mater 137:1385–1394
He P, Chen L, Shao L, Zhang H, Lü F (2019) Municipal solid waste (MSW) landfill: A source of microplastics?-Evidence of microplastics in landfill leachate. Water Res 159:38–45
Heidenreich M, Kleeberg A (2003) Phosphorus-binding in iron-rich sediments of a shallow reservoir: spatial characterization based on sonar data. Hydrobiologia 506:147–153
Heron G, Christensen TH, Tjell JC (1994) Oxidation capacity of aquifer sediments. Environ Sci Technol 28:153–158
Hossain ML, Das SR, Hossain MK (2014) Impact of landfill leachate on surface and ground water quality. J Environ Sci Technol 7:337
Huber-Humer M, Gebert J, Hilger H (2008) Biotic systems to mitigate landfill methane emissions. Waste Manage Res 26:33–46
Hussein M, Yoneda K, Zaki ZM, Amir A (2019) Leachate characterizations and pollution indices of active and closed unlined landfills in Malaysia. Environ Nanotechnol, Monitor Manage.
Ibezute AC, Erhunmwunse O (2018) Assessment of leachate characteristics and pollution index of ikhueniro dumpsite in Benin City, Edo State, Nigeria
Idowuac IA, Atherton W, Hashimb K, Kotb P, Alkhaddarb R, Aloc B, Shawa A (2019) An analyses of the status of landfill classification systems in developing countries: Sub Saharan Africa landfill experiences. Waste Manage 87:761–771
Idris A, Inanc B, Hassan MN (2004) Overview of waste disposal and landfills/dumps in Asian countries. J Mater Cycles Waste Manage 6:104–110
Ifeanyichukwu MJ (2008) New leachate treatment methods. Lund University, Lund
Imhof HK, Ivleva NP, Schmid J, Niessner R, Laforsch C (2013) Contamination of beach sediments of a subalpine lake with microplastic particles. Curr Biol 23:R867–R868
Iswa (2016) A roadmap for closing waste dumpsites the world's most polluted places. In: Association ISW (ed)
Iuga AN (2016) Waste management in the circular economy. The case of Romania. In: 20th Innovative manufacturing engineering and energy conference (IManEE 2016)
Iwegbue CM, Nwajei GE, Ogala JE, Overah CL (2010) Determination of trace metal concentrations in soil profiles of municipal waste dumps in Nigeria. Environ Geochem Health 32:415–430
Jain P, Townsend TG, Johnson P (2013) Case study of landfill reclamation at a Florida landfill site. Waste Manage 33:109–116
Jayawardhana Y, Kumarathilaka P, Herath I, Vithanage M (2016) Municipal solid waste biochar for prevention of pollution from landfill leachate. In: Environmental materials and waste. Elsevier
Jayaweera M, Gunawardana B, Gunawardana M, Karunawardena A, Dias V, Premasiri S, Thilakasiri S (2019a) Management of municipal solid waste open dumps immediately after the collapse: an integrated approach from Meethotamulla open dump, Sri Lanka. Waste Manage 95:227–240
Jayaweera M, Gunawardana B, Gunawardana M, Karunawardena A, Dias V, Premasiri S, Dissanayake J, Manatunge J, Wijeratne N, Karunarathne D (2019b) Management of municipal solid waste open dumps immediately after the collapse: an integrated approach from Meethotamulla open dump, Sri Lanka. Waste Manage 95:227–240
Joseph K (2002) Solid waste dump sites to sustainable landfills. Centre for Environmental Studies, Anna University, Chennai
Joseph K, Nagendran R, Palanivelu K, Thanasekaran K, Visvanathan C (2004) Dumpsite rehabilitation and landfill mining. CES, Anna University, Chennai. Under ARRPET available on line: http://www.swlf.ait.ac.th

Jumbe AS, Nandini N (2009) Heavy metals analysis and sediment quality values in urban lakes. Am J Environ Sci 5:678
Kanmani S, Gandhimathi R (2013) Assessment of heavy metal contamination in soil due to leachate migration from an open dumping site. Appl Water Sci 3:193–205
Kargi F, Pamukoglu MY (2003) Powdered activated carbon added biological treatment of pre-treated landfill leachate in a fed-batch reactor. Biotechnol Lett 25:695–699
Karthikeyan OP, Mehariya S, Wong JW (2017) Bio-refining of food waste for fuel and value products. In: 4th International conference on energy and environment research, ICEER 2017 Porto, Portugal. ScienceDirect, pp 14–21
Kaur S, Dhillon GS, Sarma SJ, Brar SK, Misra K, Oberoi HS (2014). Waste biomass: a prospective renewable resource for development of bio-based economy/processes. In: Biotransformation of waste biomass into high value biochemicals. Springer
Khatib IA (2010) Municipal solid waste management in developing countries: future challenges and possible opportunities. In: Integrated waste management, vol II
Kjeldsen P, Barlaz MA, Rooker AP, Baun A, Ledin A, Christensen TH (2002) Present and long-term composition of MSW landfill leachate: a review. Crit Rev Environ Sci Technol 32:297–336
Klinck B, Stuart M (1999) Human health risk in relation to landfill leachate quality. Technical report WC/99/17. British Geological Survey
Koelsch F, Fricke K, Mahler C, Damanhuri E (2005) Stability of landfills—The Bandung dumpsite disaster. Proceedings Sardinia
Koide T, Nagamori M, Wijewardane N, Watababe Y, Isobe Y, Mowjood M, Kawamoto K (2012) Landfill gases at an abandoned open dump: a case study at Udapalatha/Gampola site in the Central Province of Sri Lanka
Koliyabandara P, Cooray A, Liyanage S, Siriwardhana C (2022) Characterization of landfill leachate at the Karadiyana open dumpsite, Sri Lanka, and assessment of water pollution in its vicinity. J Natl Sci Found Sri Lanka 50
Komal Habib JHS, Christensen P (2013) A historical perspective of global warming potential from municipal solid waste management. Waste Manage 33:1926–1933
Komárek M, Tlustoš P, Száková J, Chrastný V (2007) The role of chloride salts in chemically enhanced phytoextraction of heavy metals from a contaminated agricultural soil. Bull Environ Contam Toxicol 78:176–180
Korhonen J, Nuur C, Feldmann A, Birkie SE (2018b) Circular economy as an essentially contested concept. J Clean Prod 544–552
Korhonen J, Honkasalo A, Seppälä J (2018a) Circular economy: the concept and its circular economy: the concept and its limitations. Ecol Econ 37–46
Kretschmer B, Smith C, Watkins E, Allen B et al (2013). Recycling agricultural, forestry & food wastes and residues for sustainable bioenergy and biomaterials STOA
Kross B (2002) Nitrate toxicity and drinking water standards—A review. J Prev Med 10:3–10
Kulikowska D, Klimiuk E (2008) The effect of landfill age on municipal leachate composition. Biores Technol 99:5981–5985
Kumar D, Alappat BJ (2005) Analysis of leachate pollution index and formulation of sub-leachate pollution indices. Waste Manage Res 23:230–239
Laner D, Crest M, Scharff H, Morris JW, Barlaz MA (2012) A review of approaches for the long-term management of municipal solid waste landfills. Waste Manage 32:498–512
Lange Lene CKO, Sigurjon A et al (2021) Developing a sustainable and circular bio-based economy in EU: By partnering across sectors, upscaling and using new knowledge faster, and for the benefit of climate, environment & biodiversity, and people & business. Front Bioeng Biotechnol 8
Lantz M, Svensson M, Björnsson L (2007) The prospects for an expansion of biogas systems in Sweden-incentives, barriers and potentials. Energy Policy Elsevier 35:1830–1843
Lavigne F, Wassmer P, Gomez C, Davies TA, Hadmoko DS, Iskandarsyah TYW, Gaillard JC, Fort M, Texier P, Heng MB (2014) The 21 February 2005, catastrophic waste avalanche at Leuwigajah dumpsite, Bandung, Indonesia. Geoenviron Disasters 1:10

Li X, Zhao Q (2001) Efficiency of biological treatment affected by high strength of ammonium-nitrogen in leachate and chemical precipitation of ammonium-nitrogen as pretreatment. Chemosphere 44:37–43

Logan M, Visvanathan C (2019) Management strategies for anaerobic digestate of organic fraction of municipal solid waste: Current status and future prospects. Waste Manage Res 27–39

Logeshkumaran A, Magesh N, Godson PS, Chandrasekar N (2015) Hydro-geochemistry and application of water quality index (WQI) for groundwater quality assessment, Anna Nagar, part of Chennai City, Tamil Nadu, India. Appl Water Sci 5:335–343

Lovley DR (1997) Microbial Fe(III) reduction in subsurface environments. FEMS Microbiol Rev 20:305–313

Maheshi D (2015) Environmental and economic assessment of 'open waste dump'mining in Sri Lanka. Resour Conserv Recycl 102:67–79

Maheshi D, VPS, Karel VA (2015) Environmental and economic assessment of 'open waste dump' mining in Sri Lanka. Resour Conserv Recycl 67–79

Manassaram DM, Backer LC, Moll DM (2005) A review of nitrates in drinking water: maternal exposure and adverse reproductive and developmental outcomes. Environ Health Perspect 114:320–327

Mbuligwe SE, Kaseva ME (2006) Assessment of industrial solid waste management and resource recovery practices in Tanzania. Resour Conserv Recycl 47:260–276

Merry SM, Kavazanjian E Jr, Fritz WU (2005) Reconnaissance of the July 10, 2000, Payatas landfill failure. J Perform Constr Facil 19:100–107

Menikpura SNM, Gheewala SH, Bonnet S (2012) Sustainability assessment of municipal solid waste management in Sri Lanka: problems and prospects. J Mater Cycles Waste Manag 14:181–192

Meunier N, Drogui P, Montané C, Hausler R, Mercier G, Blais J-F (2006) Comparison between electrocoagulation and chemical precipitation for metals removal from acidic soil leachate. J Hazard Mater 137:581–590

Mishra S, Tiwary D, Ohri A, Agnihotri AK (2018) Assessment of groundwater quality using WQI and GIS near the Karsara municipal landfill site, Varanasi, India. Arab J Geosci 11:252

Mohan SV, Nithila P, Reddy SJ (1996) Estimation of heavy metals in drinking water and development of heavy metal pollution index. J Environ Sci Health Part A 31:283–289

Mohit Somania MD, Gupta SK, Sreekrishnan TR, Ramana GV (2019) Comprehensive assessment of the leachate quality and its pollution potential from six municipal waste dumpsites of India. Bioresour Technol Repo 198–206

Mor S, Ravindra K, Dahiya R, Chandra A (2006) Leachate characterization and assessment of groundwater pollution near municipal solid waste landfill site. Environ Monit Assess 118:435–456

Mukherjee S, Mukhopadhyay S, Hashim MA, Sen Gupta B (2015) Contemporary environmental issues of landfill leachate: assessment and remedies. Crit Rev Environ Sci Technol 45:472–590

Mutz D, Hengevoss D, Hugi C, Gross T (2017) Waste-to-energy options in municipal solid waste management a guide for decision makers in developing and emerging countries. Deutsche Gesellschaft für Internationale Zusammenarbeit (GIZ) GmbH

Mylavarapu R (2008) Impact of phosphorus on water quality1. UF/IFAS Publ SL-275, Coop Ext Serv, IFAS, University of Florida, Iainesville, FL

Nagendran R, Selvam A, Joseph K, Chiemchaisri C (2006) Phytoremediation and rehabilitation of municipal solid waste landfills and dumpsites: a brief review. Waste Manage 26:1357–1369

Narayana T (2009) Municipal solid waste management in India: From waste disposal to recovery of resources? Waste Manage 29:1163–1166

Nava-Martínez EC, Flores-García E, Espinoza-Gomez H, Wakida FT (2012) Heavy metals pollution in the soil of an irregular urban settlement built on a former dumpsite in the city of Tijuana, Mexico. Environ Earth Sci 66:1239–1245

Naveen B, Mahapatra DM, Sitharam T, Sivapullaiah P, Ramachandra T (2017) Physico-chemical and biological characterization of urban municipal landfill leachate. Environ Pollut 220:1–12

Ness DA, Xing K (2017) Toward a resource-efficient built environment: a literature. J Ind Ecol 21:572–592

Nick Dale AG, Mcdonald B, Spacek K (2012) The value of waste: an analysis of the potential for value-added activities in waste picking communities in Southeast Asia. Full circle international

Njoku PO, Edokpayi JN, Odiyo JO (2019) Health and environmental risks of residents living close to a landfill: a case study of Thohoyandou Landfill, Limpopo Province, South Africa. Int J Environ Res Public Health 16:2125

Olafisoye OB, Adefioye T, Osibote OA (2013). Heavy metals contamination of water, soil, and plants around an electronic waste dumpsite. Polish J Environ Studies 22

Oncioiu I, Căpuşneanu S, Topor DI, Petrescu M, Petrescu AG, Toader MI (2020) The effective management of organic waste policy in Albania. Energies 13

Oonk J, Boom A (1995). Landfill gas formation, recovery and emissions. TNO Institute of Environmetal and Energy Technology

Osada T, Nemoto K, Nakanishi H, Hatano A, Shoji R, Naruoka T, Yamada M (2011) Analysis of ammonia toxicity in landfill leachates. ISRN Toxicol

Oyelami A, Aladejana J, Agbede O (2013) Assessment of the impact of open waste dumpsites on groundwater quality: a case study of the Onibu-Eja dumpsite, southwestern Nigeria. Procedia Earth Planet Sci 7:648–651

Pansuk J, Junpen A, Garivait S (2018) Assessment of air pollution from household solid waste open burning in Thailand. Sustainability 10:2553

Peter AE, Nagendra SS, Nambi IM (2019a) Environmental burden by an open dumpsite in urban India. Waste Manage 85:151–163

Peter AE, Nagendra SS, Nambi IM (2019b) Environmental burden by an open dumpsite in urban India. Waste Manage 85:151–163

Pohland F, Cross W, Gould J (1993). Metal speciation and mobility as influenced by landfill disposal practices. Metals Groundwater 411–429.

Praagh MV, Hartman C, Brandmyr E (2018) Microplastics in landfill leachates in the Nordic countries. Nordisk Ministerråd

Prasanth SS, Magesh N, Jitheshlal K, Chandrasekar N, Gangadhar K (2012) Evaluation of groundwater quality and its suitability for drinking and agricultural use in the coastal stretch of Alappuzha District, Kerala, India. Appl Water Sci 2:165–175

Press A (2016) Guatemala declares 24 missing in garbage landslide at dump [Online]. Available https://www.foxnews.com/world/guatemala-declares-24-missing-in-garbage-landslide-at-dump. Accessed 19 May 2020

Przydatek G, Kanownik W (2019) Impact of small municipal solid waste landfill on groundwater quality. Environ Monit Assess 191:169

Pujari PR, Deshpande V (2005) Source apportionment of groundwater pollution around landfill site in Nagpur, India. Environ Monit Assess 111:43–54

Qingjie G, Jun D, Yunchuan X, Qingfei W, Liqiang Y (2008) Calculating pollution indices by heavy metals in ecological geochemistry assessment and a case study in parks of Beijing. J China Univ Geosci 19:230–241

Rafiq A, Rasheed A, Arslan C, Tallat U, Siddique M (2018) Estimation of greenhouse gas emissions from Muhammad wala open dumping site of Faisalabad, Pakistan. Geology Ecol Landsc 2:45–50

Raga R, Cossu R (2014) Landfill aeration in the framework of a reclamation project in Northern Italy. Waste Manage 34:683–691

Ramakrishnan A, Blaney L, Kao J, Tyagi RD, Zhang TC, Surampalli RY (2015) Emerging contaminants in landfill leachate and their sustainable management. Environ Earth Sci 73:1357–1368

Reddy KR, Yargicoglu EN, Yue D, Yaghoubi P (2014) Enhanced microbial methane oxidation in landfill cover soil amended with biochar. J Geotech Geoenviron Eng 140:04014047

Renou S, Givaudan J, Poulain S, Dirassouyan F, Moulin P (2008) Landfill leachate treatment: review and opportunity. J Hazard Mater 150:468–493

Riediker S, Suter M-F, Giger W (2000) Benzene-and naphthalenesulfonates in leachates and plumes of landfills. Water Res 34:2069–2079

Rim-Rukeh A (2014) An assessment of the contribution of municipal solid waste dump sites fire to atmospheric pollution. Open J Air Pollution 3:53

Ritzkowski M, Stegmann R (2007) Controlling greenhouse gas emissions through landfill in situ aeration. Int J Greenhouse Gas Control 1:281–288

Röling WF, Van Breukelen BM, Braster M, Lin B, Van Verseveld HW (2001) Relationships between microbial community structure and hydrochemistry in a landfill leachate-polluted aquifer. Appl Environ Microbiol 67:4619–4629

Rügge K, Bjerg PL, Pedersen JK, Mosbæk H, Christensen TH (1999) An anaerobic field injection experiment in a landfill leachate plume, Grindsted, Denmark: 1. experimental setup, tracer movement, and fate of aromatic and chlorinated compounds. Water Resour Res 35:1231–1246

Sadasivam BY, Reddy KR (2014) Landfill methane oxidation in soil and bio-based cover systems: a review. Revi Environ Sci Bio/technol 13:79–107

Saheri S, Mir MA, Basri NEA, Begum RA, Mahmood NZB (2017). Solid waste management by considering composting potential in Malaysia toward a green country. e-Bangi 6

Saja AM, Zimar AM, Junaideen SM (2021) Municipal solid waste management practices and challenges in the southeastern coastal cities of Sri Lanka. Sustainability 13

Salem Z, Hamouri K, Djemaa R, Allia K (2008) Evaluation of landfill leachate pollution and treatment. Desalination 220:108–114

Samuding K, Abustan I, Rahman MTA, Isa MH (2009) Distribution of heavy metals profile in groundwater system at solid waste disposal site. Eur J Sci Res 37:58–66

Sanneh E, Hu AH, Chang Y, Sanyang E (2011) Introduction of a recycling system for sustainable municipal solid waste management: a case study on the greater Banjul area of the Gambia. Environ Dev Sustain 13:1065–1080

Sawyerr HO, Adeolu AT, Afolabi AS, Salami OO, Badmos BK (2017) Impact of dumpsites on the quality of soil and groundwater in satellite towns of the Federal Capital Territory, Abuja, Nigeria. J Health Pollut 7:15–22

Scheutz C, Kjeldsen P, Bogner JE, De Visscher A, Gebert J, Hilger HA, Huber-Humer M, Spokas K (2009) Microbial methane oxidation processes and technologies for mitigation of landfill gas emissions. Waste Manage Res 27:409–455

Seng B, Fujiwara T, Spoann V (2018) Households' knowledge, attitudes, and practices toward solid waste management in suburbs of Phnom Penh, Cambodia. Waste Manage Res 36:993–1000

Shakya PR, Shrestha P, Tamrakar CS, Bhattarai PK (2008) Studies on potential emission of hazardous gases due to uncontrolled open-air burning of waste vehicle tyres and their possible impacts on the environment. Atmos Environ 42:6555–6559

Singh S, Anunay G, Rohit G, Shivangi G, Vipul V (2016) Greenhouse gas emissions from landfills: a case of NCT of Delhi, India. J Climatol Weather Forecast 1–6

Sivertsen B (2006). Air pollution impacts from open air burning. WIT Trans Ecol the Environ

Soltermann F, Abegglen C, Gotz C, Von Gunten U (2016) Bromide sources and loads in Swiss surface waters and their relevance for bromate formation during wastewater ozonation. Environ Sci Technol 50:9825–9834

Soltermann F, Abegglen C, Tschui M, Stahel S, Von Gunten U (2017) Options and limitations for bromate control during ozonation of wastewater. Water Res 116:76–85

Song L, Huo J, Wang X, Yang F, He J, Li C (2016) Phosphate adsorption by a Cu(II)-loaded polyethersulfone-type metal affinity membrane with the presence of coexistent ions. Chem Eng J 284:182–193

Sundt P, Schulze P-E, Syversen F (2014) Sources of microplastic-pollution to the marine environment. Mepex for the Norwegian Environment Agency, p 86

Sunny S (2017) Delhi's Ghazipur landfill collapse: 2 dead as mountain of trash sweeps many into nearby canal. Hindustan Times, 2 Sept 2017

Taha A, El-Mahmoudi A, El-Haddad I (2004) Pollution sources and related environmental impacts in the new communities southeast Nile Delta, Egypt. Emirates J Engin Res 9:35–49

Talalaj IA, Biedka P (2016) Use of the landfill water pollution index (LWPI) for groundwater quality assessment near the landfill sites. Environ Sci Pollut Res 23:24601–24613
Tatsi A, Zouboulis A (2002) A field investigation of the quantity and quality of leachate from a municipal solid waste landfill in a Mediterranean climate (Thessaloniki, Greece). Adv Environ Res 6:207–219
Terraza H, Willumsen H (2009). Guidance note on landfill Gas Capture and utilization. Infrastructure and environment sector-technical notes
Thrane VR, Thrane AS, Wang F, Cotrina ML, Smith NA, Chen M, Xu Q, Kang N, Fujita T, Nagelhus EA (2013) Ammonia triggers neuronal disinhibition and seizures by impairing astrocyte potassium buffering. Nat Med 19:1643
Tisserant A, Pauliuk S, Merciai S, Schmidt J, Fry J, Wood R, Tukker A (2017) Solid waste and the circular economy: a global analysis of waste treatment and waste footprints. J Ind Ecol 21:628–640
Tsai FM, Bul TD, Tseng ML et al (2020) Municipal solid waste management in a circular economy: a datadriven bibliometric analysis. J Clean Prod
Turner Da WI, Kemp S (2016) Combined material flow analysis and life cycle assessment as a support tool for solid waste management decision making. J Clean Prod 234–248
Umar M, Aziz HA, Yusoff MS (2010) Variability of parameters involved in leachate pollution index and determination of LPI from four landfills in Malaysia. Int J Chem Eng
Unep (2015) Sustainable consumption and production—A handbook for policymakers, Global edn. United Nations Environment Programme
Usepa (1990) National ambient air quality standards (NAAQS table). In: Agency, U. S. E. P (ed)
Usepa (2017) Greenhouse gas emissions [Online]. Available https://www.epa.gov/ghgemissions/understanding-global-warming-potentials. Accessed 20 May
Usepa (2020) EPA's report on the environment (ROE) [Online]. United States Environmental Protection Agency. Available https://www.epa.gov/report-environment. Accessed May 01 2020
Van Der Zee D, Achterkamp M, De Visser B (2004) Assessing the market opportunities of landfill mining. Waste Manage 24:795–804
Verma M, M'hamdi N, Dkhili Z, Brar SK, Misra K (2014) Thermochemical transformation of agro-biomass into biochar: simultaneous carbon sequestration and soil amendment. In: Biotransformation of waste biomass into high value biochemicals. Springer
Visvanathan C, Tränkler J, Kuruparan P, Xiaoning Q (2002) Influence of landfill operation and waste composition on leachate control: lysimeter experiments under tropical conditions. In: 2nd Asia Pacific landfill symposium in Seoul, Korea, pp 441–447
Vrijheid M (2000) Health effects of residence near hazardous waste landfill sites: a review of epidemiologic literature. Environ Health Perspect 108:101–112
Waheed S, Siddique N, Hamid Q, Chaudhry M (2010) Assessing soil pollution from a municipal waste dump in Islamabad, Pakistan: a study by INAA and AAS. J Radioanal Nucl Chem 285:723–732
Wallingford K (1987) NIOSH (National Institute for Occupational Safety and Health) indoor air quality in office buildings. National Institute for Occupational Safety and Health, Cincinnati
Wang L-S, Hu H-Y, Wang C (2007) Effect of ammonia nitrogen and dissolved organic matter fractions on the genotoxicity of wastewater effluent during chlorine disinfection. Environ Sci Technol 41:160–165
Wang T (2020) World's biggest dump sites 2019. Available https://www.statista.com/statistics/530481/largest-dump-sites-worldwide/
WHO (2003) Background document for preparation of WHO Guidelines for drinking-water quality. World Health Organization, Geneva
WHO (2011) Guidelines for drinking-water quality. WHO Chronicle 38:104–108
WHO (2006) Monitoring and assessment of chemical quality guidelines for drinking water quality training pack. Available http://www.who.int/water_sanitation_health/dwq/S09.pdf. Accessed 24 Mar 2019

WHO (2009) Bromide in drinking-water—Background document for development of guidelines for drinking-water quality. In: Oraganization WH (ed)

Wijesekara S, Mayakaduwa SS, Siriwardana A, De Silva N, Basnayake B, Kawamoto K, Vithanage M (2014) Fate and transport of pollutants through a municipal solid waste landfill leachate in Sri Lanka. Environ Earth Sci 72:1707–1719

Wilson DC, Rodic L, Modak P, Soos R, Carpintero A, Velis K, Iyer M, Simonett O (2015). In: Global waste management outlook. UNEP

Winterstetter A, Laner D, Rechberger H, Fellner J (2015) Framework for the evaluation of anthropogenic resources: A landfill mining case study–Resource or reserve? Resour Conserv Recycl 96:19–30

Wiszniowski J, Robert D, Surmacz-Gorska J, Miksch K, Weber J (2006) Landfill leachate treatment methods: a review. Environ Chem Lett 4:51–61

Xu Q, Peng D, Li W, Dong X, Hu W, Tang M, Liu F (2017) The catastrophic landfill flowslide at Hongao dumpsite on 20 December 2015 in Shenzhen, China. Nat Hazards Earth Syst Sci 17

Yang H, Xia J, Thompson JR, Flower RJ (2017) Urban construction and demolition waste and landfill failure in Shenzhen, China. Waste Manage 63:393–396

Yousuf TB, Rahman MM (2009) Transforming an open dump into a sanitary landfill: a development effort in waste management. J Mater Cycles Waste Manage 11:277–283

Zhang Y, Guo F, Meng W, Wang X-Q (2009) Water quality assessment and source identification of Daliao river basin using multivariate statistical methods. Environ Monit Assess 152:105

Zhang DQ, Tan SK, Gersberg RM (2010) Municipal solid waste management in China: status, problems and challenges. J Environ Manage 91:1623–1633

Zhang C, Xu T, Feng H, Chen S (2019) Greenhouse gas emissions from landfills: a review and bibliometric analysis. Sustainability 11:2282

Zhaozhi Zhou YT, Dong J, Chi Y, Ni M et al (2018) Environmental performance evolution of municipal solid waste management by life cycle assessment in Hangzhou, China. J Environ Manage Elsevier 23–33

Zhou C, Fang W, Xu W, Cao A, Wang R (2014) Characteristics and the recovery potential of plastic wastes obtained from landfill mining. J Clean Prod 80:80–86

Chapter 8
Assessing Two Sanitary Landfills in the West Bank of Palestine: Current Situation and Future Obstacles

Issam A. Al-Khatib

Abstract The proper disposal of solid waste entails existence of sanitary landfills. This study aims to assess the reality of two main sanitary landfills in the West Bank of Palestine. Literature review, visits, and onsite evaluation has been conducted to the two sanitary landfills in Palestine. In addition, there were interviews with key persons in the joint services councils for solid waste, responsible for managing these landfills, as well other officials in the relevant institutions. The reality of sanitary landfills management, challenges faced, and impacts on public health and the environment need to be understood, especially in the light of unstable political conditions and difficult economic situation. Presently, the two main sanitary landfills exist in the West Bank are Zahret Al-Finjan landfill in Jenin, and Al-Minyah landfill in Bethlehem. The increasing generation of solid waste in the study area shortens life span of the landfill and therefore threatens the environment. Strategic planners and decision makers must work on establishing new sanitary landfill sites or extending existing ones with new cells. The results confirmed that the two main sanitary landfills in the West Bank face many technical problems including the absence of gas collection, leachate treatment, and odor control systems. These problems led to negative impacts on the neighboring population, agricultural lands, water resources, and air quality, especially after landfill burning from time to time. Policy and decision makers have to implement immediate interventions for a proper management and disposal of solid waste system and protection of human health and the environment of Palestine.

Keywords Domestic solid waste · Assessment · Sanitary landfills · Developing countries

I. A. Al-Khatib (✉)
Institute of Environmental and Water Studies, Birzeit University, Birzeit, Palestine
e-mail: ikhatib@birzeit.edu

A. Anouzla and S. Souabi (eds.), *Technical Landfills and Waste Management*, Springer Water, https://doi.org/10.1007/978-3-031-52633-6_8

8.1 Introduction

8.1.1 Landfills and Their Impacts

If properly maintained, sanitary landfills can be built to be an acceptable method for disposing of garbage that is not harmful to the environment (Madon et al. 2019). The cost of landfilling has, however, significantly grown as a result of new laws pertaining to landfill liners, leachate-control systems, landfill gas collection and control systems, and long-term closure requirements (Abdel-Shafy et al. 2023). Additionally, many municipalities now have to choose more remote disposal sites because there is less acceptable land for garbage sites close to surrounding urbanizing areas (Dolui and Sarkar 2021). However, the difficulty of many towns in identifying a suitable location for sanitary landfills has increased due to the Not-In-My-Back-Yard (NIMBY) mentality on the part of citizen resistance groups (Dolui and Sarkar 2021). As a result, there is more interest in the idea of recovering energy and recyclable materials from municipal solid waste (MSW) as opposed to relying on sanitary landfilling as the main long-term method of solid waste disposal (Rogoff and Screve 2011), and this interest is leading to the siting and permitting of new landfills (Nema et al. 2021).

Sanitary landfills are seen to be the most popular method of disposing of household garbage since they strive to lessen or neutralize any potential threats to the environment and public health. Ordinarily, landfills are placed in regions where dominant geographical features can act as natural barriers between the ecosystem and landfills. To keep hazardous elements from escaping, trenches are dug in the soil that has been excavated and impervious liners are created before burying the E-waste (Li et al. 2023).

However, landfills are not environmentally sound procedures because of the possibility for harmful compounds to leak into the soil and groundwater. Domestic garbage receiving landfills have been demonstrated to be a significant contributor to groundwater contamination (Yaashikaa et al. 2022). According to reports, the leachates that seep from home waste sites have much greater concentrations of trace metals, as well as dissolved and suspended organic and inorganic materials (Yaashikaa et al. 2022). These contaminants may spread through food chains and accumulate in living things before having an impact on human health. Sanitary landfills have relatively minimal health and environmental concerns, although their initial cost is greater than average (Yaashikaa et al. 2022).

As the final method of disposing of MSW, landfilling is the oldest and most widely used method of waste removal and disposal (Chelliapan et al. 2020; Koda et al. 2015). Due to their simplicity, low initial cost, large handling capacity, and low operating costs, landfills have been widely used (Li et al. 2017; Mor and Ravindra 2023). Waste management is a problem that affects all of civilization (Wolny-Koadka and Malinowski 2015).

According to other research, 95% of MSW was disposed of through landfilling globally (Ghosh et al. 2015); in the European Union (EU), the majority of member states dispose of more than 50% of their waste in landfills (Cuartas et al. 2017).

However, as urbanization and citizens' desires for better living conditions have advanced, landfill contamination has received increased attention on a global scale. This is due to the large amount of potentially harmful substances that landfills contain, some of which may jeopardize the safety of the surrounding ecosystem. All components of the environment can become contaminated by MSW dumps, which is widely recognized (Makarenko and Budak 2017). Recent research has examined the impact of landfills on groundwater. For instance, El-Salam and Abu-Zuid (2015) investigated the groundwater quality, leachate, and environmental impacts of MSW landfilling in Egypt. In Nagarajan et al. (2012) investigated the potential consequences of leachate percolation on the quality of groundwater. In a different study, Koda et al. (2017) assessed the groundwater quality at a landfill and a waste management site, paying close attention to the quantities of organic contaminants. Studies in the literature have discussed the general effects of dump sites on the neighborhood. For instance, Gworek et al. (2015) investigated the effects of mercury exposure in the groundwater, plants, and soil profiles from a 35-year-old MSW dump. Palmiotto et al. (2014) performed an integrated risk assessment for emissions of hazardous chemicals and olfactory nuisance to determine environmental quality around landfill. The negative environmental impact of landfills can be reduced by adding protection methods into the design of landfills, such as the use of suitable impermeable material for bottom and top capping (Cuartas et al. 2017; Dajić et al. 2016; Vaverkova and Adamcová 2018). These unfavorable environmental consequences must be controlled both throughout the operational phase of a landfill and after closure (referred to as the "aftercare" period) until they no longer constitute an unacceptable risk to the environment (Aryampa et al. 2022).

Numerous studies have shown that contaminants might be found in landfill leachate for municipal solid waste. According to a study by Vaverková et al. (2018), leachate samples were less harmful when rain spread the pollutants within them. Greater amounts of leachate greatly hampered plant development, with the exception of the leachate samples from June, July, and September. The toxicity of the leachate samples was lower since these three months saw the most rainfall. The water around the landfill is often not toxic. A bottom liner, a leachate collection system, and a system for treating landfill gas are all included in the construction of the landfill. As a result, leachate cannot enter the subsurface environment. The landfill has no immediate or noticeable effects on the environment or water because (i) the closed portion of the landfill is covered with an impermeable compact clay layer, geotextile membranes, nonwoven fabric, and pulper products to stop rainwater from penetrating the waste site and reaching the landfill base, (ii) the leachate from the landfill base is collected so that the leachate is entering the leachate pond, and (iii) the evapotranspiration system.

In some large cities, loans or grants have been used to construct sanitary landfills on carefully chosen sites, but typically little attention is paid to the training of a site manager and the provision of adequate resources, both material and financial, to allow a reasonable standard of operation (Dos Santos et al. 2018).

Some locations quickly turn into open dumps. A committed and knowledgeable site manager is necessary for efficient operations. It is suggested that training for this job include practical experience on well-run sites (Lavagnolo 2018).

8.1.2 Landfills Planning, Main Components, and Monitoring

In most sanitary landfill designs, preventing polluted water (leachate) from leaving the site is given first importance. Leachate from landfills has been found to be significantly produced, even in semi-arid regions. The majority of designs include expensive, painstakingly constructed impermeable barriers to prevent leachate from permeating the ground as well as drainage systems to move it to a treatment facility or storage tank. But if the tank is not properly emptied or if the plant is not operating, the leachate system actually makes the pollution worse than from an open dump because all of the leachate is concentrated in one place and natural purification mechanisms have very little chance of reducing the impact. This illustration demonstrates that even with excellent design and construction, nothing can be accomplished without good operation (Ahmed et al. 2023).

Establishing the needs for the waste site is the first step in planning for a new landfill. The location must support any auxiliary solid waste operations, such as leachate treatment, landfill gas management, and special waste services, as well as providing enough landfill capacity for the chosen design period (Arshad et al. 2023). Additionally, some locations house "material recovery facilities," which handle recyclable materials. The disposal needs of the neighborhood or communities must be calculated in order to determine the landfill's capacity (Alam and Qiao 2020).

Finding a suitable location is required when the landfill's size has been established. While various waste treatment facilities are frequently situated close to residential areas and are well-liked by the locals, it is far more challenging to get support for landfills. The declaration that a landfill is anticipated in a specific site frequently sparks fierce public opposition and the NIMBY phenomenon (not in my backyard). The permitting procedure for a new landfill or the extension of an existing one may be years behind schedule as a result of these objections (Jegadeesan et al. 2023).

An artificial use has been inserted into the landscape by building a landfill. Along with worries about directly hurting communities, competing demands for the conservation of the environment, natural resources, surface water and groundwater, and recreational uses are becoming more and more pressing (Ajibade et al. 2019).

Therefore, it becomes more and more challenging to find acceptable locations due to the growing demands imposed on future landfills. Since regional waste management is the most cost-effective, it's crucial to choose a site that's conveniently placed while keeping costs in mind. The choice of such a place is typically opposed by land-use planning and environmental considerations (Yu et al. 2023).

The site of the landfill itself acts as the lowest layer of protective barrier because no landfill liner can stop leachate migration from a landfill "over infinitely long periods of time". To lessen the danger of leaking toxins into the environment, MSW mandates

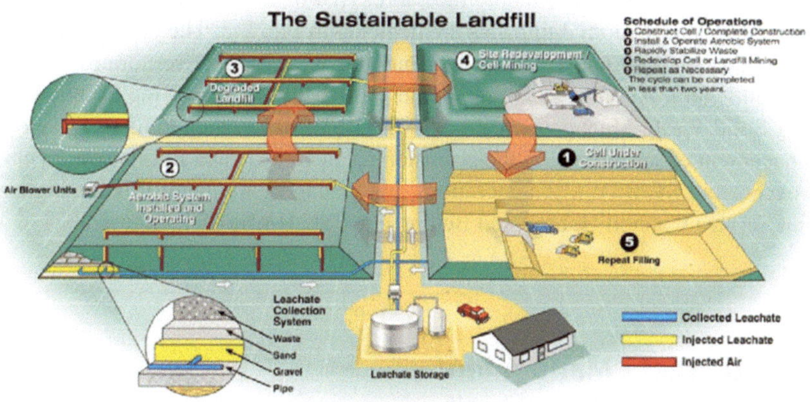

Fig. 8.1 Illustration of a sustainable landfill (Read 2023)

that the site-specific geology, hydrogeology, soil, and geotechnical conditions must be taken into account before establishing a landfill. Additionally, a distance of at least 300 m must be kept from the closest residential area in order to prevent any direct impacts on adjacent inhabitants. Geological barriers should be created by the soil beneath the dump and the surroundings (Wan et al. 2023).

Gas produced inside a landfill will flow according to a gradient of pressure and along the lines of least resistance. Unchecked gas migration can build up in sumps, basements, and sewers, which could result in disastrous results if explosions happen. Gas vents or wells must be present to stop gas migration. (Ghazi et al. 2022).

Groundwater monitoring is often done by building and sampling monitoring wells close to the dump. At the perimeter of the land, additional monitoring wells are also positioned. A range of organic and inorganic groundwater elements are sampled from these wells on a quarterly basis (Feng et al. 2020). Figure 8.1 depicts a sustainable landfill in an image.

Assessing the existing state of landfills in the West Bank of Palestine and potential challenges is the major goal of this study.

8.1.3 *Background About the Two Main Palestinian Sanitary Landfills in the West Bank*

Currently, the West Bank is home to three sanitary landfills: the Al-Minyah landfill in Bethlehem, the Jericho landfill in Jericho, and the Zahret Al-Finjan landfill in Jenin. In addition, a controlled dumpsite (CDS) was created in Beit Anan, Jerusalem, in 2018. The two landfills that will be examined in this study are Al-Minyah landfill in Bethlehem and Zahret Al-Finjan landfill in Jenin. The locations of the sanitary landfills in the West Bank, including Zahret Al-Finjan and Al-Minyah, controlled dump sites, and random dumpsites are depicted in Fig. 8.2. The Al-Minyah project is a sanitary landfill that will serve the southern West Bank governorates of Hebron and Bethlehem. Al-Minyah is one of the principal products of the Southern West Bank Solid Waste Management Project (SWBSWMP), which aims to enhance the management of solid waste in the governorates of Hebron and Bethlehem. In charge of overseeing SWBSWMP's operations and managing the landfill is the higher-up Joint Service Council Hebron and Bethlehem (JSC H&B) (Nammoura 2021).

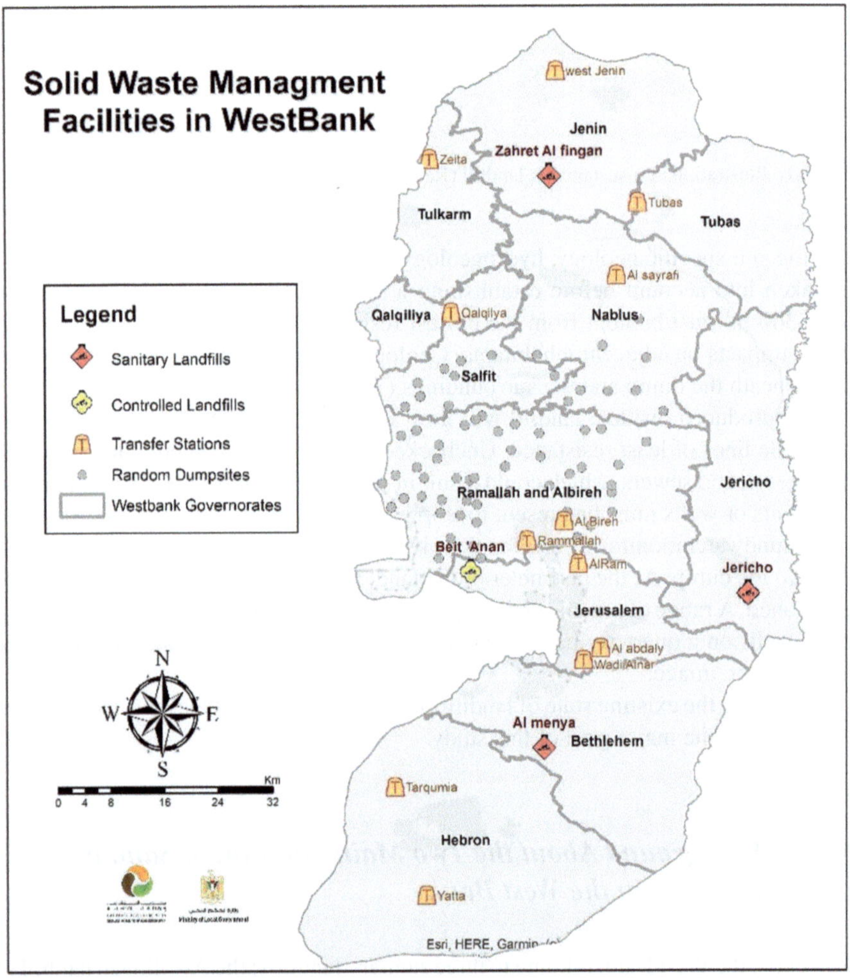

Fig. 8.2 Locations of the Zahret Al-Finjan and Al-Minyah landfills in the West Bank (MoLG and JICA 2023)

In the Bethlehem Governorate, Al-Minyah Landfill is situated around 17 km southeast of Bethlehem City and 25 km northeast of Hebron City. 27 km south of Jerusalem, in the center of the West Bank's southern region.

The World Bank has set around $12 million towards the building of the Al-Minyah landfill and the closure of dumps. About $2.25 million was spent on the land parcels designated for the dump, of which $2.25 million was provided by the Palestinian Authority.

The Zahrat Al Finjan area is situated between the towns of Arraba and Ajja in the Jenin Governorate's northeastern region of the West Bank, 18 km south of Jenin City, 26 km from Tubas, 23 km north of Nablus, 24 km east of Tulkarem, and 50 km northeast of Qalqilya (El-Kelani et al. 2017). It occupies a land area of around 240,000 m^2, of which 90,000 m^2 are used for waste cells.

The landfill, which was created to dispose of municipal solid waste from the governorates of Jenin and Tubas, was fully constructed and put into operation in June 2007. It currently receives waste from Tulkarem and portions of the governorates of Nablus, Ramallah, and Al-Bireh.

8.2 Research Methodology

The methodology for data collection in this study included a combination of interviews with stakeholders, site visits, and document analysis. The purpose of this methodology was to gather comprehensive and accurate data related the Zahret Al-Finjan landfill in Jenin and Al-Minyah landfill in Bethlehem.

Semi-structured interviews were conducted with the executive directors of the joint services councils responsible for these two sanitary landfills, in addition to the technical engineers responsible for the technical aspects of these landfills. The interview questions were designed to elicit information about the challenges, successes, and future plans for solid waste management in the two landfills.

In addition to interviews, site visits were conducted to the two landfills. The purpose of these visits was to gather visual and first-hand information about the infrastructure, operations, and environmental impacts of two landfills. During site visits, detailed notes were taken, and photographs were captured to supplement the qualitative data collected through interviews. Data about waste generation rates and components were collected from secondary sources.

All data collected through interviews, site visits, and document analysis were carefully reviewed, categorized, and analysed using both qualitative and quantitative methods. The findings were then presented (Figs. 8.3 and 8.4).

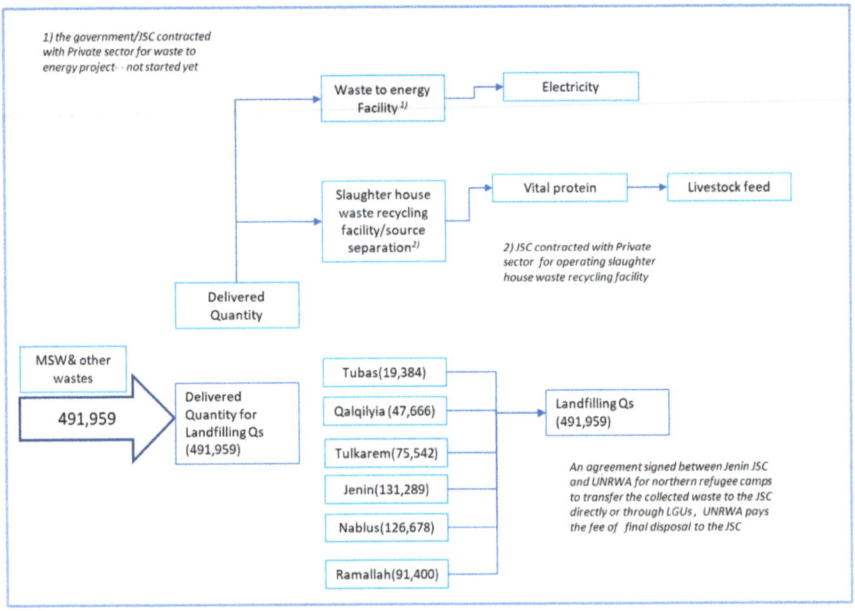

Fig. 8.3 Waste Stream for ZF SLF-Jenin JSC for the Year 2022 (unit: t/year) (MoLG and JICA 2023)

8.3 Results and Discussion

8.3.1 Waste Streams, Composition and Recycling Activities for the Two Sanitary Landfills

The Zahret Al-Finjan sanitary landfill (ZF SLF) receives solid waste from Tulkarm, Qalqilyah, Nablus, Ramallha, Jenin, and Tubas governorates with an annual amount of 491,959 tons. The 100% of delivered collected amount is disposed to ZF SLF as shown in Fig. 8.1.

The waste at ZF SLF had the typical waste characteristics of most developing countries, such as large organic fractions, which contribute to the production of leachate and landfill gasses with the additional problem of the presence of an unpleasant odor. The overall waste composition consisted of three main parts in the form of organic matter (42.04%), plastics (21.37%) paper and cardboard (13.9%). The remainder accounted for 22.69 % of the total and contained metal (3.3%), glass (1.97%), building materials (2.82%), fine material (2.82%) and others (Al-Khatib and Mahmoud 2022).

The total disposed waste to Al-Minyah sanitary landfill (SLF) is around 470,464 t/year which is around 1288.94 t/day. The 99 % of collected amount is disposed to the SLF. JSC estimated that the recycling for collected separated inorganic recyclable materials is around 0.3%. For organic materials the JSC will produce compost.

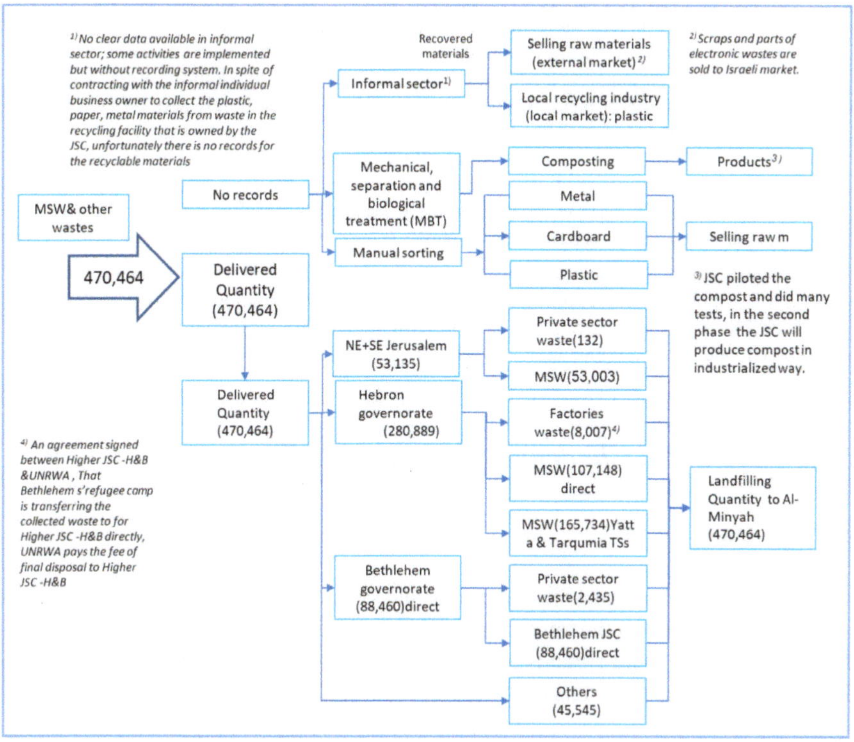

Fig. 8.4 Regional Waste Stream for Al-Minyah SLF for the Year 2022 (unit: t/year) (MoLG and JICA 2023)

The waste of Al-Minyah SLF had the typical waste characteristics of most developing countries, such as large organic fractions, which contribute to the production of leachate and landfill gasses with the additional problem of the presence of an unpleasant odor. The waste composition consisted of three main parts in the form of organic matter (51.29%), paper (13.96%), and plastics (13.44%). The remainder accounted for 21.31% of the total and contained metal (2.37%), glass (0.99%), household hazardous (1.36%), building materials (0.86%), fine material (2.84%) and others (Al-Khatib and Mahmoud 2022).

As can be noticed from Figs. 8.2 and 8.3, most of the collected waste from the different areas is disposed in the two landfills with the absence of the 3 Rs (Reduction, Reuse, and Recycling) practices.

There have been minimal efforts to sort and recycle garbage at two sanitary landfills. Only 1% of total solid trash, according to the estimate, is currently recycled. According to Musleh and Al-Khatib (2010), plastic recycling made up around a fifth of all recycling in Palestine in 2010. When recovered or reused materials are taken into account, this ratio rises to 3%. However, recycling and composting have a remarkable potential to not only help tackle the problem of the growing amounts of solid waste, but also to improve cost recovery and provide new job possibilities. This is especially true when you consider that the bulk of the solid waste produced in Palestine is made up of recyclables and biodegradables (Atallah 2020).

At the moment, ZF SLF is receiving funding from the business sector to process chicken flesh and animal slaughterhouse waste into high-protein animal feed. Palestine's recycling and reuse industry is still quite tiny and largely unorganized. Glass, plastic, and paper/cardboard recycling is a part of this process, which produces raw materials for both Israeli and international companies in addition to the local one (mostly from the Hebron governorate). The recycling of metal is also occurring, although it is largely unaccounted for in the municipal garbage stream due to its acquisition by roaming trucks that collect it from homes and establishments. There are some effective cases of garbage separation and recycling, like the one at the Al Minyah Landfill, where recyclable materials including plastic, metal, cardboard, and glass are transferred there along with organic waste that is converted into subpar compost that is used at the landfill.

The majority of recyclable material is mixed with other municipal waste at the two sanitary landfills, which restricts recycling there. Official documents mention recycling regulations and the 3 Rs (reduce, recycle, re-use) idea; nevertheless, separation at source is not effectively enforced, and waste minimization is essentially nonexistent.

8.3.2 *Technical Challenges at the Two SLFs*

8.3.2.1 Daily Cover

The daily cover is not placed in a scientific and systematic manner in the two landfills, and this contributes to the spread of stray dogs, strange birds, insects and rodents, and the waste flies and spreads around the two landfills. The reason for this is the lack of soil or other materials necessary to cover the waste adequately on daily basis.

8.3.2.2 Leachate

The two sanitary landfills currently are unable to handle the large quantities of leachate produced, especially in the winter, as the pond designated for collecting the leachate does not fit.

the resulting quantity, and thus the leachate overflows onto agricultural lands, causing great damage, as this was confirmed by people who live around the Zahret Al-Finjan landfill, as well as officials in the Ministry of Health and the Ministry of Agriculture in Jenin Governorate, which leads to pollution of the agricultural lands surrounding the landfill, and the appearance of unpleasant odors in population centers near the landfill area, and this was confirmed by citizens who were interviewed.

8.3.2.3 Gas Collection System

Since the beginning of the operation of the Zahret Al-Finjan landfill, a system for collecting gas from the landfill has not been installed, despite the existence of a design for that system. With the accumulation of waste in the landfill over time, several problems occurred, the most important of which is the burning of waste repeatedly. The last of which was during the year 2023, when the landfill was exposed to the burning of waste twice, the first time the fire continued for 8 consecutive days, despite the great efforts made to extinguish the fire.

8.3.2.4 Groundwater Monitoring Wells

In spite of establishing of groundwater monitoring wells near the each of the two sanitary landfills, they have been damaged and neglected, and is no longer fit for use, and thus there is no monitoring plan for the wells and constituent lists, and the absence of quality control requirements.

8.3.3 Environmental and Health Impacts of the Two Landfills

8.3.3.1 Zahret Al-Finjan Landfill

In a study conducted by the Rai Consult and the Universal Institute for Applied and Health Research (UIAHR) (2022), and through the interviews conducted with representatives from the Ministry of Health and Ministry of Agriculture, and from the results of a study conducted by Salah et al. (2019), it was found that Zahret al-Finjan landfill has different health and environmental impacts that can be summarized as follows.

Unpleasant odors are the result of the decomposition of the organic components of the waste in Zahret al-Finjan landfill.

When the landfill catches fire, air pollution levels increase and become unbearable, as happened recently during August 2022. The main causes of the fire are the lack of a gas collection system, gas leakage from landfills, and the toxic gases emissions from the landfill including methane, hydrogen sulfide, carbon monoxide, and carbon dioxide, ammonia, and others.

Zahret al-Fanjan landfill contributes to the spread of flies, mosquitoes, bird droppings, rodents, and spread of stray dogs, due to the presence of food waste. Consequently, these insects can carry microbes and transmit them to humans near landfill. The root cause of the problem is the failure to cover the waste with an appropriate layer of soil at the end of the work.

As confirmed by farmers, Zahret al-Finjan landfill contributes to soil pollution in neighboring lands, because the garbage leachate that is overflowing into surrounding agricultural lands leads to soil degradation and a change in the physical properties of soil, and thus a lack of fertility. It also plays a significant role in forcing farmers to abandon their agricultural land, due to the unpleasant odors. It is worth noting that various tests on soil were conducted by the Ministry of Agriculture, and there was no pollution, but the constant leakage of leachate into the agricultural land transformed it to be contaminated. The Ministry doesn't conduct periodic soil testing, and therefore it is necessary to have periodic tests on the agricultural lands adjacent to the landfill to ensure its safety.

The presence of Zahrat Al-Finjan landfill has greatly contributed to reducing the value of surrounding lands and decreasing their prices, as people prefer to keep away from landfill, due to the stinking smell felt from a distance, insects gathering, and the unattractive view.

Rainfall on the garbage leachate that leaks from Zahrat Al-Finjan landfill, contributes to mixing the dissolved and suspended waste with running or stagnant surface water, which harmfully affects the quality of water, and leads to the spread of viral diseases such as hepatitis and typhoid. It has also been polluted with heavy metals, such as mercury, cadmium, and others, which may result in the pollution of springs. In addition, citizens living in the areas surrounding the landfill are reluctant to construct rainwater harvesting wells due to the spread of various pollutants, depriving them of an important source of water which is much needed during summer.

8.3.3.2 Al-Minyah Landfill

The people of the villages of Kaisan, Minya, and the town of Tuqu, south of Bethlehem, protested against the authority's decision to expand the Minya garbage dump, which serves all municipal bodies and councils in the Hebron and Bethlehem governorates, and constitutes the main dump for all Israeli settlements established on the lands of the two governorates. Safa (A Palestinian Press Agency) conducted a dialogue with one of the demonstrators, which can be summarized as follows: A citizen from the village of Minya mentioned that the landfill, as it stands, constitutes a source of diseases and environmental pollution that the villagers suffer from. He pointed out the spread of the virus among the children of the villages of Kisan and Al-Minya and the town of Tuqu' due to environmental pollution, according to medical diagnoses. He said, "The current landfill has become a fait accompli and the authority and the occupation benefit from it, but we reject the expansion, which will be at the expense of the health of our children, our lands, and our crops, which constitute the only source of livelihood for the people of the village."

He added that the government purchased about 215 new donoms (1 donom = 1000 m^2) with the aim of expanding the landfill, so the distance separating the farthest house from the landfill would not exceed only 100 m. He explained that the current landfill was built without the slightest consideration for international standards and specifications that guarantee the safety of neighboring residents, pointing out that residents suffer from heavy flies, bad odors, the spread of diseases, and environmental pollution. He stated that the councils of Minya, Kaysan, and the town of Tuqu' submitted letters to the government rejecting the expansion decision, but the letters were not considered by the government. Then he said that "The dump threatens the lives of about 50,000 people, and I and my children will not be a victim of the government's project."

He called on the competent authorities to consider the humanitarian situation experienced by the people of the villages adjacent to the dump and stop the expansion decision, which threatens to double and increase the damage.

8.4 Conclusions and Recommendations

More than ever, household separation is required to prevent contamination of recyclables when mixed with other municipal waste. More crucially, though, the current economic system's consumption patterns need to alter, and people need to be made aware of the importance of reducing waste. We cannot recycle our way out of the plastic and solid waste problem, as has been stated in several sections of our Plastic Atlas. Palestinians may genuinely look back on the ecologically friendly methods and rich traditional knowledge of their forefathers, who lived a sustainable lifestyle.

Additionally, by using straightforward techniques like composting for gardening at the household level or on larger sizes, it is possible to significantly reduce the amount of organic waste, which makes up the largest portion of the total municipal garbage at the two landfills. Furthermore, plastic pollution is a worldwide issue that is destroying entire ecosystems and getting worse by the day.

Palestine must participate in the global effort to address the plastic crisis despite political restrictions. Governmental agencies, civil society groups, and environmental and social activists must work together to raise community awareness of the problem, imploring residents to refrain from using plastic in their daily activities and to reduce waste in general.

Manufacturers and traders must also be pushed to stop using plastic and switch to more eco-friendly materials. They must also be held accountable for their part in the plastic market oversupply.

The installation of a leachate collection system and treatment plant is urgently needed at the two sanitary landfills because leachate is escaping from the site and the leachate system actually makes the pollution worse than from an open dump because all the leachate is concentrated in one area and natural purification systems have very little chance of reducing the pollution impact.

Tragic effects, like as the latest incident at ZF SLF, are caused by the obscene gas collection systems at both landfills, which drive the uncontrolled migrating gas to accumulate in sewers, sumps, and basements. At the two SLFs, gas collection devices must be installed in order to prevent gas migration and the need for gas vents or wells.

Since groundwater serves as the primary source of drinking water in the two landfill areas, the Ministry of Health and the Ministry of Local Government must work together through municipalities to periodically monitor the quality of the water to ensure that it is fit for consumption and other domestic uses. Additionally, it is crucial to pay closer attention to the elements of landfills in Palestine, particularly as numerous studies are beginning to be conducted as a first step in the establishment of the Ramon landfill in the Ramallah and Al-Bireh Governorate to service the central West Bank.

References

Abdel-Shafy HI, Ibrahim AM, Al-Sulaiman AM, Okasha RA (2023) Landfill leachate: sources, nature, organic composition, and treatment: an environmental overview. Ain Shams Eng J 102293

Ahmed DN, Ibrahim MO, Abdul-Kareem MB, Shoshan AAA, Faisal AA (2023) Novel low-cost composite sorbent for remediating synthetic sanitary landfill leachates: batch and column study. Alex Eng J 64:205–218

Ajibade FO, Olajire OO, Ajibade TF, Nwogwu NA, Lasisi KH, Alo AB, Owolabi TA, Adewumi JR (2019) Combining multicriteria decision analysis with GIS for suitably siting landfills in a Nigerian state. Environ Sustain Ind 3:100010

Alam O, Qiao X (2020) An in-depth review on municipal solid waste management, treatment and disposal in Bangladesh. Sustain Cities Soc 52:101775

Al-Khatib IA, Mahmoud N (2022) Waste composition study for west bank, Palestine. In: Technical report submitted to ministry of local government (MoLG) and the Japan international cooperation agency (JICA). Institute of Environmental and Water Studies, Birzeit University, Palestine

Arshad M, Hasan MA, Al Mesfer MK, Al Alwan BA, Qureshi MN, Eldirderi M (2023) Sustainable landfill sites selection using geospatial information and AHP-GDM approach: a case study of Abha-Khamis in Saudi Arabia. Heliyon 9(6)

Aryampa S, Maheshwari B, Sabiiti EN, Bateganya NL, Olobo C (2022) Understanding the impacts of waste disposal site closure on the livelihood of local communities in Africa: a case study of the Kiteezi landfill in Kampala, Uganda. World Dev Perspect 25:100391

Atallah N (2020) Palestine: solid waste management under occupation. In: Germany: The Heinrich-B€ oll-Stiftung in a nutshell. Ramallah, Palestine. Retrieved on 30 Sep 2023 from: https://ps.boell.org/en/2020/10/07/palestine-solid-waste-management-under-occupation

Chelliapan S, Arumugam N, Din MFM, Kamyab H, Ebrahimi SS (2020) Anaerobic treatment of municipal solid waste landfill leachate. In: Bioreactors. Elsevier, pp 175–193

Cuartas M, López A, Pérez F, Lobo A (2017) Analysis of landfill design variables based on scientific computing. Waste Manage 71:287–300

Dajić A, Mihajlović M, Jovanović M, Karanac M, Stevanović D, Jovanović J (2016) Landfill design: need for improvement of water and soil protection requirements in EU landfill directive. Clean Technol Environ Policy 18(3):753–764

Dolui S, Sarkar S (2021) Identifying potential landfill sites using multicriteria evaluation modeling and GIS techniques for Kharagpur city of West Bengal, India. Environ Challenges 5:100243

Dos Santos IFS, Gonçalves ATT, Borges PB, Barros RM, da Silva Lima R (2018) Combined use of biogas from sanitary landfill and wastewater treatment plants for distributed energy generation in Brazil. Resour Conserv Recycl 136:376–388

El-Kelani RJ, Shadeed SM, Hasan AFR, Ghodieh AM, Burqan MA (2017) Geospatial implications assessment of Zahrat Al Finjan solid waste landfill, North of West Bank, Palestine. IUG J Nat Stud 25(2):01–09

El-Salam MMA, Abu-Zuid GI (2015) Impact of landfill leachate on the groundwater quality: a case study in Egypt. J Adv Res 6(4):579–586

Feng SJ, Zhao Y, Zhang XL, Bai ZB (2020) Leachate leakage investigation, assessment and engineering countermeasures for tunneling underneath a MSW landfill. Eng Geol 265:105447

Ghazi FMG, Abbaspour M, Rahimpour MR (2022) Transport phenomena in gas membrane separations. In: Current trends and future developments on (Bio-) membranes. Elsevier, pp 193–208

Ghosh P, Gupta A, Thakur IS (2015) Combined chemical and toxicological evaluation of leachate from municipal solid waste landfill sites of Delhi, India. Environ Sci Pollut Res 22(12):9148–9158

Gworek B, Dmuchowski W, Gozdowski D, Koda E, Osiecka R, Borzyszkowski J (2015) Influence of a municipal waste landfill on the spatial distribution of mercury in the environment. PLoS One 10(7)

Jegadeesan C, Somanathan A, Jeyakumar RB (2023) Sanitary landfill leachate treatment by aerated electrochemical Fenton process. J Environ Manage 337:117698

Koda E, Osinski P, Sieczka A, Wychowaniak D (2015) Areal distribution of ammonium contamination of soil-water environment in the vicinity of old municipal landfill site with vertical barrier. Water 7(6):2656–2672

Koda E, Miszkowska A, Sieczka A (2017) Levels of organic pollution indicators in groundwater at the old landfill and waste management site. Appl Sci 7(6):638

Lavagnolo MC (2018) 13.3—landfilling in developing countries. In: Cossu R, Stegmann R (eds) Solid waste landfilling. Elsevier, pp 773–796. ISBN 9780128183366, https://doi.org/10.1016/B978-0-12-407721-8.00036-X

Li J, Wang C, Du L, Lv Z, Li X, Hu X, Niu Z, Zhang Y (2017) Did municipal solid waste landfill have obvious influence on polychlorinated dibenzo-p-dioxins and polychlorinated dibenzofurans (PCDD/Fs) in ambient air: a case study in East China. Waste Manage 62:169–176

Li M, Zheng F, Wang J, Jia D, Mao X, Li P, Zhen YQ, Yu Y (2023) Energy-saving production of high value-added foamed glass ceramic from blast furnace slag and hazardous wastes containing heavy metal ions. J Clean Prod 383:135544

Madon I, Drev D, Likar J (2019) Long-term risk assessments comparing environmental performance of different types of sanitary landfills. Waste Manage 96:96–107

Makarenko N, Budak O (2017) Waste management in Ukraine: municipal solid waste landfills and their impact on rural areas. Ann Agrarian Sci 15(1):80–87

Ministry of Local Government (MoLG) and Japan International Cooperation Agency (JICA) (2023) Data Book on Solid Waste Management in Palestine (2022). Ramallah, Palestine

Mor S, Ravindra K (2023) Municipal solid waste landfills in lower-and middle-income countries: environmental impacts, challenges and sustainable management practices. Proc Saf Environ Protect 174:510–530. ISSN 0957-5820, https://doi.org/10.1016/j.psep.2023.04.014

Musleh R, Al-Khatib A (2010) An assessment of solid waste sorting and recycling in the northern and southern West Bank, and identification of suitable pilot projects for implementation in Hebron and Bethlehem Governorates. European Commission and International Management Group (IMG): Jerusalem Office, 42 Mt. of Olives, Sheikh Jarrah, Jerusalem

Nagarajan R, Thirumalaisamy S, Lakshumanan E (2012) Impact of leachate on groundwater pollution due to non-engineered municipal solid waste landfill sites of erode city, Tamil Nadu, India. Iran J J Environ Health Sci Eng 9(1):35

Nammoura A (2021) Converting municipal solid waste into energy—Al-Minya landfill as a case study. Graduation project II, energy engineering and environment department, faculty of engineering and information technology, An-Najah National University, Nablus, Palestine

Nema A, Zacharia KMB, Kumar A, Singh E, Varma VS, Sharma D (2021) Challenges and opportunities associated with municipal solid waste management. Curr Dev Biotech Bioeng 231–258

Palmiotto M, Fattore E, Paiano V, Celeste G, Colombo A, Davoli E (2014) Influence of a municipal solid waste landfill in the surrounding environment: toxicological risk and odor nuisance effects. Environ Int 68:16–24

Rai Consult and the Universal Institute for Applied and Health Research (UIAHR) (2022) Case study: sanitary landfills between importance and risks, Zahrat Al Finjan Lanfill in Jenin as a case study. part of a youth-led research study on the most significant environmental hazards in the West Bank and Gaza Strip, under the supervision of MA'AN development center and funded by ActionAid Palestine, September 2022

Read AD (2023) Illustration of a sustainable landfill. Retrieved on 21 Sep 2023 from: https://www.researchgate.net/figure/Illustration-of-A-Sustainable-Landfill-Environmental-Control-Systems-Inc-2000_fig1_266573145

Rogoff MJ, Screve F (2011) Chapter 1—introduction and overview. In: Rogoff MJ, Screve F (eds) Waste-to-energy, 2nd ed. William Andrew Publishing, pp 1–7. ISBN 9781437778717, https://doi.org/10.1016/B978-1-4377-7871-7.10001-2

Salah MM, Al-Sari' MI, Al-Khatib IA, Kontogianni S (2019) Local residents' perception of landfill impacts in Palestine: the case of Zahrat Al-Finjan landfill. J Mater Cycles Waste Manag https://doi.org/10.1007/s10163-019-00959-6

Vaverkova MD, Adamcová D (2018) Case study of landfill reclamation at Czech landfill site. Environ Eng Manag J (EEMJ) 17(3)

Vaverková MD, Adamcová D, Zloch J, Radziemska M, Boas Berg A, Voběrková S, Maxianová A (2018) Impact of municipal solid waste landfill on environment–a case study. J Ecol Eng 19(4):55–68

Wan Y, Dong Z, Cai Y, Xue Q, Liu K, Liu L, Guo D (2023) Geomembrane leaks detection and leakage correlation factor analysis of composite liner systems for fifty-five (55) solid waste landfills in China. Environ Technol Innov 32:103308

Wolny-Koładka K, Malinowski M (2015) Assessment of the microbiological contamination of air in a municipal solid waste treatment company. Ecol Chem Eng A 22(2):175–183

Yaashikaa PR, Kumar PS, Nhung TC, Hemavathy RV, Jawahar MJ, Neshaanthini JP, Rangasamy G (2022) A review on landfill system for municipal solid wastes: insight into leachate, gas emissions, environmental and economic analysis. Chemosphere 136627

Yu TT, Chen CY, Wu TH, Chang YC (2023) Application of high-dimensional uniform manifold approximation and projection (UMAP) to cluster existing landfills on the basis of geographical and environmental features. Sci Total Environ 167013

Chapter 9
Industrial Solid Wastes and Environment: An Overview on Global Generation, Implications, and Available Management Options

Snigdha Nath, Konthoujam Khelchandra Singh, Sumpam Tangjang, and Subhasish Das

Abstract Increasing industrial waste generation is a matter of global concern and should be encountered solemnly and critically. The implications of different industrial wastes on the environment and human health have become more critical and pertinent around the world over the recent years. The conceivable leaching and accumulation of the toxic contaminants from these industrial wastes into the environmental matrices concomitantly affect human health and other living organisms. Over recent years, various researchers have studied the control of industrial waste pollution from the perspective of evaluating proper management strategies, terminal disposal methods, and other waste utilization techniques. This warrants an analysis of the various treatment and disposal methods currently available in industrial waste management. Hence, the current study was initiated with the goal to provide an overview of the global scenario in industrial waste generation, their environmental and health implications, and the different industrial waste management techniques currently employed.

Keywords Industrial wastes · Treatment · Disposal

S. Nath · K. K. Singh · S. Das (✉)
Department of Environmental Science, Pachhunga University College,
Mizoram University (A Central University), Aizawl, Mizoram 796001, India
e-mail: dassubhasish@pucollege.edu.in

K. K. Singh
Department of Forestry and Environmental Science, Manipur University (A Central University), Imphal, Manipur 795003, India

S. Tangjang
Department of Botany, Rajiv Gandhi University (A Central University), Doimukh, Arunachal Pradesh 791112, India

A. Anouzla and S. Souabi (eds.), *Technical Landfills and Waste Management*,
Springer Water, https://doi.org/10.1007/978-3-031-52633-6_9

9.1 Introduction

The improving standards of living and economic growth have contributed to an increased consumption, which has aggravated the issue of waste generation from industries on a global scale (Grazhdani 2016). Recent reports illustrate that countries in the European (EU) conglomerate produce on an average about 482 kg per capita of industrial wastes annually (Minelgaitė and Liobikienė 2019). Industrial wastes significantly vary from industry to industry depending on the raw materials used, production outlets, and manufacturing processes (Arockiam JeyaSundar et al. 2020; Soliman and Moustafa 2020). Industrial wastes encompass all types of industrial process end-products or by-products (Soliman and Moustafa 2020). A major concern regarding industrial wastes is its management and disposal as humongous quantities of such wastes get disposed in the environment (El Zrelli et al. 2019). Hence, there is an urgent need to reduce the environmental load caused by industrial wastes as it concomitantly leads to social, environmental, and health problems (Lee et al. 2022). According to UNEP UAP (1991), 650×10^6 t of industrial sewage sludges is dumped in the Mediterranean Sea annually. Disposal of industrial wastes not only impacts the marine environment, but also impacts the terrestrial environment with deposition of toxic contaminants from these wastes in the environmental matrices (Bergmann et al. 2015; Gaur et al. 2020). In 2017, the total amassment of industrial wastes amounted to be about 60–70 billion tons in China, however, the all-inclusive utilization rate was only around 60% (Zhang et al. 2021).

Industrial wastes generated from the different processing activities or sources have several generalized environmental and health implications (Azmi et al. 2018). As estimated, about 10–15% of the net industrial wastes generated in the world are reckoned to be non-biodegradable and harmful (Krishnan et al. 2021). The conceivable leaching of toxic contaminants from industrial wastes into the environment is a matter of serious concern (He and Kappler 2017). Groundwater and surface water contamination caused by leachates, as well as, air pollution due to burning of industrial solid wastes are some of the immediate environmental threats (Matinde et al. 2018). In marine ecosystems, the radioactive elements and heavy metals from industrial wastes are known to potentially cause coastal erosion, contamination, and modification (El Zrelli et al. 2019). It is also very difficult and exorbitant to eliminate industrial pollutants after their entry into surface water, thus resulting in their further accumulation (Hairom et al. 2021). Industrial wastes also have negative impacts in the terrestrial environment, for instance, processes such as composting and landfilling emit hazardous gases in the environment (Gaur et al. 2020). These hazardous gases react with atmospheric water forming chemical compounds (examples: NH_4NO_3, HNO_3, and H_2SO_4), which later precipitate down as acid rain (Ramírez-García et al. 2019). The different toxic constituents found in the industrial wastes have also been reported for inducing genotoxicity in living organisms (Ericson and Larsson 2000). In human beings, these toxic constituents are mainly observed to alter hormone levels, induce nephromegaly, and hydronephrosis (Singh and Chandra 2019). Additionally, bioaccumulation of toxic heavy metals in aquatic organisms leads to their further biomagnification in the higher trophic levels. This concomitantly induces negative impacts on the food chain (Nasyitah Sobihah et al. 2018; Rajeshkumar et al. 2018).

In the recent past, various researchers have studied the control of industrial waste pollution from the perspective of evaluating management strategies, terminal disposal methods, and utilization techniques (Zhang et al. 2021). Technological advancement has greatly developed the means in subsequent treatment and disposal of industrial wastes from varied sectors. For instance, different waste-treatment routes like incineration, gasification, combustion, pyrolysis, compaction, etc., have radically transformed the waste sector (Bhattacharya and Kim 2016). Additionally, concepts like waste-to-energy have come into the limelight in the recent years. However, the associated high cost of operation, requirement of specialized infrastructure, and skilled manpower have greatly restricted the adoption of these techniques evenly among different nations (Das et al. 2019). In this context, traditional waste recycling methods like composting and vermicomposting also could extend a sustainable way towards mitigating the industrial wastes' menace. These environment friendly integrated waste management techniques utilize minimal infrastructure, cost, and do not require any skilled labor for execution (Das et al. 2019). Although, a lot of work has been documented in industrial waste management, but there are several grey areas which warrant in-depth analysis of the existing industrial waste management techniques. Hence, the current study was initiated with the following objectives: (1) To evaluate the global generation scenario and current developments in industrial waste management, (2) To understand the environmental foot-print of industrial wastes and their health implications (3) To analyze and compare the available methods in industrial waste management.

9.2 Global Generation Scenario of Different Types of Industrial Waste

Industrial wastes can be broadly categorized into two types: (1) hazardous wastes (examples: flammable, corrosive, toxic materials, etc.) and (2) non-hazardous wastes (examples: agro-industrial wastes, plastic, paper, glass, other organic wastes, etc.) (Millati et al. 2019; Arockiam JeyaSundar et al. 2020). In comparison, the volume of non-hazardous wastes produced in the world is substantially greater than the volume of hazardous wastes (Millati et al. 2019). About only 1.5 (~ 4.5 MT) and 1.1% (~ 11.6 MT) of the total industrial wastes generated was classified as hazardous in India and China, respectively (Pappu et al. 2007; Duan et al. 2008).

Considered as the major support in the economic upliftment of developing countries, the agro-based industries have been growing extensively with growing global population. Thus, a major portion of non-hazardous industrial wastes comes from this sector alone (Gaur et al. 2020). Agro-industrial wastes are mainly remnants produced during harvesting, processing, and storage of raw agricultural products (Obi et al. 2016; Mo et al. 2018). Globally, about 1.3 billion tons of agro-industrial wastes (i.e.,

1/3rd of the overall agricultural output) are generated (Ravindran et al. 2018). In the European (EU) conglomerate, about 367 million tons of wastes are generated from the agro-industrial processes annually (Correddu et al. 2020). According to Food and Agriculture Organization (2013), the estimated global generation of non-edible agro-industrial wastes is about 250 million tons (Heredia-Guerrero et al. 2017). As per the reports of Madurwar et al. (2013), the amount of agro-industrial wastes produced in India annually was over 350 million tons.

Similar to agro-industries, various food processing industries like dairy industry, meat and poultry industry, fruit and vegetable industry, edible oil industry, etc., also contribute to the generation of a substantial amount of industrial wastes (Gaur et al. 2020; Sharma et al. 2020). The food processing industries in European Union dispose of waste of around 88 million tons annually (Scherhaufer et al. 2018). Fruit juice processing industries, for examples, apple juicing industry produces 10.91% of pomace after its extraction, papaya juicing industry generates 15% wastes as peels and seeds, and mango processing industry generates 42% wastes as pulps, peels, and seeds. Altogether, around 6 MMT of fruit and vegetable wastes are generated from caning and frozen food industries (Sagar et al. 2018).

Paper industry is considered as a major polluting industry which generates extensive amounts of hazardous wastes (Chakraborty et al. 2019). The wastes from paper industries mainly consist of various inorganic and organic substances including lime mud, ashes, slaker grits, green liquor dregs, etc. (Monte et al. 2009). A major solid waste that is produced from the various paper-making processes is paper sludge (Zhang and Sun 2018). In China, the production of dry paper sludge was about 0.26 Gt/a in the year 2012 (Fang et al. 2015). In Europe, wastes of about 11 million tons were generated from paper production. Likewise, in 2005, 7.7 million tons of solid wastes were generated from the production of 47.3 million tons of recycled paper (Monte et al. 2009).

Textile industry is also a fast-growing industrial sector, contributing to global industrial waste generation (Sivaram et al. 2018). Textile wastes are mainly categorized into three types: (1) production wastes, (2) pre-consumer wastes, and (3) post-consumer wastes (Yalcin-Enis et al. 2019). The textile sector's waste generation was predicted to be risen by 63% by 2030 (Tedesco and Montacchini 2020). In 2007, the total generation of textile wastes in the US was about 11.9 million tons, while the recovery accounted only 15.9% of the total textile wastes (Wang 2010). Annually, estimated volumes of about 10.5 million tonnes, 350,000 tonnes, and 287,000 tonnes of post-consumer textile wastes are disposed of in the landfills of USA, UK, and Turkey, respectively (Karaosman et al. 2017). US Environmental Protection Agency (USEPA) reported that textile wastes occupy almost 5% of all landfill spaces (Radhakrishnan 2014).

The mining industry goes through complex processes such as metal extraction, mineral beneficiation, smelting, refining, etc., which also resultantly produce considerable amounts of industrial solid wastes, liquid wastes, and gaseous emissions (Matinde et al. 2018). In India, annually substantial volumes of wastes are generated from different mines, for instance, production of manganese, coal, bauxite, and iron ores generate wastes of 4, 75, 1.2, and 55 million tonnes, respectively (Deshpande and Shekdar 2005). According to statistics, China discharges over 0.5 billion tons of tailings produced from mining annually (Lu and Cai 2012).

Petroleum refinery is another major industrial sector associated with the production of fuels, lubricants, and various intermediates of petrochemicals (Jain et al. 2020). A large volume of water is required during the crude-oil extraction processes, which resultantly generates a substantial volume of wastewater (~ 0.4–1.6 times of the amount of produced crude oil) (Coelho et al. 2006; Jain et al. 2020). The liquid effluents generated during the refining processes undergo through depuration processes for their treatment, subsequently producing oil refinery sludges (ORS), a potential waste product having high amounts of petroleum-derived hydrocarbons and heavy metals (Marin et al. 2005; Martí et al. 2009). About more than one billion tons of petroleum sludge is produced each year globally (Hu et al. 2013). The oil refinery sludges produced from the refining processes corresponds up to 1/3rd and 1/4th of the amount of used crude oil (Said et al. 2021). According to the reports of US EPA, the refineries in US produced 30,000 tons of ORS on an average annually (Hu et al. 2013). Annually, about 3×10^6 tons of oil refinery sludge are discarded by the petroleum industries in China (Wang et al. 2012). The total amount of ORS generated in India is about 28,220 tons per year from petroleum industries (Bhattacharyya and Shekdar 2003).The waste generation scenario of some of the major industries around the world have been summarized (Table 9.1).

9.3 Environmental Impacts of Industrial Waste

Due to the non-biodegradable properties of industrial wastes, disposal of these wastes into the environment have accelerated the pollution of freshwater bodies and clean water resources, as well as, terrestrial ecosystems (Azmi et al. 2018; Hairom et al. 2021). One of the major polluters of the environment are the wastes generated from paper and pulp industries, characterized by high concentrations of toxic compounds such as chlorophenols, organic acids, phosphorus, sulfur, and heavy metals (Haq et al. 2016). These hazardous wastes are disposed of into lakes and rivers at a rate of 20–100 m^3 per Mg of the products (Lindholm-Lehto et al. 2015). In textile industries, about 4% of the total global nitrogen and phosphorus-based fertilizers are utilized in the production of cotton, which concomitantly pollutes the clean water resources, further escalating into eutrophication and algal blooms if integrated into the flowing rivers (Ütebay 2020). The wastes from textile industries are exceedingly alkaline and consist of higher concentrations of total dissolved solids (TDS), as well as, toxic compounds such as vat dyes, nitrates, hydrogen peroxides, heavy metals, and certain auxiliary chemicals, which results in adverse environmental complications if not treated appropriately prior to their disposal (Paul et al. 2012; Parisi et al. 2015; Holkar et al. 2016; Sivaram et al. 2018). Radioactive wastes, as well as, wastes such as silver foils, fly ash, synthetic fibers, etc., from different industrial plants relentlessly affect the environment (Azmi et al. 2018). Phosphogypsum (PG), an environmentally hazardous industrial radiochemical by-product, is disposed of in the Mediterranean

Table 9.1 Different types of industrial wastes and their global generation

Sl. No	Industry	Waste type	Production volume	Country	References
1	Agro-industry	Bagasse	90 million tonnes	India	Pappu et al. (2007)
2	Agro-industry	Rice husk	20 million tonnes	India	Pappu et al. (2007)
3	Battery industry	Lithium-ion battery (LIB) waste	355,000 tonnes	China	Sun et al. (2021)
4	Construction industry	Construction wastes	14.5 million tonnes	India	Pappu et al. (2007)
5	Dairy industry	Dairy processed sludges	19.7 kg sludge/m^3	Ireland	Shi et al. (2021)
6	Dairy industry	Dairy processed sludges	31 kg organic waste/tonnes product	Australia	Shi et al. (2021)
7	Mining industry	Tailings	0.5 billion tonnes	China	Lu and Cai (2012)
8	Mining industry	Tailings	~ 80 million tonnes	India	Agrawal et al. (2004)
9	Mining industry	Iron ore wastes	55 million tonnes	India	Deshpande and Shekdar (2005)
10	Mining industry	Coal wastes	60 million tonnes	India	Pappu et al. (2007)
11	Paper and pulp industry	Paper sludge	0.26 Gt/a	China	Fang et al. (2015)
12	Steel industry	Steel slags	> 100 million tonnes	China	Gao et al. (2020)
13	Tannery industry	Fleshing wastes	70,000–100,000 tonnes	India	Selvaraj et al. (2019)
14	Tannery industry	Offal (Fleshing wastes)	150,000 tonnes	India	Ravindran et al. (2014)
15	Textile industry	Post-consumer wastes	350,000 tonnes	UK	Karaosman et al. (2017)
16	Textile industry	Disposable textile products	10 million tonnes	America	Ütebay (2020)
17	Textile industry	Disposable textile products	20 million tonnes	China	Ütebay (2020)
18	Petroleum refineries	Refinery sludges	30,000 tonnes	US	Hu et al. (2013)
19	Petroleum refineries	Refinery sludges	3 million tonnes	China	Wang et al. (2012)
20	Petroleum refineries	Refinery sludges	28,220 tonnes	India	Bhattacharyya and Shekdar (2003)

marine environment (Periáñez 2002; El Zrelli et al. 2018). Recent reports suggest that solid industrial wastes consisting of by-produced phosphogypsum from the Tunisian Chemical Group of Gabes have severely affected the coastal habitats of central Gulf of Gabes (Kateb et al. 2018).

Wastes from agro-based and food processing industries also cause negative effects on the environment, for example, emission of greenhouse gases (GHGs), and inefficient usage of water and land cover (Salihoglu et al. 2018; Sharma et al. 2020). Reports have shown that wastes from olive oil producing industries degrade the quality of soil, as well as, induce toxicity on the arthropods and bacteria living in the soil (Gaur et al. 2020). In addition, the by-products from dairy industries consist of milk residues which are harmful to the environment. The large amount of whey which is released as wastes during cheese production is a potential environmental pollutant (Sar et al. 2022). These wastes or by-products are rich in fats, and have high BOD and COD levels because of the high content of lactose and milk-based proteins (Kumar Awasthi et al. 2022; Sar et al. 2022). Greenhouse gases (GHGs) such as CO_2 and CH_4 emitted from brewery, food, and beverage industries may contribute to global warming. Different industries such as battery, chemical, iron, steel, petroleum, paper and pulp, tannery and textiles emit SO_X and NO_s into the environment, resultantly producing acid rains (Ofoezie and Sonibare 2004).

Soil heavy metal pollution caused by different industries is another environmental problem of concern, which resultantly degrades the soil productivity as higher concentrations of heavy metals penetrate into the pedosphere (Wei et al. 2016; Yang et al. 2018). Soils of large industrial sectors are more susceptible to greater heavy metal pollution (Alloway 2013). Changes in the soil organic content, as well as, in its acid–base conditions further facilitate the mobility of these heavy metals, further contaminating groundwater and surface water resources (Barsova et al. 2019). On a global scale, more than 30,000 and 800,000 tons of Cr and Pb, respectively, have been released into the environment, consequently causing heavy metal pollution (Yang et al. 2018). Also, solid wastes from food processing industries which have been treated with high concentration of salts have detrimental effects on soil structure (Gaur et al. 2020). Deleterious effects of high salt concentration in soil include dispersion of soil particles, alteration of soil permeability, and alteration of soil pore size, which in turn affect agricultural processes (Loehr 1978). Studies have also accumulated showing the adverse effects of air pollution in agriculture from different industries, especially crop yield reduction by 30–60% (Lindhjem et al. 2007; Wang et al. 2020).

Various constituents of oil refinery sludges (ORS) from petroleum industries such as the aliphatic and aromatic hydrocarbons, resins, asphaltenes, heavy metals, etc., are potentially hazardous compounds for the environment (Jasmine and Mukherji 2015; Vdovenko et al. 2015). The petroleum hydrocarbons (PHCs) present in these sludges may enter into terrestrial ecosystems through leaching and alter the different properties of soil (Robertson et al. 2007) (Fig. 9.1). These contaminated soils are found to be nutrient deficient, resultantly inhibiting seed germination, as well as, restricting growth in plants (Al-Mutairi et al. 2008). PHCs in ORS also inhibit the soil enzyme activity and soil microbial activity (Suleimanov et al. 2005). As these wastes

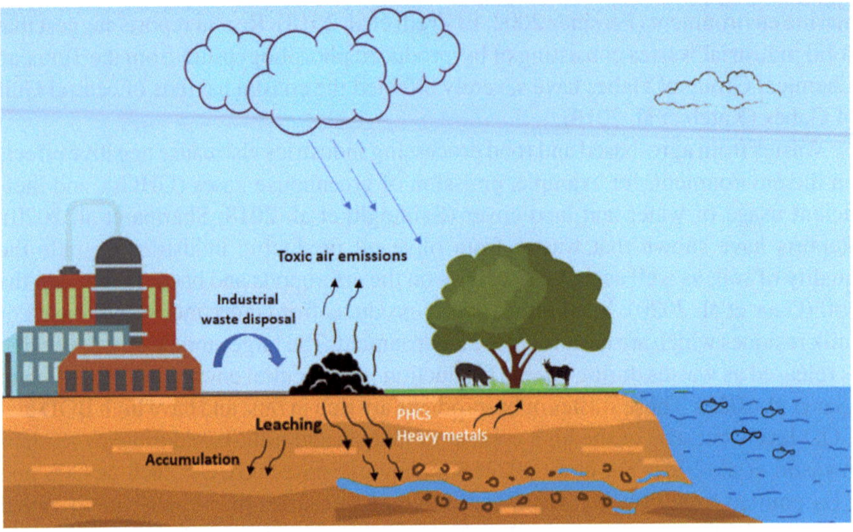

Fig. 9.1 Fate and pathways of toxic constituents of industrial waste into the environment

are highly viscous, they form a continuous cover on the soil surface, thus resulting in low water retention capacity, hydraulic conductivity, and moisture (Johnson and Affam 2019). These PHCs also penetrate through the soil profile into groundwater, consequently affecting its related aquatic ecosystems (Hu et al. 2013) (Table 9.2).

9.4 Health Implications of Industrial Waste

The hazardous wastes generated from the different industries abundantly consist of harmful dyes, chemicals, compounds, heavy metals, etc., wherein the heavy metals are considered toxically most potent to human health (Soliman and Moustafa 2020). The different compounds found in wastes of paper and pulp industries were reported for being genotoxic (Ericson and Larsson 2000). Wastes from steel and battery manufacturing industries consist of different concentrations of lead (Pb) released into the water bodies, exposure to which causes nervous system damage, mental retardation, kidney disease, anemia, and cancer in humans (Ata et al. 2019; Soliman and Moustafa 2020). Cadmium (Cd) is another potentially toxic metal discharged into the environment from industries such as petroleum, metal plating, and cement. Exposure to even low concentrations of Cd(II) has deleterious effects in human health, such as cancer, respiratory diseases, damage to liver, kidneys, bone, etc. (Iqbal et al. 2016) (Table 9.3). Exposure to high concentrations of chromium (Cr) causes irritation at the site of contact and nasal mucosa, and skin ulcers (Sathwara et al. 2007). Of all the oxidized states of chromium, Cr(VI) is the most toxic as it is highly water soluble, and is found to carcinogenic for the stomach and lungs (Soliman and Moustafa 2020).

Table 9.2 Types, sources, and environmental impacts of wastes from major industrial sectors

Industrial sector	Waste type	Source/process/ activity	Environmental impacts
Agro	Dust, acid, gypsum	Mixing, chemical reactions, utilities	Pollution of air and water, contamination of soil, GHG émissions
Battery	Dust, mist, fume, acid, wastewater, plastic	Currying, bobbing, mixing, packaging	Acid rain, pollution of air and water, soil contamination
Brewery	Fume, mist, spent grain, wastewater	Preparation, brewing, washing, packaging, utilities	Odours, pollution of air and water
Cement	Dust, fume, slurry, sludge, wastewater	Extraction, grinding, calcination, packaging, utilities	Air pollution, water pollution, soil contamination
Chemical and allied	Dust, fume, mist, smog, acid, waste water, gypsum, sludge	Reactions, washing, utilities, packaging, treatment discharges	Pollution of air and water, contamination of soil
Dye	Organic vapour, wastewater	Nitration, chlorination, sulphonation of aromatic rings	Pollution of air and water, bad odours
Food and beverage	Fume, mist, dust, wastewater, oily wastes, solid wastes	Milling, mixing, grinding, packaging, utilities	Odours, water and soil contamination, GHG emissions
Iron and steel	Dust, fume, wastewater, slags, sludges	Rolling, smelting, foundry process	Air contamination, water pollution
Metal work and plating	Fume, wastewater, sludges	Cutting, anoding, cleaning, rust proofing, stripping operation	Air contamination, water pollution, soil degradation
Paper and pulp	Dust, foul gases, saw dust, clarifier sludge, wastewater	Digestion, bleaching, mixing, utilities	Acid rain, air and water contamination, soil contamination
Petrochemical	Fume, sludge, spent catalyst	Nitration, alkylation, oxidation, polymerization	Odours, pollution of air and water, soil contamination
Petroleum exploration and production	Dust, fume, drilling fluids, oily wastewater	Washing, cutting, drilling, utilities	Pollution of air and water, soil contamination, impaired photosynthesis

(continued)

Table 9.2 (continued)

Industrial sector	Waste type	Source/process/activity	Environmental impacts
Petroleum refining	Gaseous emissions, wastewater, oily sludges, spent catalysts	Crude distillation, hydrocarbon processing, thermal/catalytic cracking, heat exchange operations	Odours, acid rain, pollution of air and water, soil contamination
Pharmaceutical	Dust, fume, wastewater	Intermediates, mixing, grinding, washing, packaging	Air pollution, water pollution, soil contamination
Plastic and synthetic	Fume, wastewater, plastic products	Moulding, washing, cooling, mould preparation	Pollution of air and water, soil contamination
Tannery	Fume, sludges, wastewater	Tanning, hide washing, treatment, screening, shaving, drying	Odours, colours, acid rain, pollution of air, soil and water
Textile	Gaseous wastes, wastewater, solid waste	Weaving, scouring, dyeing, printing	Odours, colours, air and water contamination, soil contamination

Source UNEP UAP (1991), Faboya (1997), Ofoezie and Sonibare (2004)

Also, exposure to mercury (Hg) can cause serious damage to the nervous system, which is mainly found in the wastes of paper industry and battery manufacturing industries (Alluri et al. 2007).

Wastes from mining industries are also found to be potentially toxic to human health when not disposed of appropriately (Covre et al. 2022). Bioaccumulation of toxic heavy metals in fishes leads to their further biomagnification in higher trophic levels, which resultantly damages the circulatory and central nerve system of human beings (Nasyitah Sobihah et al. 2018; Rajeshkumar et al. 2018). The textile wastes consist of harmful phenolic compounds whose accumulation in the animals and human beings causes health complications (Thasneema et al. 2021). These phenolic compounds can easily get into the human organ systems through absorption from the skin and gastrointestinal tract, after which they undergo metabolism to form harmful reactive intermediates, especially quinone moieties (Soto-Hernández et al. 2017). Reports have shown that Bisphenol A and certain alkylphenols alter the development of mammary glands, resultantly exerting harmful effects on the endocrine system (Munoz-de-Toro et al. 2005). Similarly, chlorophenol has been reported to damage liver, kidneys, lungs, skin, and digestive tract in humans (Anku et al. 2017). Chlorinated solvents from textile industries are also suspected to be carcinogens and harmful to aquatic animals (Sivaram et al. 2018). Industries such as textile, leather, rubber, furniture, and cement use formaldehyde as a raw material in manufacturing

Table 9.3 Health effects of some of toxic metals from industrial sources

Toxic heavy metal	Industrial sources	Health effects	References
Cd	Paint, dyes, fertilizer industries	Lung cancer, pneumonitis, proteinuria, etc.	Singhal et al. (1976), Waalkes et al. (1996)
Cr	Tanneries and steel industries	Lung cancer, pulmonary diseases, renal failure, haemolysis, etc.	Wang et al. (2006), Velma et al. (2009)
As	Mining industry	Hypopigmentation, cancer, nephalopathy, nausea, diarrhoea	Tchounwou et al. (2003, 2004)
Cu	Mining and fertilizer industries	Cardiovascular diseases, nausea, inflammation, Wilson disease, etc.	Tchounwou et al. (2012), Lakherwal (2014)
Hg	Coal industries	Nephritic syndromes, vomiting, diarrhoea, hypersensitivity, etc.	Lund et al. (1991), Clarkson and Magos (2006)
Pb	Battery industries	Anaemia, abdominal diseases, osteoporosis, neurologic degeneration, etc.	Hertz-Picciotto (2000)

of various products, which subsequently is released into the environment as wastewater effluent (Duong et al. 2011). As an extremely volatile compound, exposure to even low concentrations of formaldehyde can have deleterious effects on health (Wen et al. 2011). Irritation of eyes, nose and throat, skin allergies, and allergic asthma are caused as a result of long exposure to formaldehyde (Kim and Kim 2005). The International Agency for Research on Cancer (IARC) has also regarded formaldehyde to be a carcinogen by Tang et al. (2009), Duong et al. (2011) (Fig. 9.2).

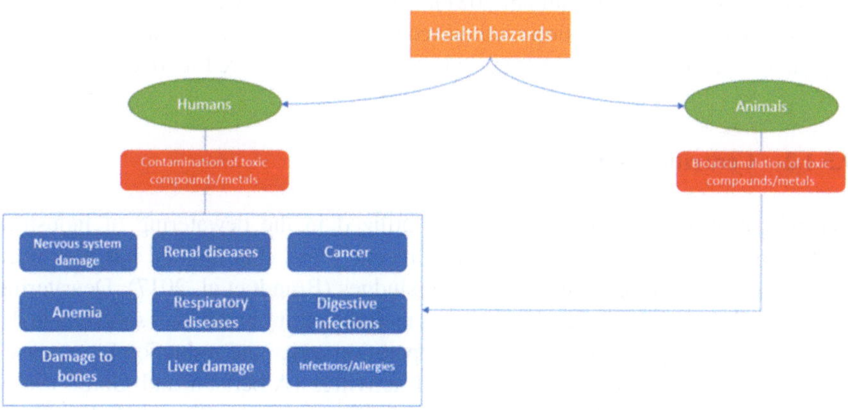

Fig. 9.2 Generalized health implications of industrial wastes

9.5 Available Treatment and Disposal Methods for Industrial Wastes

Industrial waste management is required and warrants the collection and treatment of wastes produced from all levels of processing. It is also highly required that the different industries collect the wastes from its origin phase to its final treatment and disposal (Vaishnavi et al. 2023). In industrial waste management, several techniques have recently developed which have been mainly utilized to convert waste into energy, recover or remediate different constituents of wastes, as well as, improve the route efficiency and other disposal techniques (Singh 2023).

9.5.1 Treatment Methods in Industrial Waste Management

9.5.1.1 Solvent Extraction

Among the various treatment methods, solvent extraction is a widely used conventional method for toxic metals separation, solid–liquid separation, and recovery of other beneficial components. This process primarily consists of a selective extraction and concentration of metal ions into an organic phase from an aqueous phase (organic solvent). This is then followed by a subsequent stripping of the extracted metals back into the aqueous phase (Mansur 2011). In industrial waste treatment, this method is primarily utilized for the removal of phenols, creosols and other phenolic acids (Kiezyk and Mackay 1971). This method does not necessarily degrade wastes, however, separates hazardous constituents from sludges and sediments. Thus, reducing the volume of the hazardous wastes that must be treated or disposed of (USEPA 1990). Different industries such as food, pharmaceutical, petrochemical, ore processing, etc., utilize the solvent extraction method for extraction of beneficial components (Gasser and Rahman 2021). For instance, Taiwo and Otolorin (2009) reported that solvent extraction method recovered about 67.5% hydrocarbons from oily sludges of petroleum industries, out of which 86.7% were found to be aromatic.

9.5.1.2 Centrifugation

Centrifugation method has always been utilized in the dewatering of industrial sludges in industrial waste treatment. In recent years, this method has also been successfully used for coagulant industrial sludges (Brandt et al. 2017). Dewatering of industrial sludges has several advantages including minimization of solid residuals, improvement in operation, reduction of operation costs and reduction in expenses for storage, processing and other related activities (Albertson and Guidi 1969). Reducing the volume of industrial wastes by the centrifugation or sludge dewatering method improves the overall economics (Wang et al. 2007). In centrifugation process, constituents of the industrial sludges are separated from one another with the help of strong centrifugal force generated by a high-speed rotating instrument, based on the difference in their densities. To increase the efficacy of this process, several sludge pre-treatment methods are also incorporated (Huang et al. 2014). In the treatment and management of petroleum sludges, Cambiella et al. (2006) reported an efficient water–oil separation process (92–96%) by centrifugation with the addition of small amount of a coagulant salt $CaCl_2$.

9.5.1.3 Surfactant Enhanced Oil Recovery (SEOR)

SEOR is a widely used treatment method in petroleum industries, utilizes the potential effectiveness of different surface-active compounds which help in increasing the biodegradation rates of hydrophobic organic compounds. To meet the global demand for crude oil, it is necessary to recover oil which could potentially be lost during the different production processes, wherein the SEOR method plays a very crucial role (Al Sabagh and Mohamed 2022; Nadir et al. 2022). This is primarily achieved by increasing their total aqueous solubility (Cort et al. 2002). For instance, Abdel Azim et al. (2011) investigated the breakdown potential of three nonyl phenol ethoxylates-based ($n = 9, 11, 13$) demulsifier systems on petroleum sludges. The results showed that nonyl phenol ethoxylates ($n = 13$) based demulsifier was the most effective in completely breaking down the petroleum sludge.

9.5.1.4 Pyrolysis

In this treatment method, materials are thermally decomposed at very high temperatures (ranging from 500 to 1000 °C) and in the absence of oxygen. The output products such as the char, liquid, or gas depend on the operational conditions (Jafarinejad 2017). This method has a few advantages compared to the conventional combustion method such as it requires lower process temperature and it emits lower air contaminants like polybrominated diphenyl ethers (PBDEs). This method is also convenient for a wide variety of domestic and industrial residues (Czajczyńska et al. 2017). In petroleum industry waste management, about 70–84% of the oil was separated from the oil refinery sludges of a tanker through pyrolysis by maintaining temperatures ranging from 460 to 650 °C (Schmidt and Kaminsky 2001). This technique has also been used a waste-to-energy conversion method in some of the major oil industries (Hai et al. 2021).

9.5.1.5 Gasification

Gasification is a thermochemical conversion technique in which carbon-based materials are converted into gaseous product at high temperatures with the aid of some gasification agents. With the aid of different heterogenous reactions, the gasification agents allow the feedstocks to be rapidly converted into gas (Yun 2018). In gasification, industrial solid wastes are subjected to very high heat (above 600 °C) in an oxygen deficient environment. Oxygen levels are normally kept low to avoid immediate combustion of the substances. This resultantly decomposes the carbon-based fraction of the solid wastes into synthetic gas (syngas) and a solid residue (slag, ash, or char) (Belgiorno et al. 2003; Tangri and Wilson 2017). The by-product of gasification technology, i.e., syngas can be potentially used as energy source. The chars which are produced as a result of gasification have been utilized for removal of toxic metals (Dias et al. 2018).

9.5.1.6 Microwave Irradiation

Microwave heating is a more effective method and provides faster processing rates compared to other heating methods because of its volumetric heating effects (Abdulbari et al. 2011). However, the effectiveness of microwave heating depends upon various factors such as its applied power, process duration, surfactants used, and other sludge properties (Fortuny et al. 2007; Hu et al. 2013). It is a useful technique in industrial waste treatment, wherein this technique can be utilized in several ways like combining of microwave with oxidants. This resultantly provides a higher reaction temperature which increases the degradation rate of pollutants (Shaheen 2018). Since recent years, microwave irradiation has primarily been acting as an alternative to conditional heating methods in industrial sludge management (Zhen et al. 2017). For instance, microwave irradiation method separated 146 barrels of oil and 42 barrels of water from 188 barrels of water–oil emulsion, showing its higher separation efficiency (Fang and Lai 1995).

9.5.1.7 Ultrasonic Irradiation

Ultrasonic irradiation method is mainly used in disintegration of activated sludges from industries like paper, pulp, and sugar. Reports have shown that utilization of low-frequency ultrasonication was extremely effective in decreasing the bacterial population in industrial sludges, and also the results showed reduction in particle sizes and chemical oxygen demand (COD) of the treated industrial wastes (Singh et al. 2010). In petroleum industries, this method is extensively used wherein the application of ultrasonic waves in a treatment chamber helps produce compression and rarefactions. This resultantly decreases the water–oil mixture stability and makes it possible to separate solid–liquid suspensions in high concentration (Song et al. 2012; Li et al. 2013). Xu et al. (2009) recorded a separation rate of 55.6% of oil from petroleum sludge through ultrasonic cavitation achieved with the application of frequency of 28 kHz. Studies have also showed that an oil recovery rate of 80% can be achieved by the application of ultrasonic irradiation for 10 min at a frequency of 20 kHz and power of 66 W (Zhang et al. 2012).

9.5.1.8 Electrokinetic Method

In this treatment method, a direct current (DC) of low intensity is applied across two electrodes on a permeable medium. As a result, electrons move to the higher concentration region from the lower concentration region through the permeable medium. Recovery of oil from petroleum sludges by this method is a three-stage mechanism, wherein initially segregation of the colloidal aggregates takes place in an electrical field. Consequently, the colloidal particles of the sludge and the separated liquid are relocated to the anode area and the cathode area respectively (Yang et al. 2005). Elektorowicz and Habibi (2005) reported that the application of electrokinetics in oily sludge treatment removed the water content of the sludge and light hydrocarbon by 63 and 43% respectively. Reports have also accumulated showing efficient metal recovery by this method from mining industrial wastes (Peppicelli et al. 2018). Reports have also shown the efficiency of this method in reduction of heavy metal content from industrial wastes such as mining, and iron and steel. Also, adequate strategies could be adopted depending on the sludge characteristics to increase the output efficiency of this method (Pazos et al. 2009).

9.5.1.9 Froth Floatation

In this technique, fine solids are separated from an aqueous solution by the use of surface chemistry-based unit operation. This method utilizes air bubbles in an aqueous slurry to help catch oil droplets/small solids, which later is accumulated in a froth layer (Hu et al. 2013). In petroleum industries, studies have shown that froth floatation of oil sludge at optimum conditions can yield a recovery of 55% of oil (Ramaswamy et al. 2007). Froth floatation method has also been used in mining industries (Chase 1958).

9.5.2 Available Disposal Methods in Industrial Waste Management

9.5.2.1 Incineration

Incineration is a thermal method based on the application of high temperature (870–1200 °C) for decomposing organic materials. The efficiency of this method owing to the application of proper incinerators is extremely high for degradation of toxic constituents like PCBs and dioxins (Fernández et al. 2022). In petroleum industries, the oily sludges generated from the refineries go through complete combustion. This takes place in the presence of surplus air and auxiliary fuel. Mainly two incinerator types: a) fluidized bed (730–760 °C) and b) rotary kiln (temp: 980–1200 °C) are used for combustion (Scala and Chirone 2004). Of the two incinerator types, fluidized bed is more effective (Zhou et al. 2009). For instance, (Sankaran et al. 1998) investigated the efficiency of fluidized bed on three oily sludge wastes, reporting a combustion and an incineration efficiency of 98 and 99% respectively for all the three types. Incineration can also be utilized in disposing of wastes from rubber industries (Kulkarni et al. 2018).

9.5.2.2 Stabilization/Solidification

Solidification/stabilization treatment is a disposal method used to treat hazardous wastes and in the remediation/site restoration of contaminated lands (Wilk 2004). This method mainly aims to prevent the leaching of contaminates wastes into the environment by encapsulating or sealing it with the use of a binder (either through physical or chemical mean). This helps in converting the wastes into products such as construction materials or non-hazardous waste for landfill disposal (Johnson and Affam 2019). Solidification mainly generates a resistant solid matrix which capture contaminants and stabilization transforms them into less toxic forms. Mainly cement-based stabilization/solidification is used to encapsulate inorganic constituents (Leonard and Stegemann 2010; Johnson and Affam 2019).

9.5.2.3 Oxidation Treatment

In this method, industrial waste contaminants are mainly degraded with the help of chemical and oxidation agents. In this process, the organic constituents in the industrial waste are oxidized into H_2O and CO_2 or into non-hazardous materials with the addition of reagents (Ferrarese et al. 2008). Studies have shown that a Fenton type reagent reduced the concentration of phenols, PAHs, and other contaminants in petroleum waste contaminated soil (Mater et al. 2006). Zhao et al. (2019) investigated the treatment of oil sludges through two-stage wet air oxidation and reported that this method removed 93.1% oil from the oily sludge and reduce the volume of such sludge by 85.4%.

9.5.2.4 Composting/Vermicomposting

Integrated techniques such as composting and vermicomposting are emerging as proficient ones in industrial waste management owing to the rapidity of their process and production of quality compost (Tian et al. 2012; Goswami et al. 2014; Karwal and Kaushik 2020). Composting includes the accelerated biodegradation of organic materials by thermophilic microorganisms under controlled conditions (Lung et al. 2001). On the other hand, vermicomposting is an alternative composting technique in which organic wastes are biodegraded with the synergistic action of earthworms and mesophilic microorganisms and are then transformed into a valuable finished clean product called vermicompost (Wang et al. 2009; Bhat et al. 2013; Lim et al. 2016).

Various reports have accumulated showing the effectiveness of vermicomposting in industrial waste management, for instance, paper mill sludge (Kaur et al. 2010), textile industry waste (Bhat et al. 2013), sugar industry waste (Sen and Chandra 2007), tannery industry waste (Vig et al. 2011), agro mill waste (Suthar 2010), etc. (Table 9.4).

Table 9.4 Different treatment and disposal methods in industrial waste management with corresponding advantages and disadvantages

Treatment/disposal methods	Application level	Advantages	Disadvantages
Solvent extraction	Field	Simple and fast	Low efficiency
Centrifugation	Field	Clean, efficient, low energy consumption	Costly for installation and environmental concerns
Surfactant enhanced oil recovery	Field	Requires less time and less expensive	Environmentally toxic
Pyrolysis	Field	Simple and easy method	High energy required, high maintenance cost and high operating cost
Microwave irradiation	Field	Requires less time and efficient	High energy required and expensive
Freeze/thaw	Laboratory	Suitable for cold regions	Not suitable for other regions
Ultrasonic irradiation	Laboratory	Fast and efficient	Very expensive
Electrokinetic	Laboratory	Fast and efficient	Only small-scale applications
Froth flotation	Laboratory	Simple and low energy consumption	Very low efficiency
Incineration	Field	Fast and efficient	Costly and environmentally detrimental
Composting/vermicomposting	Field	Efficient and environment friendly	Slow and labour intensive
Stabilization/solidification	Laboratory	Fast and efficient	Only for specific industrial sludges
Oxidation	Laboratory	Fast and efficient	Costly and environmentally detrimental

Source Johnson and Affam (2019)

9.6 Conclusion

As summarized in the review, the generation of hazardous industrial wastes cannot be completely avoided at present times due to existing global industrialization and consumerism. In this review, global waste generation scenario from some major industrial sectors and their impacts on health and environment was summarized. In addition, an evaluation of the prevailing treatment and disposal strategies in industrial waste management was made to see the pros and cons of these existing treatment and disposal methods. In this regard, we can conclude that several treatment and disposal methods have successfully evolved to manage industrial sludges. However, some treatment and disposal methods, despite being effective, are cost-expensive, time consuming and limited when it comes to their application in large scale.

References

Abdel Azim AAA, Abdul-Raheim ARM, Kamel RK, Abdel-Raouf ME (2011) Demulsifier systems applied to breakdown petroleum sludge. J Pet Sci Eng 78:364–370. https://doi.org/10.1016/j.petrol.2011.07.008

Abdulbari Hayder A, Abdurahman NH, Rosli YM, Azhari NH (2011) Demulsification of Triton X-100, low-sulfur wax residue (LSWR), sorbitan monooleate (Span 83) and sodium dodecyl sulfate (SDS)-stabilized petroleum emulsions with a microwave separation method. Res J Chem Environ 15:3

Agrawal A, Sahu KK, Pandey BD (2004) Solid waste management in non-ferrous industries in India. Resour Conserv Recycl 42:99–120. https://doi.org/10.1016/j.resconrec.2003.10.004

Al Sabagh AM, Mohamed A (2022) Improving the heavy oil recovery by surfactants from wastes. In: Enhanced oil recovery-selected topics. IntechOpen

Albertson OE, Guidi EE (1969) Centrifugation of waste sludges. J Water Pollut Control Fed 12:607–628

Alloway BJ (2013) Sources of heavy metals and metalloids in soils BT. In: Alloway BJ (ed) Heavy metals in soils—trace metals and metalloids in soils and their bioavailability. Springer, Dordrecht, pp 11–50

Alluri HK, Ronda SR, Settalluri VS et al (2007) Biosorption: an eco-friendly alternative for heavy metal removal. Afr J Biotechnol 6:2924–2931. https://doi.org/10.5897/ajb2007.000-2461

Al-Mutairi N, Bufarsan A, Al-Rukaibi F (2008) Ecorisk evaluation and treatability potential of soils contaminated with petroleum hydrocarbon-based fuels. Chemosphere 74:142–148. https://doi.org/10.1016/j.chemosphere.2008.08.020

Anku WW, Mamo MA, Mamo MA et al (2017) Phenolic compounds compounds in in water: water—sources, phenolic toxicity and treatment methods toxicity and treatment methods. Phenolic Compd Nat Sour Import Appl 21:419–443

Arockiam JeyaSundar PGS, Ali A, Zhang Z (2020) Waste treatment approaches for environmental sustainability. INC

Ata S, Tabassum A, Bibi I et al (2019) Lead remediation using smart materials: a review. Zeitschrift Fur Phys Chem 233:137–1409. https://doi.org/10.1515/zpch-2018-1205

Azmi AA, Jai J, Zamanhuri NA, Yahya A (2018) Precious metals recovery from electroplating wastewater: a review. IOP Conf Ser Mater Sci Eng 358:1757. https://doi.org/10.1088/1757-899X/358/1/012024

Barsova N, Motuzova G, Kolchanova K et al (2019) The effect of humic substances on Cu migration in the soil profile. Chem Ecol 35:86–101. https://doi.org/10.1080/02757540.2018.1540613

Belgiorno V, De Feo G, Della Rocca C, Napoli R (2003) Energy from gasification of solid wastes. Waste Manag 23:1–15. https://doi.org/10.1016/S0956-053X(02)00149-6

Bergmann M, Gutow L, Klages M (2015) Marine anthropogenic litter. Mar Anthropog Litter 165:1–447. https://doi.org/10.1007/978-3-319-16510-3

Bhat SA, Singh J, Vig AP (2013) Vermiremediation of dyeing sludge from textile mill with the help of exotic earthworm *Eisenia fetida* Savigny. Environ Sci Pollut Res 20:5975–5982. https://doi.org/10.1007/s11356-013-1612-2

Bhattacharya SS, Kim K-H (2016) Utilization of coal ash: is vermitechnology a sustainable avenue? Renew Sustain Energy Rev 58:1376–1386

Bhattacharyya JK, Shekdar AV (2003) Treatment and disposal of refinery sludges: Indian scenario. Waste Manag Res 21:249–261. https://doi.org/10.1177/0734242X0302100309

Brandt MJ, Johnson KM, Elphinston AJ, Ratnayaka DD (2017) Chapter 13—energy use, sustainability and waste treatment. In: Brand MJ, Johnson, KM, Elphinston, AJ, Ratnayaka, DD (eds) Twort's water supply, 7th ed, pp 553–580

Cambiella A, Benito JM, Pazos C, Coca J (2006) Centrifugal separation efficiency in the treatment of waste emulsified oils. Chem Eng Res Des 84:69–76. https://doi.org/10.1205/cherd.05130

Chakraborty D, Dahiya S, Amulya K, et al (2019) Valorization of paper and pulp waste: opportunities and prospects of biorefinery

Chase ES (1958) Flotation treatment of sewage and industrial wastes. Sewage Ind Waste 30:783–791

Clarkson TW, Magos L (2006) The toxicology of mercury and its chemical compounds. Crit Rev Toxicol 36:609–662

Coelho A, Castro AV, Dezotti M, SantAnna GL (2006) Treatment of petroleum refinery sourwater by advanced oxidation processes. J Hazard Mater 137:178–184. https://doi.org/10.1016/j.jhazmat.2006.01.051

Correddu F, Lunesu MF, Buffa G et al (2020) Can agro-industrial by-products rich in polyphenols be advantageously used in the feeding and nutrition of dairy small ruminants? Animal 10:1442

Cort TL, Song M-S, Bielefeldt AR (2002) Nonionic surfactant effects on pentachlorophenol biodegradation. Water Res 36:1253–1261. https://doi.org/10.1016/S0043-1354(01)00320-7

Covre WP, Ramos SJ, Pereira WV et al (2022) Impact of copper mining wastes in the Amazon: properties and risks to environment and human health. J Hazard Mater 421:126688. https://doi.org/10.1016/j.jhazmat.2021.126688

Czajczyńska D, Anguilano L, Ghazal H et al (2017) Potential of pyrolysis processes in the waste management sector. Therm Sci Eng Prog 3:171–197. https://doi.org/10.1016/j.tsep.2017.06.003

Das S, Lee S-H, Kumar P et al (2019) Solid waste management: scope and the challenge of sustainability. J Clean Prod 228:658–678. https://doi.org/10.1016/j.jclepro.2019.04.323

Deshpande PV, Shekdar VA (2005) Sustainable waste management in the Indian mining industry. Waste Manag Res 23:343–355. https://doi.org/10.1177/0734242X05056994

Dias D, Lapa N, Bernardo M et al (2018) Cr(III) removal from synthetic and industrial wastewaters by using co-gasification chars of rice waste streams. Bioresour Technol 266:139–150

Duan H, Huang Q, Wang Q et al (2008) Hazardous waste generation and management in China: a review. J Hazard Mater 158:221–227. https://doi.org/10.1016/j.jhazmat.2008.01.106

Duong A, Steinmaus C, McHale CM et al (2011) Reproductive and developmental toxicity of formaldehyde: a systematic review. Mutat Res Rev Mutat Res 728:118–138. https://doi.org/10.1016/j.mrrev.2011.07.003

El Zrelli R, Rabaoui L, Ben Alaya M et al (2018) Seawater quality assessment and identification of pollution sources along the central coastal area of Gabes Gulf (SE Tunisia): evidence of industrial impact and implications for marine environment protection. Mar Pollut Bull 127:445–452. https://doi.org/10.1016/j.marpolbul.2017.12.012

El Zrelli R, Rabaoui L, Ben Alaya M et al (2019) Decadal effects of solid industrial wastes on the coastal environment: Gulf of Gabes (Tunisia, Southern Mediterranean Sea) as an example. Estuar Coast Shelf Sci 224:281–288. https://doi.org/10.1016/j.ecss.2019.04.021

Elektorowicz M, Habibi S (2005) Sustainable waste management: recovery of fuels from petroleum sludge. Can J Civ Eng 32:164–169. https://doi.org/10.1139/l04-122

Ericson G, Larsson Å (2000) DNA adducts in perch (*Perca fluviatilis*) living in coastal water polluted with bleached pulp mill effluents. Ecotoxicol Environ Saf 46:167–173. https://doi.org/10.1006/eesa.1999.1892

Faboya OOP (1997) Industrial pollution and waste management. Dimens Environ Probl Niger 24:12–25

FAO (2013) Food wastage footprint: impacts on natural resources. FAO, Rome

Fang CS, Lai PMC (1995) Microwave heating and separation of water-in-oil emulsions. J Microw Power Electromagn Energy 30:46–57. https://doi.org/10.1080/08327823.1995.11688257

Fang S, Yu Z, Lin Y et al (2015) Thermogravimetric analysis of the co-pyrolysis of paper sludge and municipal solid waste. Energy Convers Manag 101:626–631. https://doi.org/10.1016/j.enconman.2015.06.026

Fernández MD, Sánchez-Arguello PS, García-Gómez CBT (2022) Soil pollution remediation. Elsevier, Amsterdam

Ferrarese E, Andreottola G, Oprea IA (2008) Remediation of PAH-contaminated sediments by chemical oxidation. J Hazard Mater 152:128–139. https://doi.org/10.1016/j.jhazmat.2007.06.080

Fortuny M, Oliveira CBZ, Melo RLFV et al (2007) Effect of salinity, temperature, water content, and pH on the microwave demulsification of crude oil emulsions. Energy Fuels 21:1358–1364. https://doi.org/10.1021/ef0603885

Gao D, Wang FP, Wang YT, Zeng YN (2020) Sustainable utilization of steel slag from traditional industry and agriculture to catalysis. Sustainability 12:1–9. https://doi.org/10.3390/su12219295

Gasser MS, Rahman ROA (2021) Sustainability of solvent extraction techniques in pollution prevention and control. Handb Adv Approaches Towar Pollut Prev Control 147:33–66

Gaur VK, Sharma P, Sirohi R et al (2020) Assessing the impact of industrial waste on environment and mitigation strategies: a comprehensive review. J Hazard Mater 398:123019. https://doi.org/10.1016/j.jhazmat.2020.123019

Goswami L, Sarkar S, Mukherjee S et al (2014) Vermicomposting of tea factory coal ash: metal accumulation and metallothionein response in *Eisenia fetida* (Savigny) and *Lampito mauritii* (Kinberg). Bioresour Technol 166:96–102. https://doi.org/10.1016/j.biortech.2014.05.032

Grazhdani D (2016) Assessing the variables affecting on the rate of solid waste generation and recycling: an empirical analysis in Prespa Park. Waste Manag 48:3–13. https://doi.org/10.1016/j.wasman.2015.09.028

Hai A, Bharath G, Rambabu K et al (2021) Pyrolysis of different date palm industrial wastes into high-quality bio-oils: a comparative study. Clean Technol Environ Policy 23:55–64

Hairom NHH, Soon CF, Mohamed RMSR et al (2021) A review of nanotechnological applications to detect and control surface water pollution. Environ Technol Innov 24:102032. https://doi.org/10.1016/j.eti.2021.102032

Haq I, Kumar S, Kumari V et al (2016) Evaluation of bioremediation potentiality of ligninolytic *Serratia liquefaciens* for detoxification of pulp and paper mill effluent. J Hazard Mater 305:190–199. https://doi.org/10.1016/j.jhazmat.2015.11.046

He J, Kappler A (2017) Recovery of precious metals from waste streams. Microb Biotechnol 10:1194–1198. https://doi.org/10.1111/1751-7915.12759

Heredia-Guerrero JA, Heredia A, Domínguez E et al (2017) Cutin from agro-waste as a raw material for the production of bioplastics. J Exp Bot 68:5401–5410. https://doi.org/10.1093/jxb/erx272

Hertz-Picciotto I (2000) The evidence that lead increases the risk for spontaneous abortion. Am J Ind Med 38:300–309

Holkar CR, Jadhav AJ, Pinjari DV et al (2016) A critical review on textile wastewater treatments: possible approaches. J Environ Manage 182:351–366. https://doi.org/10.1016/j.jenvman.2016.07.090

Hu G, Li J, Zeng G (2013) Recent development in the treatment of oily sludge from petroleum industry: a review. Elsevier, Amsterdam

Huang Q, Han X, Mao F et al (2014) A model for predicting solid particle behavior in petroleum sludge during centrifugation. Fuel 117:95–102. https://doi.org/10.1016/j.fuel.2013.09.002

Iqbal M, Iqbal N, Bhatti IA et al (2016) Response surface methodology application in optimization of cadmium adsorption by shoe waste: a good option of waste mitigation by waste. Ecol Eng 88:265–275. https://doi.org/10.1016/j.ecoleng.2015.12.041
Jafarinejad S (2017) Solid-waste management in the petroleum industry
Jain M, Majumder A, Ghosal PS, Gupta AK (2020) A review on treatment of petroleum refinery and petrochemical plant wastewater: a special emphasis on constructed wetlands. J Environ Manag 272:111057. https://doi.org/10.1016/j.jenvman.2020.111057
Jasmine J, Mukherji S (2015) Characterization of oily sludge from a refinery and biodegradability assessment using various hydrocarbon degrading strains and reconstituted consortia. J Environ Manag 149:118–125. https://doi.org/10.1016/j.jenvman.2014.10.007
Johnson OA, Affam AC (2019) Petroleum sludge treatment and disposal: a review. Environ Eng Res 24:191–201. https://doi.org/10.4491/EER.2018.134
Karaosman H, Brun A, Morales-Alonso G (2017) Vogue or vague: sustainability performance appraisal in luxury fashion supply chains. Environ Footprints Eco-Des Prod Process 24:301–330. https://doi.org/10.1007/978-981-10-2917-2_14
Kateb A El, Stalder C, Rüggeberg A et al (2018) Impact of industrial phosphate waste discharge on the marine environment in the Gulf of Gabes (Tunisia). PLoS One 13: https://doi.org/10.1371/journal.pone.0197731
Karwal M, Kaushik A (2020) Co-composting and vermicomposting of coal fly-ash with press mud: changes in nutrients, micro-nutrients and enzyme activities. Environ Technol Innov 18:100708
Kaur A, Singh J, Vig AP et al (2010) Cocomposting with and without *Eisenia fetida* for conversion of toxic paper mill sludge to a soil conditioner. Bioresour Technol 101:8192–8198
Kiezyk PR, Mackay D (1971) Waste water treatment by solvent extraction. Can J Chem Eng 49:747–752
Kim S, Kim HJ (2005) Comparison of standard methods and gas chromatography method in determination of formaldehyde emission from MDF bonded with formaldehyde-based resins. Bioresour Technol 96:1457–1464. https://doi.org/10.1016/j.biortech.2004.12.003
Krishnan S, Zulkapli NS, Kamyab H et al (2021) Current technologies for recovery of metals from industrial wastes: an overview. Environ Technol Innov 22:101525. https://doi.org/10.1016/j.eti.2021.101525
Kulkarni GS (2018) 1—Introduction to polymer and their recycling techniques. In: Thomas S, Rane AV, Kanny K et al (eds) Plastics design library. William Andrew Publishing, Norwich, pp 1–16
Kumar Awasthi M, Paul A, Kumar V et al (2022) Recent trends and developments on integrated biochemical conversion process for valorization of dairy waste to value added bioproducts: a review. Bioresour Technol 344:126193. https://doi.org/10.1016/j.biortech.2021.126193
Lakherwal D (2014) Adsorption of heavy metals: a review. Int J Environ Res Dev 4:41–48
Leonard SA, Stegemann JA (2010) Stabilization/solidification of petroleum drill cuttings: leaching studies. J Hazard Mater 174(1–3):484–491
Lee D, Kim J, Park H-S (2022) Characterization of industrial hazardous waste generation in South Korea using input-output approach. Resour Conserv Recycl 183:106365. https://doi.org/10.1016/j.resconrec.2022.106365
Li J, Song X, Hu G, Thring RW (2013) Ultrasonic desorption of petroleum hydrocarbons from crude oil contaminated soils. J Environ Sci Heal A 48:1378–1389. https://doi.org/10.1080/10934529.2013.781885
Lim SL, Lee LH, Wu TY (2016) Sustainability of using composting and vermicomposting technologies for organic solid waste biotransformation: recent overview, greenhouse gases emissions and economic analysis. J Clean Prod 111:262–278. https://doi.org/10.1016/j.jclepro.2015.08.083
Lindhjem H, Hu T, Ma Z et al (2007) Environmental economic impact assessment in China: problems and prospects. Environ Impact Assess Rev 27:1–25. https://doi.org/10.1016/j.eiar.2006.08.004
Lindholm-Lehto PC, Knuutinen JS, Ahkola HSJ, Herve SH (2015) Refractory organic pollutants and toxicity in pulp and paper mill wastewaters. Environ Sci Pollut Res 22:6473–6499. https://doi.org/10.1007/s11356-015-4163-x

Loehr RC (1978) Hazardous solid waste from agriculture. Environ Health Perspect 27:261–273. https://doi.org/10.1289/ehp.7827261

Lu Z, Cai M (2012) Disposal methods on solid wastes from mines in transition from open-pit to underground mining. Proced Environ Sci 16:715–721. https://doi.org/10.1016/j.proenv.2012.10.098

Lund B-O, Miller DM, Woods JS (1991) Mercury-induced H_2O_2 production and lipid peroxidation in vitro in rat kidney mitochondria. Biochem Pharmacol 42:S181–S187

Lung AJ, Lin C-M, Kim JM et al (2001) Destruction of Escherichia coli O157: H7 and *Salmonella enteritidis* in cow manure composting. J Food Prot 64:1309–1314

Madurwar MV, Ralegaonkar RV, Mandavgane SA (2013) Application of agro-waste for sustainable construction materials: a review. Constr Build Mater 38:872–878. https://doi.org/10.1016/j.conbuildmat.2012.09.011

Mansur MB (2011) Solvent extraction for metal and water recovery from industrial wastes and effluents. Rem Rev Esc Minas 64:51–55

Marin JA, Hernandez T, Garcia C (2005) Bioremediation of oil refinery sludge by landfarming in semiarid conditions: influence on soil microbial activity. Environ Res 98:185–195. https://doi.org/10.1016/j.envres.2004.06.005

Martí MC, Camejo D, Fernández-García N et al (2009) Effect of oil refinery sludges on the growth and antioxidant system of alfalfa plants. J Hazard Mater 171:879–885. https://doi.org/10.1016/j.jhazmat.2009.06.083

Mater L, Sperb RM, Madureira LAS et al (2006) Proposal of a sequential treatment methodology for the safe reuse of oil sludge-contaminated soil. J Hazard Mater 136:967–971. https://doi.org/10.1016/j.jhazmat.2006.01.041

Matinde E, Simate GS, Ndlovu S (2018) Mining and metallurgical wastes: a review of recycling and re-use practices. J South Afr Inst Min Metall 118:825–844

Millati R, Cahyono RB, Ariyanto T et al (2019) Agricultural, industrial, municipal, and forest wastes: an overview. Elsevier, Amsterdam

Minelgaitė A, Liobikienė G (2019) Waste problem in European Union and its influence on waste management behaviours. Sci Total Environ 667:86–93. https://doi.org/10.1016/j.scitotenv.2019.02.313

Mo J, Yang Q, Zhang N et al (2018) A review on agro-industrial waste (AIW) derived adsorbents for water and wastewater treatment. J Environ Manag 227:395–405. https://doi.org/10.1016/j.jenvman.2018.08.069

Monte MC, Fuente E, Blanco A, Negro C (2009) Waste management from pulp and paper production in the European Union. Waste Manag 29:293–308. https://doi.org/10.1016/j.wasman.2008.02.002

Munoz-de-Toro M, Markey C, Wadia PR et al (2005) Perinatal exposure to Bisphenol A alters peripubertal mammary gland development in mice. Endocrinology 146:4138–4147. https://doi.org/10.1210/en.2005-0340

Nadir N, Shahruddin S, Othman J (2022) Surfactant evaluation for enhanced oil recovery: phase behavior and interfacial tension. Open Chem 20:1110–1120

Nasyitah Sobihah N, Ahmad Zaharin A, Khairul Nizam M et al (2018) Bioaccumulation of heavy metals in maricultured fish, *Lates calcarifer* (Barramudi), *Lutjanus campechanus* (red snapper) and *Lutjanus griseus* (grey snapper). Chemosphere 197:318–324. https://doi.org/10.1016/j.chemosphere.2017.12.187

Obi FO, Ugwuishiwu BO, Nwakaire JN (2016) Agri wastes. Niger J Technol 35:957–964

Ofoezie EI, Sonibare J (2004) Health and environmental consequences of industrial wastes and toxic chemicals. Environtropica 1:161–175

Pappu A, Saxena M, Asolekar SR (2007) Solid wastes generation in India and their recycling potential in building materials. Build Environ 42:2311–2320. https://doi.org/10.1016/j.buildenv.2006.04.015

Parisi ML, Fatarella E, Spinelli D et al (2015) Environmental impact assessment of an eco-efficient production for coloured textiles. J Clean Prod 108:514–524. https://doi.org/10.1016/j.jclepro.2015.06.032

Paul SA, Chavan SK, Khambe SD (2012) Studies on characterization of textile industrial waste water in Solapur city. Int J Chem Sci 10:635–642

Pazos M, Alcántara MT, Cameselle C, Sanromán MA (2009) Evaluation of electrokinetic technique for industrial waste decontamination. Sep Sci Technol 44:2304–2321

Peppicelli C, Cleall P, Sapsford D, Harbottle M (2018) Changes in metal speciation and mobility during electrokinetic treatment of industrial wastes: implications for remediation and resource recovery. Sci Total Environ 624:1488–1503. https://doi.org/10.1016/j.scitotenv.2017.12.132

Periáñez R (2002) The enhancement of 226Ra in a tidal estuary due to the operation of fertilizer factories and redissolution from sediments: experimental results and a modelling study. Estuar Coast Shelf Sci 54:809–819. https://doi.org/10.1006/ecss.2001.0858

Radhakrishnan S (2014) Natural plant fibers: production, processing, properties and their sustainability parameters. Roadmap Sustain Text Cloth 15:41–62. https://doi.org/10.1007/978-981-287-065-0

Rajeshkumar S, Liu Y, Zhang X et al (2018) Studies on seasonal pollution of heavy metals in water, sediment, fish and oyster from the Meiliang Bay of Taihu Lake in China. Chemosphere 191:626–638. https://doi.org/10.1016/j.chemosphere.2017.10.078

Ramaswamy B, Kar DD, De S (2007) A study on recovery of oil from sludge containing oil using froth flotation. J Environ Manag 85:150–154. https://doi.org/10.1016/j.jenvman.2006.08.009

Ramírez-García R, Gohil N, Singh V (2019) Chapter 21—recent advances, challenges, and opportunities in bioremediation of hazardous materials. In: Pandey VC, Bauddh KBT (eds) Elsevier, pp 517–568

Ravindran B, Contreras-Ramos SM, Wong JWC et al (2014) Nutrient and enzymatic changes of hydrolysed tannery solid waste treated with epigeic earthworm *Eudrilus eugeniae* and phytotoxicity assessment on selected commercial crops. Environ Sci Pollut Res 21:641–651. https://doi.org/10.1007/s11356-013-1897-1

Ravindran R, Hassan SS, Williams GA, Jaiswal AK (2018) A review on bioconversion of agro-industrial wastes to industrially important enzymes. Bioengineering 5:1–20. https://doi.org/10.3390/bioengineering5040093

Robertson SJ, McGill WB, Massicotte HB, Rutherford PM (2007) Petroleum hydrocarbon contamination in boreal forest soils: a mycorrhizal ecosystems perspective. Biol Rev 82:213–240. https://doi.org/10.1111/j.1469-185X.2007.00012.x

Sagar NA, Pareek S, Sharma S et al (2018) Fruit and vegetable waste: bioactive compounds, their extraction, and possible utilization. Compr Rev Food Sci Food Saf 17:512–531. https://doi.org/10.1111/1541-4337.12330

Said OB, Cravo-Laureau C, Armougom F et al (2021) Enhanced pilot bioremediation of oily sludge from petroleum refinery disposal under hot-summer Mediterranean climate. Environ Technol Innov 24:102037. https://doi.org/10.1016/j.eti.2021.102037

Salihoglu G, Salihoglu NK, Ucaroglu S, Banar M (2018) Food loss and waste management in Turkey. Bioresour Technol 248:88–99. https://doi.org/10.1016/j.biortech.2017.06.083

Sankaran S, Pandey S, Sumathy K (1998) Experimental investigation on waste heat recovery by refinery oil sludge incineration using fluidised-bed technique. J Environ Sci Heal A Toxic/hazard Subst Environ Eng 33:829–845. https://doi.org/10.1080/10934529809376764

Sar T, Harirchi S, Ramezani M et al (2022) Potential utilization of dairy industries by-products and wastes through microbial processes: a critical review. Sci Total Environ 810:152253. https://doi.org/10.1016/j.scitotenv.2021.152253

Sathwara NG, Patel KG, Vyas JB et al (2007) Chromium exposure study in chemical based industry. J Environ Biol 28:405–408

Scala F, Chirone R (2004) Fluidized bed combustion of alternative solid fuels. Exp Therm Fluid Sci 28:691–699. https://doi.org/10.1016/j.expthermflusci.2003.12.005

Scherhaufer S, Moates G, Hartikainen H et al (2018) Environmental impacts of food waste in Europe. Waste Manag 77:98–113. https://doi.org/10.1016/j.wasman.2018.04.038

Schmidt H, Kaminsky W (2001) Pyrolysis of oil sludge in a fluidised bed reactor. Chemosphere 45:285–290. https://doi.org/10.1016/S0045-6535(00)00542-7

Selvaraj S, Jeevan V, Rao Jonnalagadda R, Nishad Fathima N (2019) Conversion of tannery solid waste to sound absorbing nanofibrous materials: a road to sustainability. J Clean Prod 213:375–383. https://doi.org/10.1016/j.jclepro.2018.12.144

Sen B, Chandra TS (2007) Chemolytic and solid-state spectroscopic evaluation of organic matter transformation during vermicomposting of sugar industry wastes. Bioresour Technol 98:1680–1683. https://doi.org/10.1016/j.biortech.2006.06.007

Shaheen YA (2018) Utilization of microwave for industrial wastewater treatment

Sharma P, Gaur VK, Kim S-H, Pandey A (2020) Microbial strategies for bio-transforming food waste into resources. Bioresour Technol 299:122580. https://doi.org/10.1016/j.biortech.2019.122580

Shi W, Healy MG, Ashekuzzaman SM et al (2021) Dairy processing sludge and co-products: a review of present and future re-use pathways in agriculture. J Clean Prod 314:128035. https://doi.org/10.1016/j.jclepro.2021.128035

Singh D (2023) Advances in industrial waste management. Waste management and resource recycling in the developing world. Elsevier, Amsterdam, pp 385–416

Singh AK, Chandra R (2019) Pollutants released from the pulp paper industry: aquatic toxicity and their health hazards. Aquat Toxicol 211:202–216. https://doi.org/10.1016/j.aquatox.2019.04.007

Singh MK, Kesari KK, Behari J (2010) Ultrasonic irradiation of activated industrial sludge. Int J Environ Pollut 43:52–65

Singhal RL, Merali Z, Hrdina PD (1976) Aspects of the biochemical toxicology of cadmium. In: Federation proceedings, pp 75–80

Sivaram NM, Gopal PM, Barik D (2018) Toxic waste from textile industries. Elsevier, Amsterdam

Soliman NK, Moustafa AF (2020) Industrial solid waste for heavy metals adsorption features and challenges: a review. J Mater Res Technol 9:10235–10253. https://doi.org/10.1016/j.jmrt.2020.07.045

Song W, Li J, Zhang W et al (2012) An experimental study on the remediation of phenanthrene in soil using ultrasound and soil washing. Environ Earth Sci 66:1487–1496. https://doi.org/10.1007/s12665-011-1388-y

Soto-Hernández M, Tenango MP, García-Mateos R (eds) (2017) Phenolic compounds: natural sources, importance and applications. BoD–Books on Demand

Suleimanov RR, Gabbasova IM, Sitdikov RN (2005) Changes in the properties of oily gray forest soil during biological reclamation. Izv Akad Nauk Ser Biol 32:109–115

Sun W, Shahrajabian MH, Cheng Q (2021) Organic waste utilization and urban food waste composting strategies in china-a review. Not Sci Biol 13:1–16

Suthar S (2010) Pilot-scale vermireactors for sewage sludge stabilization and metal remediation process: comparison with small-scale vermireactors. Ecol Eng 36:703–712. https://doi.org/10.1016/j.ecoleng.2009.12.016

Taiwo EA, Otolorin JA (2009) Oil recovery from petroleum sludge by solvent extraction. Pet Sci Technol 27:836–844. https://doi.org/10.1080/10916460802455582

Tang X, Bai Y, Duong A et al (2009) Formaldehyde in China: production, consumption, exposure levels, and health effects. Environ Int 35:1210–1224. https://doi.org/10.1016/j.envint.2009.06.002

Tangri N, Wilson M (2017) Waste gasification and pyrolysis: high risk, low yield processes for waste management. Technol Risk Anal Rep

Tchounwou PB, Patlolla AK, Centeno JA (2003) Invited reviews: carcinogenic and systemic health effects associated with arsenic exposure—a critical review. Toxicol Pathol 31:575–588

Tchounwou PB, Centeno JA, Patlolla AK (2004) Arsenic toxicity, mutagenesis, and carcinogenesis: a health risk assessment and management approach. Mol Cell Biochem 255:47–55

Tchounwou PB, Yedjou CG, Patlolla AK, Sutton DJ (2012) Heavy metal toxicity and the environment. Mol Clin Environ Toxicol Environ Toxicol 3:133–164

Tedesco S, Montacchini E (2020) From textile waste to resource: a methodological approach of research and experimentation. Sustain 12:1–12. https://doi.org/10.3390/su122410667

Thasneema KK, Dipin T, Thayyil MS et al (2021) Removal of toxic heavy metals, phenolic compounds and textile dyes from industrial waste water using phosphonium based ionic liquids. J Mol Liq 323:114645. https://doi.org/10.1016/j.molliq.2020.114645

Tian Y, Chen L, Gao L et al (2012) Composting of waste paint sludge containing melamine resin and the compost's effect on vegetable growth and soil water quality. J Hazard Mater 243:28–36

UNEP UAP (1991) UNEP/GEMS environment library. Freshw Pollut 36:128

USEPA (1990) Solvent extraction treatment. Eng Bull 27:116–139

Ütebay B (2020) Textile wastes: status and perspectives. In: Çelik P (ed) IntechOpen, Rijeka

Vaishnavi J, Jabez Osborne W, Samuel J (2023) Chapter 11—Microorganism in waste valorization and its impact on the environment and economy. Samuel J, Kumar A, Singh JBT(eds). Elsevier, Amsterdam, pp 191–205

Vdovenko S, Boichenko S, Kochubei V (2015) Composition and properties of petroleum sludge produced at the refineries. Chem Chem Technol 9:257–260

Velma V, Vutukuru SS, Tchounwou PB (2009) Ecotoxicology of hexavalent chromium in freshwater fish: a critical review. Rev Environ Health 24:129–146

Vig AP, Singh J, Wani SH, Singh Dhaliwal S (2011) Vermicomposting of tannery sludge mixed with cattle dung into valuable manure using earthworm *Eisenia fetida* (Savigny). Bioresour Technol 102:7941–7945. https://doi.org/10.1016/j.biortech.2011.05.056

Waalkes MP, Misra RR, Chang LW (1996) Toxicology of metals. CRC Press, Boca Raton

Wang Y (2010) Fiber and textile waste utilization. Waste Biomass Valoriz 1:135–143. https://doi.org/10.1007/s12649-009-9005-y

Wang X-F, Xing M-L, Shen Y et al (2006) Oral administration of Cr(VI) induced oxidative stress, DNA damage and apoptotic cell death in mice. Toxicology 228:16–23

Wang LK, Chang S-Y, Hung Y-T et al (2007) Centrifugation clarification and thickening. Biosolids Treat Process 14:101–134

Wang LK, Hung Y-T, Li KH (2009) Vermicomposting process. Biological treatment processes. Springer, New York, pp 715–732

Wang X, Wang Q, Wang S et al (2012) Effect of biostimulation on community level physiological profiles of microorganisms in field scale biopiles composed of aged oil sludge. Bioresour Technol 111:308–315. https://doi.org/10.1016/j.biortech.2012.01.158

Wang Z, Wei W, Zheng F (2020) Effects of industrial air pollution on the technical efficiency of agricultural production: evidence from China. Environ Impact Assess Rev 83:106407. https://doi.org/10.1016/j.eiar.2020.106407

Wei L, Wang K, Noguera DR et al (2016) Transformation and speciation of typical heavy metals in soil aquifer treatment system during long time recharging with secondary effluent: depth distribution and combination. Chemosphere 165:100–109. https://doi.org/10.1016/j.chemosphere.2016.09.027

Wen Q, Li C, Cai Z et al (2011) Study on activated carbon derived from sewage sludge for adsorption of gaseous formaldehyde. Bioresour Technol 102:942–947. https://doi.org/10.1016/j.biortech.2010.09.042

Wilk CM (2004) Solidification/stabilization treatment and examples of use at port facilities. In: Ports 2004: port development in the changing world, pp 1–10

Xu N, Wang W, Han P, Lu X (2009) Effects of ultrasound on oily sludge deoiling. J Hazard Mater 171:914–917. https://doi.org/10.1016/j.jhazmat.2009.06.091

Yalcin-Enis I, Kucukali-Ozturk M, Sezgin H (2019) Risks and management of textile waste. Nanosci Biotechnol Environ Appl 319:29–53. https://doi.org/10.1007/978-3-319-97922-9_2

Yang L, Nakhla G, Bassi A (2005) Electro-kinetic dewatering of oily sludges. J Hazard Mater 125:130–140. https://doi.org/10.1016/j.jhazmat.2005.05.040

Yang Q, Li Z, Lu X et al (2018) A review of soil heavy metal pollution from industrial and agricultural regions in China: pollution and risk assessment. Sci Total Environ 642:690–700. https://doi.org/10.1016/j.scitotenv.2018.06.068

Yun Y (2018) Gasification for low-grade feedstock. BoD–Books on Demand

Zhang L, Sun X (2018) Influence of sugar beet pulp and paper waste as bulking agents on physical, chemical, and microbial properties during green waste composting. Bioresour Technol 267:182–191. https://doi.org/10.1016/j.biortech.2018.07.040

Zhang J, Li J, Thring RW et al (2012) Oil recovery from refinery oily sludge via ultrasound and freeze/thaw. J Hazard Mater 203–204:195–203. https://doi.org/10.1016/j.jhazmat.2011.12.016

Zhang X, Zhou M, Li J et al (2021) Analysis of driving factors on China's industrial solid waste generation: Insights from critical supply chains. Sci Total Environ 775:145185. https://doi.org/10.1016/j.scitotenv.2021.145185

Zhao Y, Yan X, Zhou J et al (2019) Treatment of oily sludge by two-stage wet air oxidation. J Energy Inst 92:1451–1457. https://doi.org/10.1016/j.joei.2018.08.006

Zhen G, Lu X, Kato H et al (2017) Overview of pretreatment strategies for enhancing sewage sludge disintegration and subsequent anaerobic digestion: current advances, full-scale application and future perspectives. Renew Sustain Energy Rev 69:559–577

Zhou L, Jiang X, Liu J (2009) Characteristics of oily sludge combustion in circulating fluidized beds. J Hazard Mater 170:175–179. https://doi.org/10.1016/j.jhazmat.2009.04.109

Chapter 10
Micro/Nanoplastics Pollution from Landfill Sites: A Comprehensive Review on the Formation, Distribution, Effects and Potential Mitigation

Md. Mostafizur Rahman, Mohammad Toha, and Sadia Sikder

Abstract In recent years, landfills have become well-known worldwide as sanctuaries for various pollutants. Due to various reasons, landfills receive a massive amount of plastic daily. Then, this plastic waste is degraded, making it one of the most significant sources of microplastics (MPs) in landfill environments. Later, this enormous amount of MPs is dispersed over the landfills and contaminates nearby leachate, soil, groundwater, and aquatic bodies, posing a severe ecotoxicological risk. Besides, MPs can carry antibiotic-resistance genes and hazardous pollutants due to their affinity for organic and inorganic particles. Thus, landfills can be hubs for emerging diseases and destroy human health. Nevertheless, despite the large MP production and its significant ecotoxicological effects, not much deep attention has been drawn to this issue. Furthermore, some technologies, such as biochemical, flocculation, sedimentation, membrane filtration, coagulation, and oxidation photo-electrolysis, have been used recently to remove MPs from landfill environments, which have less efficiency than some of the newly invented technologies. Therefore, this book chapter provides a significant idea regarding the formation, detection, and distribution of MPs in surrounding environments. This book chapter also highlights these MPs' associated environmental and health risks. In addition, this book chapter evaluates upgrading removal technologies of MPs from landfills. Overall, this book chapter focuses on the importance of adequately managing landfills, monitoring MP pollution, and implementing upgrading MP removal technologies. Furthermore, more research is needed regarding these issues for the betterment of Mother Nature.

Keywords Landfills · Groundwater · Leachate · Health effects · Soil

Md. M. Rahman (✉) · M. Toha · S. Sikder
Department of Environmental Science, Bangladesh University of Professionals, Mirpur, Cantonment, Dhaka 1216, Bangladesh
e-mail: rahmanmm@juniv.edu

Md. M. Rahman
Laboratory of Environmental Health and Ecotoxicology, Department of Environmental Sciences, Jahangirnagar University, Savar, Dhaka 1342, Bangladesh

A. Anouzla and S. Souabi (eds.), *Technical Landfills and Waste Management*, Springer Water, https://doi.org/10.1007/978-3-031-52633-6_10

10.1 Introduction

Globally, large amounts of plastics have emerged in our environment due to the development of various easy manufacturing processes, population growth, increasing prosperity, extensive use of MPs, and a lack of recycling since 1950, which is already an issue of emerging concern as this increasing trend of MPs has been ubiquitously found in all of our environmental compartments, has potential bioaccumulation ability, and negatively affects terrestrial and aquatic life. In the past sixty years, the yearly production of plastics climbed from 500,000 tonnes in 1960 to 367 million tonnes in 2020 and is anticipated to reach up to 53 million metric tonnes per year by 2030, with an estimated 60% of them being dumped into the marine environment (Myszka et al. 2023, Kitahara and Nakata 2020). Globally, the average per capita waste generation value is 0.74 kg/day, where high-income countries generate 51% of total waste, indicating a rapidly higher waste generation value than low- and middle-income countries (World Bank UN report 2022).

After that, the ultimate destination of this tremendous number of wastes is landfills. Generally, landfilling is a convenient way to dispose of municipal solid waste (MSW), where 21–42% of waste is plastic (Shen et al. 2022). In addition, it has been predicted that landfills will receive 12,000 million metric tons of plastic waste in 2050 (Bharath et al. 2022). So, this significant number of plastic waste produced and disposed of in landfills shows the dire situation for future environmental balance. The wastes in landfills constantly break through various biological, chemical, and physical processes and create more complex tiny pollutants that are < 5 mm and < 100 nm, respectively, giving birth to the issue of microplastics (MPs) and nano plastics (NPs) (Yu et al. 2022; Singh et al. 2023). Microplastics come from both primary (direct) and secondary (indirect) sources. The primary sources of microplastics are introduced directly into the environment as plastic beads, which are used as raw materials in the plastic manufacturing sector and personal care and hygiene products. In addition, microplastics in the form of fibres and small pieces can be released indirectly from secondary sources by breaking larger plastic particles caused by photocatalytic breakdown, oxidation, and mechanical weathering (Patchaiyappan et al. 2021).

However, due to the degradation process, liquid wastes are created from landfills known as landfill leachate (LL) (Kumar et al. 2021, 2023). Furthermore, various environmental factors, MSW composition, and landfill age are the major factors for Landfill leachate composition. Landfill leachates contain various types of MPs, heavy metals, refractory organic pollutants, harmful bacteria, and antibiotic-resistance genes that harm our ecosystem (Shen et al. 2022; Rahman et al. 2022). This landfill leachate pollutes the surrounding landfills, specifically sediments and groundwater, due to leaching and surface water through runoff (Wan et al. 2022). According to recent studies, various types and concentrations of MPs have already been identified in LL and its nearby area (Yu et al. 2022; Xu et al. 2020). Polyethylene terephthalate (PETE), low-density polyethylene (LDPE), high-density polyethylene (HDPE), polystyrene (PS or styrofoam), polyethylene (PE), polyvinyl, and chloride (PVC or vinyl) with various forms like films, pellets, foam, fragments, fibers, lines, and strands have already been detected in LL and groundwater where most of the MPs are white, black, yellow, red, and green in color (Praagh et al. 2018; Kilponen 2016; Sholokhova et al. 2023).

Landfills are the sacred places of various emerging pollutants, specifically MPs, antibiotics and heavy metals, with a tremendous ecotoxicological risk (Rahman et al. 2022). Landfills act as a source and sinks of MPs and are distributed throughout the landfill area by various processes. Despite the detrimental effects of landfill MPs and rapid dispersion rate, very few studies have been conducted regarding these issues. Most are ecotoxicological risk-related (Bharath et al. 2022; Rahman et al. 2022). The quantification levels of MPs, sources, pathways, distribution patterns, and removal of MPs from landfills are yet to be discussed comprehensively. Therefore, it is essential to comprehensively study MPs' occurrence, fate, and removal techniques from landfills.

Consequently, this study aims to discuss the formation of MPs in landfills, the detection of MPs, and their characteristics. This study also highlighted the distribution pathways of MPs in landfills and present removal techniques of MPs from landfills. Lastly, this study also aims to raise awareness among researchers globally about the need to do extensive studies on developing emerging micropollutants and their detrimental effects on mother nature.

10.2 Formation of Microplastics (MPs) in Landfills Environment

Landfills continuously receive a massive amount of plastic from various sources, making them one of the most significant sources of microplastic (MPs) production. From beginning to end, landfills go through the aerobic biodegradation stage, the conversion stage from aerobic to anaerobic condition, the anaerobic acidification stage, the methanogenesis stage, and the stable stage to generate MPs (1). In landfills, plastics are degraded by biological, physical, and chemical actions where long-chain polymers are converted into short-chain polymers under high-temperature conditions known as abiotic or thermal degradation (Yu et al. 2022; Rahman et al. 2022; Shi et al. 2022; Kumar et al. 2021). After that, these short-chain abiotic products are simultaneously degraded through several biotic processes (Shen et al. 2022). In these degradation processes, light, temperature, moisture, and some microbes significantly influence the conversion of macroplastics into microplastics rapidly (Yu et al. 2022). This formation process of MPs in landfill environments is shown in Fig. 10.1. Plastic fragmentation is created through the above factors, especially moisture, heat, and biological enzymes that can weaken and shorten the macroplastics (Bharath et al. 2022; Singh et al. 2023). Due to the increases in oxygen, water, and temperature availability, aerobic bacteria can get a favourable situation and increment the temperature from 35 to 70 °C (Shen et al. 2022). As a result, cracks and fractures started to be created in macroplastics, known as the oxidative degradation process.

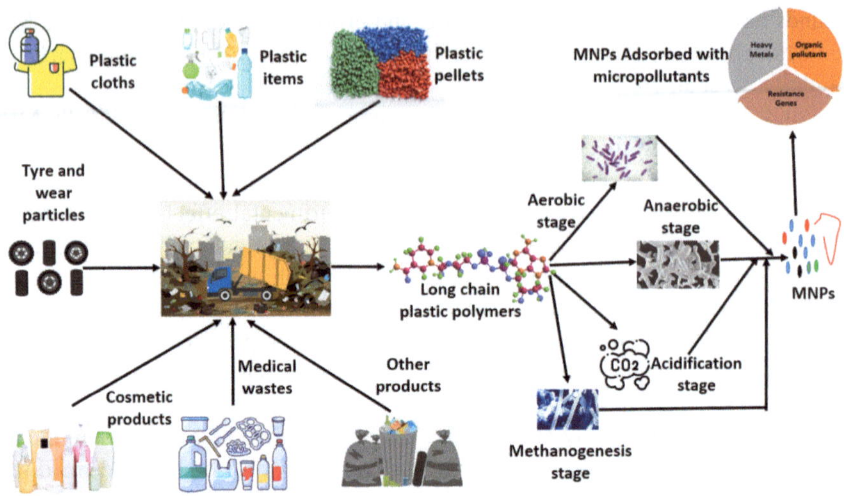

Fig. 10.1 Formation of microplastics (MPs) in landfill environments

Later, thermal and mechanical shear stress is added to it, creating small fragments of plastics, considered a pre-stage of microplastics (MPs) (Yu et al. 2022; Rahman et al. 2022). In addition, incineration plant ash, Cu, Fe, Cr, and other heavy metals have a great potential to accelerate thermal degradation (Shen et al. 2022).

One kg of incineration plant ash contains 171 MP particles (Singh et al. 2023; Shen et al. 2022). Photodegradation of macroplastics occurs due to the higher energy of UV-A and UV-B irradiation. However, different macroplastics contain different components of polymers and have various degradation processes. Polyethylene (PE) is not involved in the photodegradation process due to the absence of light-absorbing particles in their polymeric components, whereas Polyethylene terephthalate (PET) is capable of joining the photodegradation process (Singh et al. 2023; Bharath et al. 2022). The hydrolysis reaction of the transition and acidification stage can decay the fragments of plastic particles resulting from thermal and photodegradation. Finally, some microbes can further degrade and consume these plastic particles, creating microplastics or nanoplastics. Lastly, these MPs are moved into the landfill environment's surroundings, creating an unfavourable situation (Rahman et al. 2022). Thus, it is indispensable to identify the level of MP contamination and their ecotoxicological risk in landfill surroundings.

10.3 Distribution of Microplastics (MPs) in Landfill Environments

The environment of landfills is very complex and dynamic, with many pollutants. In landfills, MPs are distributed throughout the surrounding area due to surface runoff, atmospheric factors, and leaching (Rahman et al. 2022) (Fig. 10.2 and Table 10.1). Moreover, MPs can penetrate the soil pores and move into bedrock zones, making the groundwater contaminate. As a consequence, scientists have already found a large number of MPs from the different compartments of landfills. In Finland, the number of MPs was found to be 0.8–1.1 particles/L from the leachate of landfills, whereas the value was 4.5 particles/L for Iceland. The concentration of MPs was 4–13 particles/L in the landfills in Shanghai, China (Shen et al. 2022).

Furthermore, the abundance of MPs was 291 ± 91 particles/L in other cities in China (Xu et al. 2020). A recent study investigated that the concentration of MPs depends on the age of landfills, geographical area, environmental factors, and physical and chemical characteristics of wastes (Yu et al. 2022; Wan et al. 2022). In addition, the density of polymers significantly impacts the abundance of MPs on landfill leachate. MPs polymers PVC have a greater density (1.38 g/cm^3) than PET (1.20 g/cm^3). As a result, PVC settled down in the basin of landfills and showed lower concentration than PET.

In Bangladesh, numerous polymers (LDPE, HDPE, and CA) have already been found in the soil samples of landfills. According to a recent study, 570–14,200 particles/kg of MPs were found in the soil samples of landfills in South China. However, the presence of MPs in the groundwater of landfills has also been detected in various nations. In Mexico, the amount of MPs in groundwater samples was found to be

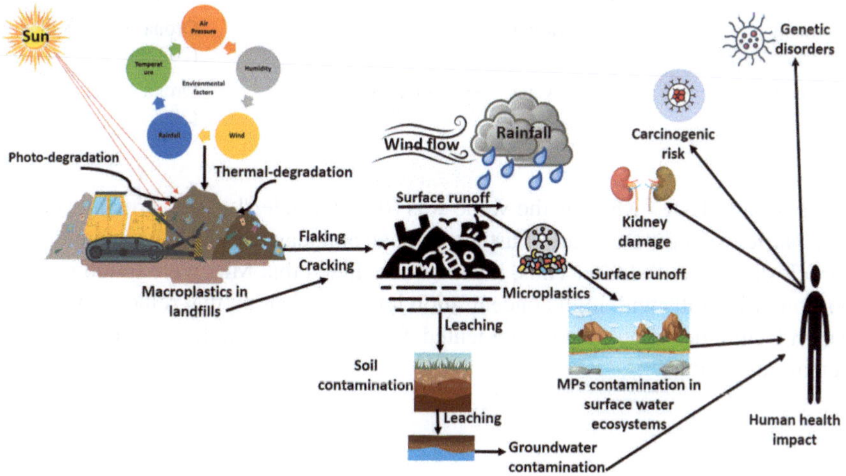

Fig. 10.2 Distribution of microplastics (MPs) in landfill environments

Table 10.1 Distribution of microplastics (MPs) in various landfill environments

Location	Sample type	Concentration of MPs	References
Bushehr port landfill, Iran	Leachate	63–92 (particles/L)	Mohammadi et al. (2022)
Helsinki, Finland	Leachate	0.80–1.10 (particles/L)	Kilponen (2016)
Mojokerto, Indonesia	Leachate	9 (particles/L)	Trihadiningrum et al. (2023)
Kaunas, Lituania	Leachate	55 (particles/g)	Sholokhova et al. (2023)
Guangdong Province, China	Leachate	3–25 (particles/L)	Wan et al. (2022)
Norway and Iceland	Leachate	0–4.50 (particles/L)	Praagh et al. (2018)
Gulf, Thailand	Leachate	13.50–27.50 (particles/Kg)	Puthcharoen and Leungprasert (2019)
Thailand	Leachate	8.80–9.93 (particles/L)	Nayahi et al. (2022)
Shanghai, China	Leachate	71.3–653 (items/g plastics)	Huang et al. (2022)
Istanbul, Turkey	Leachate	147–196 (particles/L)	Kara et al. (2022)
Suzhou, China	Leachate	235 (particles/L)	Sun et al. (2021)
Bosnia and Herzegovina	Leachate	0.3–2.2 mg/L	Narevski et al. (2021)
Novi Sad, Serbia	Leachate	0.64–2.16 mg/L	Narevski et al. (2021)
Shanghai, China	Leachate	291 (particles/L)	Xu et al. (2020)
Kahrizak landfill, Iran	Soil	863 ± 681 (particles/kg)	Shirazi et al. (2023)
South China	Soil	570–14,200 (particles/kg)	Wan et al. (2022)
South India	Groundwater	80 (particles/L)	Bharath et al. (2022)
Mexico	Groundwater	18 ± 7 (particles/L)	Kumar et al. (2023)
Germany	Groundwater	0.90 (particles/L)	Orona-Navar et al. (2022)
Denmark	Groundwater	0–0.80 (particles/L)	Orona-Navar et al. (2022)

(18 ± 7 particles/L) whereas the value was (0.70 particles/L) for Germany and (0–0.90 particles/L) for Denmark. Furthermore, an abundance of MPs was found (0.80 particles/L) in South India. Thus, it can be concluded that MPs were found in every portion of landfills with high concentration levels, indicating that removal and mitigation techniques should be implemented as soon as possible for the betterment of humans and nature.

10.4 Detection of Microplastics (MPs) in Landfill Environments

Despite the massive number of microplastics (MPs) already found in landfill environments, no standard methods for detecting MPs have yet to be discovered. So, it is very challenging to identify and quantify the contamination level of MPs in landfill environments. However, four steps: sampling, pretreatment, identification, and quantification are found for detecting the MPs in existing research (Shen et al. 2022). Autosamplers, containers, pumping, and trawling are usually used in sampling collection steps. Pretreatment is the most critical step, where several techniques and procedures are applied for detecting MPs (Myszka et al. 2023; Kitahara and Nakata 2020). In pretreatment steps, density separation separates the MPs from the collecting samples. In this process, several types of salt solutions, NaCl (1.2 g/cm^3), NaI (1.7 g/cm^3), $ZnCl_2$ (1.6 g/cm^3), and (K (HCOO)) are used based on polymer density. In addition, polypropylene (PP) and polyethene (PE) have a similar density of water (1 g/cm^3) (Kiran et al. 2022). Consequently, no salt solutions are required for these two polymers. On the other hand, most of the polymer's densities are in between (1.2–1.9 g/cm^3). Problematically, despite NaI and (K (HCOO)) having a higher recovery rate, these two salt solutions are costly and not eco-friendly. Salt solutions of NaCl and $ZnCl_2$ are widely recognized for their excellent recovery rate, inexpensive, and eco-friendly (Myszka et al. 2023). Thus, the scientific community preferred to use NaCl and $ZnCl_2$ solutions for density separation (Kiran et al. 2022). Another important pretreatment step is digestion. Based on the types of samples, oxidative, acid, alkali, and enzymatic digestion are usually used in pretreatment steps. After that, filter paper with 0.45 mm size and vacuum pump is used for filtration (Shen et al. 2022; Myszka et al. 2023). Then, the filter paper goes through the drying procedure. In identification and quantification procedures, this dried filter paper is used to identify MPs' physical and chemical characteristics (Kiran et al. 2022). Stereomicroscopes with various models are generally used to assess MPs' physical characteristics (colour, shape, size). On the other hand, Fourier Transform infrared spectroscopy (FTIR), Scanning electron microscopy (SEM), and energy dispersive X-ray spectroscopy (EDS) are used for identifying the chemical composition of MPs (Bharath et al. 2022).

10.5 Type of Microplastic Polymers in a Landfill Environment

Geographical location, waste composition, and seasonal variations significantly impact the types of polymers. According to numerous studies, various types of polymers were detected from different landfill environments. In China, PE, PP, PVC, PS, ABS, PET, PUR, EVA, PA, PMMA, PC and other polymers were detected in various landfills leachate (Wan et al. 2022; Huang et al. 2022; Sun et al. 2021). As china, various types of polymers were detected in Iran (PP, PS, PC, Nylon), Indonesia (PE, PP, PVC, PS, PET, PUR, PMMA), Thailand (PE, PP, PS, PC, PVC, ABS), Bangladesh (LDPE, HDPE, CA), and Finland (PE, PP, PVC, PS, PET) (Mohammadi et al. 2022; Trihadiningrum et al. 2023; Puthcharoen and Leungprasert 2019; Afrin et al. 2020; Orona-Navar et al. 2022). In addition, PE, PP, PS, and PET had the highest percentages, 80 and 70%, in China and Iran, respectively (Singh et al. 2023). Thus, since all types of microplastic polymers have been detected all over the regions, even where we all are aware of the harmful effects of these polymers, it is necessary to focus on proper waste management programs worldwide.

10.6 Environmental Impact and Health Risk of Microplastics (MPs)

At present, the impact of MPs on our environment and human health is being observed at extreme levels. As our environment and the development of human civilization are inextricably linked, it is our responsibility to be aware of the impact of MPs. As a result, the environmental impact and health risks of MPs have been discussed in this section.

10.6.1 Environmental Impact

Small pieces of plastic waste are accumulated in aquatic bodies in several ways, for example- landfill leachate vents and spillage from leachate ponds. Rain and wind also help to transfer MPs from landfill sources to nearby habitations (Rahman et al. 2022). As all water bodies have a final destination to sea, a significant course is created for the plastic rush from land to sea, including micro and mesoplastics. Contaminating the groundwater and being bioaccumulated in animal bodies, these plastics find an entryway to human life. The impacts are vital on every element of the environment (Kumar et al. 2023). The impact of MPs has already been found in Lycopersicum plants and birch tree roots. The concentration level of MPs in Chinese cabbage was 2.5–20 gm/kg soil (Yang et al. 2021). This amount of MPs can reduce crop production by increasing urease activity and reducing the chlorophyll concentration of plants (Bharath et al. 2022). Microplastics in soil can inhibit the movement of earthworms and microarthropods.

Moreover, these organisms can be highly contaminated by taking these MPs (Kiran et al. 2022). As a result, these MPs can quickly enter our food chain. The substantial contamination of MPs can alter the alkalinity and pH of the soil. After that, these MPs can damage bacterial activity and sometimes act as a carrier of antibiotic-resistance genes (Kumar et al. 2023; Kiran et al. 2022). In recent research, it has been found that the capability of MPs to carry various pollutants like heavy metals, antibiotic-resistant genes, and organic and inorganic pollutants was (6–70) times greater than the surrounding environment (Atugoda et al. 2021). Thus, it is imperative to extend more scientific work and awareness programs regarding these issues for the betterment of our globe, especially in developing nations, as they suffer the most due to the lack of appropriate mitigation techniques and strategies.

10.6.2 Health Risk

After the dispersion of MPs in various trophic levels, they can quickly enter our human body by inhalation, ingestion, and dermal exposure. Firstly, consuming contaminated water and seafood is the primary reason (Afrin et al. 2022). The plankton consumes microplastic particles, and later on, fish feed on these. Eventually, they end up being served on our dishes. Mussels are said to be especially inclined to microplastic contamination. Other seafood like shrimp, fish, and bottled water carry many MPs (Prata et al. 2020). Once ingested, they freely proceed to lymphatic tissues by means of translocation, redistribution, and retention (Xu et al. 2022). As it is proven that a significant quantity of nano-plastic is present in aquatic organisms, the potential toxicity of NPs in the food chain is also evident. Another notable entryway of MPs to human bodies is inhalation. Industrial operations, domestic activities, dried sludge, and degradation of plastic waste release MPs into the atmosphere (Wu et al. 2022; Xiang et al. 2022). As an outshot of inhaling these, respiratory diseases, cell toxicity and other auto-immune diseases might occur. Workers in the polymer industry are under severe exposure to occupational hazards, which might finally develop into cancer (Shi et al. 2022; Prata et al. 2020).

Nevertheless, more accurate research is yet to be done to get the right measure of potential threats of aerial inhalation of MPs by humans, seeing that lab experiments on animals do not render copper-bottomed data owing to differences in species, body functioning, exposer trackway, etc. MPs can also enter through dermal exposure. Cosmetics, scrubs, and washes may contain MPs, and when they come in contact with skin, these products carry the potential to transfer MPs to human bodies (Rahman et al. 2022). Urea, hydroxyl acids, and glycerol are essential components of body lotion, which tweak the nanoparticle's potentiality to perforate the skin barrier (Xu et al. 2022). PS and PE particles in human colon fibroblast cells range from nearly 50–500 μm if ingested through passive translocation (Wu et al. 2022; Shi et al. 2022). Acute oral exposure to positively charged PS nanoparticles changes iron transmission in the intestine and cellular intake. MNPs of measurement < 2.5 μm, on getting into the gastrointestinal tract utilizing presorption, lead to inflammation precipitated through the inert nature of the MNPs (Kumar et al. 2023). Inhaled MNPs can show various respiratory disorders, such as granulomas with fibre inclusions (chronic pneumonia, extrinsic allergic alveolitis), diffused interstitial fibrosis, and immediate bronchial (asthma-like) reactions. Artificial small particles have bio-persistence and toxicity that may create lung cancers in the human body (Xiang et al. 2022; Bharath et al. 2022). In the bargain, the toxicity of small-sized PS NPs was stated to cause low cellular feasibility, cell cycle arrest during the S phase, stimulated transcription of inflammatory genes, modification in protein expression retrieving to the cell cycle, and pro-apoptosis. Aging, polymerization, treating, and UV irradiations in sum end in the discharge of free radicals of MNPs, initiating oxidation of target tissues and the cytotoxic effect of MNPs were confirmed by in vitro human brain and epithelial cells (Natesan et al. 2021). It was also emphasized that IBD's harshness was proportionate to MNP exposure, which may have helped in IBD severity or vice versa. Drinking water exposure to 3 μm PS MNP particles extends to 4 g yearly for humans. Likewise, just about 66.4% of the water used for human consumption in France originates from groundwater sources, compared to 50% in the United States (Harvey and Watts 2018). Hence, groundwater MNPs are anticipated to be exposed to humans. Studies say bottled water exposes Americans to 90,000 particles (Harvey and Watts 2018). Studies found that microplastics can pass in the fetus through the placenta, and children may perhaps swallow the particles via breast milk as well (Thornton Hampton et al. 2022). Plastic feeding bottles and teething toys increase children's microplastic exposure (Issac and Kandasubramanian 2021).

Lastly, the distribution of MPs has already been found in every trophic level in ecosystems that have the potential to change the biogeochemical cycle and make our mother nature more complex. Furthermore, this interaction of MPs in ecosystems also has a detrimental impact on human health. Despite several consequences on human health and ecosystems, a large number of comprehensive studies are yet to be discovered. Thus, exploring more comprehensively the sources, fates, and impacts of MPs on our environment is essential.

10.7 Removal and Mitigation Strategies of Microplastics (MPs)

If not properly managed, the microplastics present in landfill leachate pose a significant environmental threat. This is due to the potential release of microplastics and harmful substances into the ecosystem, endangering local ecology. Various methods are available for treating landfill leachate, such as batch-activated sludge, membrane biological reaction, flocculation, combined filtration (ultrafiltration, nanofiltration, and reverse osmosis), and constructed wetlands (Shen et al. 2022). However, the effectiveness of these techniques in removing microplastics from landfill leachate has yet to be thoroughly investigated. Two promising approaches for microplastic removal are physical membrane filtration and membrane bioreactors. These methods not only minimize excess sludge generation but also show potential in efficiently eliminating microplastics. Another effective method involves advanced oxidation processes. Utilizing compounds like BiOCl-X, these processes generate hydroxyl radicals that rapidly degrade microplastics, suggesting their viability for photocatalytic degradation. While leachate treatment processes can address microplastics

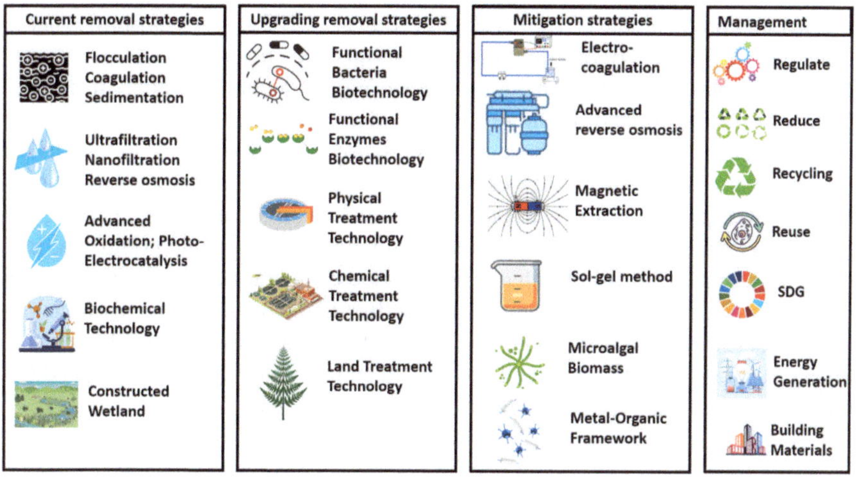

Fig. 10.3 Removal, mitigation, and management strategies of microplastics (MPs)

and nanoplastics, it is crucial to assess their potential impact on other pollutants (Shen et al. 2022; Kumar et al. 2023). Despite the available techniques, research on microplastics and nanoplastics removal in current leachate treatment methods still needs to be completed. Thus, it is indispensable to promote and implement the upgrading removal strategies. Removal, mitigation, and management strategies of microplastics (MPs) are shown in Fig. 10.3.

10.7.1 Biotechnology

Certain microorganisms, including bacteria and fungi, can convert hydrocarbons within plastic waste into carbon dioxide (CO_2), water (H_2O), and biomass. They can also break down larger carbon chains into acetic acid, which can be further converted into methane by methanogens. The digestive bacteria in mealworms play a crucial role in breaking down and mineralizing polystyrene (PS) plastics (Shen et al. 2022). Many bacterial genera can utilize monomer styrene as a carbon source, leading to the degradation of styrene into tricarboxylic acid cycle precursors (Kabir et al. 2023; Singh et al. 2023). Research involving Ideonella sakaiensis has shown PETase, an enzyme produced by this bacterium, effectively degrades PET plastics (Shen et al. 2022). The degradation process requires the presence of oxygen. In natural settings, certain fungi actively facilitate the biodegradation of polyethene (PE) microplastics. Leachate generated from household waste landfills contains complex components, complicating the biodegradation of plastics and microplastics. While numerous studies have demonstrated microbial degradation of plastics, the process and rate are influenced by factors like plastic particle size, type of plastic, microbial species, and initial biomass (Hou et al. 2021; Sun et al. 2021). Various plastics necessitate the involvement of different microorganisms and enzymes. Biofilm formation is pivotal in plastic degradation, and strategies employing biotechnology to enhance biofilm formation might prove more effective. Microplastics in landfill leachate might pose greater ecotoxicity than those in other environments due to exposure to diverse leachate components, including persistent, bioaccumulative, and toxic substances.

Moreover, plastic additives could harm microbial communities (Silva et al. 2021; Shen et al. 2022). The intricate mixture of microplastics, organic pollutants, inorganic ions, and biofilms in leachate increases the challenges of employing biotechnology for microplastics treatment, given the fluctuating treatment processes, potential microorganism toxicity, and enzyme inhibition (Shen et al. 2022; Sharma et al. 2023). Furthermore, the complexity of implementing biotechnology at scale adds to the operational challenges of leachate treatment equipment.

10.7.2 Physical and Chemical Treatment Technology

The issue of landfill leachate leakage is significant, and the composition of landfill leachate is intricate and subject to considerable variation. While physical and chemical techniques can address microplastics in landfill leachate, information on their effective removal remains limited. Following biological treatment, advanced steps are necessary for leachate treatment, including ultrafiltration, nanofiltration, and reverse osmosis (Sun et al. 2021; Kumar et al. 2023). A combined approach involving membrane bioreactors, nanofiltration, and reverse osmosis can treat landfill leachate, ensuring compliance with discharge standards (Hou et al. 2021). In a study conducted by Wang et al. (2020), it was observed that biochar filters exhibited a removal efficiency of over 95% for microplastics with a ten μm particle size.

Furthermore, modified materials achieved removal rates exceeding 96% for plastics like PE and PA (Shen et al. 2022). Emerging water treatment methods such as photocatalysis, advanced oxidation, and electrochemical catalysis have recently gained attention. However, the intense colouration of landfill leachate could impede the degradation and oxidation of microplastics. In such cases, electrocoagulation has proven to be a practical approach for removing pollutants from leachate.

10.7.3 Land Treatment Technology

The constructed wetland technique efficiently captures and breaks down contaminants present in water using a combination of vegetation and microorganisms, ultimately achieving the desired standard for landfill leachate discharge (Shen et al. 2022). Aquatic vascular plants play a significant role in gathering, stabilizing, and capturing plastics, microplastics, and nanoplastics within the surrounding environment. Upon introduction into aquatic settings, microplastics tend to float on the water's surface and interact with floating vegetation (Shen et al. 2022; Singh et al. 2023). The ability of nanoplastics to be taken up by vascular plants hinges on their size, and the process of transpiration plays a pivotal role in disseminating nanoplastics. Plant stabilization stands as a bioremediation method that harnesses the capabilities of plant roots and leaves to stabilize, break down, and bind pollutants. This approach has successfully removed diverse pollutants, including dissolved contaminants and engineered nanoparticles (Shen et al. 2022).

10.7.4 Mitigation Approaches

Various mitigation approaches have already been invented to control landfill leachate in this modern era. Thus, some mitigation approaches have been discussed in this section.

1. Membrane filtration methods, including nanofiltration, ultrafiltration, and microfiltration, effectively eliminate MNPs. Nanofiltration (NF), characterized by 1–10 nm pore sizes, has been successfully used alongside reverse osmosis, driven by solar power, to treat groundwater (Lofty et al. 2022). This combination achieved a remarkable removal efficiency of 95–99% in batch operations. Membrane fouling remains a significant concern, where particles interacting with the membrane can hinder pore function and reduce removal efficiency (Gandhi et al. 2022; Shaik et al. 2022).
2. Reverse osmosis is a water purification method used in treatment plants to filter out impurities, contaminants, heavy metals, and salts (Mengesha and Sahu 2022). Conventional reverse osmosis techniques have some fouling and capacity-related limitations. A promising technique called RO-UF (reverse osmosis combined with ultrafiltration) has emerged to address this. This approach boasts high separation efficiency and employs external power to separate desired elements from liquid samples (Khanzada et al. 2022). Notably, a pilot study by Loganathan et al. successfully demonstrated the feasibility of RO-UF, combined with zero-liquid discharge crystallization, for treating deep groundwater aquifers.
3. Coagulation, valued for its cost-effectiveness and high efficiency, employs iron or aluminium-based salts. Coagulants create flocs by encapsulating particles (Hanif et al. 2021; Kumar et al. 2023). Aluminium-based methods are preferred due to superior removal efficiency and fewer minor blockages, increasing surface area (Govindaraj et al. 2022). Despite its effectiveness in neutralizing the positive zeta potential charge on MNPs, coagulation has not systematically achieved the separation of particles smaller than 1 μm. As this process is a bit costly and can create a massive amount of residual sludge, some bio-coagulants have already gained attention—for landfill leachate treatment (Hanif et al. 2021). They have demonstrated notable reductions in pollutants from liquid samples and present a cost-effective alternative approach.

4. Electrocoagulation is favoured for its safety and effective removal of MNPs by generating coagulants electrically. This process involves applying an electric field to create micro-coagulants through electron generation at the anode. Electrode polarity reversal prevents oxide layer buildup, maintaining separation efficiency. pH plays a crucial role, as neutral pH promotes coagulant production, facilitating the formation of aluminium hydroxide to separate MNPs. The operational cost relies on current density. Coagulant presence destabilizes pollutants, leading to micro-floc formation and contaminant removal. An up-flow EC reactor achieved significant efficiency (84–100%) in removing pollutants (Rahat et al. 2021).
5. The density separation method segregates MNPs based on density, allowing heavier particles to settle while lighter ones remain suspended and are separated from the solution. Salt solutions are preferred for their affordability, effective particle recovery, and minimal environmental impact (Zhao et al. 2022). Sodium chloride (NaCl) has drawbacks like a lower recovery rate (90%) and increased vulnerability to errors. Conversely, sodium iodide (density: 1.6 g/cm^3) offers a 99% recovery capacity for separating high molecular weight MNPs (Grbic et al. 2019).
6. Magnetic extraction is a technique employed for MNP removal from water sources. This method utilizes external magnetic fields, magnetic seeds, and acid to improve separation. These techniques' efficiency in removing pollutants is (up to 98.3%) (Kumar et al. 2023). Notably, the MNPs were rendered hydrophobic by coating them with Cetyl trimethoxy silane before extraction (Kumar et al. 2023). The obtained data highlights the preference for magnetic extraction in removing small MNPs due to its superior removal efficiency.
7. Various factors such as chemical composition, surface area, sorption capacity, and hydrophilicity influence the physicochemical properties of nanomaterials. To address the challenges associated with recovery rate and toxicity risks, bio-based nanomaterials like chitosan-based aerogels and plant-derived sponges are extensively explored for microplastic (MP) remediation in aqueous media (Goh et al. 2022). Nano zero-valent iron (nZVI) is effective for groundwater remediation, though agglomeration and hydroxide corrosion can diminish efficiency. In-situ groundwater remediation was achieved using biodegradable Janus particles alongside ZVI nanoparticles. Applying nano-Fe_3O_4 demonstrated a removal rate exceeding 80% (Li et al. 2022; Kumar et al. 2023).
8. Biochar, Sol–gel, and Metal–Organic frameworks methods could also be a part of possible mitigation approaches (Abuwatfa et al. 2021; Singh et al. 2021; Kumar et al. 2023; Jagodzinska et al. 2022; Chen et al. 2020). In addition, melamine-based foam is a stable substrate to load zirconium MOFs, achieving over 95% MNP removal efficiency with enhanced recyclability (Jagodzinska et al. 2022).
9. Microalgae serve various purposes, such as biodiesel production, bioremediation, and nutritional supplements. They effectively interact with pollutants, altering aggregate buoyancy and causing differential sedimentation rates, aiding in MNP separation. Anionic polysaccharides on the algal surface contribute to

MNP removal through electrostatic interaction with positively charged MNPs (Kumar et al. 2023). This technique offers advantages without chemical, electrical, or mechanical processes. However, efficiency may vary based on the algal surface's physiological or topographical characteristics.

10. Advanced oxidation processes (AOP) involving nanohybrids and complex operational requirements enable MNP removal via oxidizing radicals (Kang et al. 2019). Moreover, peroxymonosulfates (PMS) with cobalt, iron, and manganese activators generate reactive oxygen species like sulfate radicals. Magnetic carbon nanotubes (M-CNTs) have emerged recently, employing magnetic force for MNPs (PE, PA, PET) removal and enabling reuse through thermal treatment without compromising their magnetic or separation efficiency (Rajendran et al. 2022).

Lastly, these upgrading removal techniques and mitigation approaches could be potential strategies for removing MPs from landfill leachates. However, every pollutant is dynamic, and their patterns change continuously due to various factors. So, researchers need to focus on new inventions daily to keep pace with the dynamic pattern change of these pollutants.

10.8 Conclusions

In this modern era, plastics are becoming one of the most essential materials in our daily lives. As a result, the world is witnessing excessive plastic production, and the ultimate destination of these plastics is landfills. Due to the limited capacity and mismanagement of these plastics in landfills, various plastics with large numbers are created in landfills through thermal and photodegradation processes. However, these MPs can carry various toxic elements and disperse throughout the landfill's surrounding area. However, microplastic (MP) pollution in landfills has yet to be discovered correctly. Various removal and mitigation strategies are present to eradicate MPs from landfill environments. Problematically, landfill environments vary from country to country as they depend greatly on environmental factors. So, exploring the behavior, formation, and interaction mechanisms of MPs in landfill environments is indispensable to implementing proper removal and mitigation strategies.

On the other hand, landfills and their leachate tend to change their behavior with time. Consequently, standardized methods are yet to be discovered to correctly identify and quantify the MPs in landfill environments. So, standardized methods should be explored in future studies for detecting the characterization of MPs. The negative impacts of this MP's pollution should be appropriately addressed, and policymakers are responsible for adopting strict rules and regulations to reduce the detrimental impacts of this MP's pollution. Besides, the movement of MP pollution from lower to higher trophic levels and the potential co-contamination of MPs must be addressed.

Lastly, due to a lack of proper treatment, developing nations face more severe problems than developed nations. Thus, cost-effective hybrid MP reduction technologies should be implemented for the betterment of our planet.

References

Abuwatfa WH, Al-Muqbel D, Al-Othman A, Halalsheh N, Tawalbeh M (2021) Insights into the removal of microplastics from water using biochar in the era of COVID-19: a mini review case studies in chemical and environmental. Engineering 4:100151. https://doi.org/10.1016/jcscee 2021100151

Afrin S, Uddin MK, Rahman MM (2020) Microplastics contamination in the soil from Urban Landfill site, Dhaka, Bangaladesh. Heliyon 6:11

Afrin S, Rahman MM, Hossain MN, Uddin MK, Malafaia G (2022) Are there plastic particles in my sugar? A pioneering study on the characterization of microplastics in commercial sugars and risk assessment. Sci Total Environ 837:155849. https://doi.org/10.1016/j.scitotenv.2022.155849

Alam R, Sheob M, Saeed B, Khan SU, Shirinkar M, Frontistis Z, Basheer F, Farooqi IH (2021) Use of electrocoagulation for treatment of pharmaceutical compounds in water/wastewater: a review exploring opportunities and challenges. Water 13(15):2105. https://doi.org/10.3390/w13 152105

Atugoda T, Vithanage M, Wijesekara H, Bolan N, Sarmah A, Bank M, You S, Ok Y (2021) Interactions between microplastics, pharmaceuticals and personal care products: implications for vector transport. Environ Int 149:6367. https://doi.org/10.1016/j.envint.2020.106367

Bharath KM, Muthulakshmi AL, Usha N (2022) Microplastic contamination around the landfills: distribution, characterization and threats—a review. Curr Opin Environ Sci Health 31:100422

Chen YJ, Chen Y, Miao C, Wang YR, Gao GK, Yang RX, Zhu HJ, Wang JH, Li SL, Lan YQ (2020) Metal–organic framework-based foams for efficient microplastics removal. J Mater Chem A 8(29):14644–14652

Gandhi TP, Sampath PV, Maliyekkal SM (2022) A critical review of uranium contamination in groundwater: treatment and sludge disposal. Sci Total Environ 825:153947. https://doi.org/10.1016/j.scitotenv.2022.153947

GodvinSharmila V, Shanmugavel SP, Tyagi VK, Banu JR (2023) Microplastics as emergent contaminants in landfill leachate: source, potential impact and remediation technologies. J Environ Manag 343:118240

Goh PS, Kang HS, Ismail AF, Khor WH, Quen LK, Higgins D (2022) Nanomaterials for microplastic remediation from aquatic environment: why nano matters? Chemosphere 299:134418. https://doi.org/10.1016/j.chemosphere.2022.134418

Govindaraj V, Manoharan K, Govindarajan M (2022) Cost-effective treatment for polluted groundwater using eco-friendly bio-coagulants

Grbic J, Nguyen B, Guo E, You JB, Sinton D, Rochman CM (2019) Magnetic extraction of microplastics from environmental samples. Environ Sci Tech Lett 6(2):68–72. https://doi.org/10.1021/acsestlett8b00671

Hanif MA, Ibrahim N, Dahalan FA, Ali UFM, Hasan M, Jalil AA (2021) Microplastics and nanoplastics: recent literature studies and patents on their removal from aqueous environment. Sci Total Environ 115:152115. https://doi.org/10.1016/jscitotenv2021152115

Harvey F, Watts J (2018) Microplastics found in human stools for the first time. Guardian 22:10082

Hou L, Kumar D, Yoo CG, Gitsov I, Majumder ELW (2021) Conversion and removal strategies for microplastics in wastewater treatment plants and landfills. Chem Eng J 406:126715

Huang Q, Cheng Z, Yang C, Wang H, Zhu N, Cao X, Lou Z (2022) Booming microplastics generation in landfill: an exponential evolution process under temporal pattern. Water Res 223:119035

Issac MN, Kandasubramanian B (2021) Effect of microplastics in water and aquatic systems. Environ Sci Pollut Res 28(16):19544–19562. https://doi.org/10.1007/s11356-021-13184-2

Jagodzinska K, Jonsson PG, Yang W (2022) Pyrolysis and in-line catalytic decomposition of excavated landfill waste to produce carbon nanotubes and hydrogen over Fe-and Ni-based catalysts—investigation of the catalyst type and process temperature. Chem Eng J 446:136808. https://doi.org/10.1016/jcej2022136808

Kabir MS, Wang H, Luster-Teasley S, Zhang L, Zhao R (2023) Microplastics in landfill leachate: sources, detection, occurrence, and removal. Environ Sci Ecotechnol 12:100256

Kale SK, Deshmukh AG, Dudhare MS, Patil VB (2015) Microbial degradation of plastic: a review. J Biochem Technol 6(2):952–961

Kang J, Zhou L, Duan X, Sun H, Ao Z, Wang S (2019) Degradation of cosmetic microplastics via functionalized carbon nanosprings. Matter 1(3):745–758

Kara N, Sari Erkan H, Onkal Engin G (2022) Characterization and removal of microplastics in landfill leachate treatment plants in Istanbul, Turkey. Anal Lett 14:1–14

Khanzada NK, Jassby D, An AK (2022) Conductive reverse osmosis membrane for electrochemical chlorine reduction and sustainable brackish water treatment. Chem Eng J 435:134858. https://doi.org/10.1016/jcej2022134858

Kilponen J (2016) Microplastics and harmful substances in urban runoffs and landfill leachates: possible emission sources to marine environment. Bachelor Thesis. Lahti University of Applied Sciences

Kiran BR, Kopperi H, Venkata Mohan S (2022) Micro/nano-plastics occurrence, identification, risk analysis and mitigation: challenges and perspectives. Rev Environ Sci Bio/technol 21(1):169–203

Kitahara KI, Nakata H (2020) Plastic additives as tracers of microplastic sources in Japanese road dusts. Sci Total Environ 736:139694

Kumar M, Chen H, Sarsaiya S, Qin S, Liu H, Awasthi MK, Taherzadeh MJ (2021) Current research trends on micro-and nano-plastics as an emerging threat to global environment: a review. J Hazard Mater 409:124967

Kumar V, Singh E, Singh S, Pandey A, Bhargava PC (2023) Micro-and nano-plastics (MNPs) as emerging pollutant in ground water: environmental impact, potential risks, limitations and way forward towards sustainable management. Chem Eng J 459:141568

Li G, Li J, Zhang S, Hou X, Liu X, Yu Q, Li M (2022) In-situ growing of metal-organic frameworks on iron mesh as a recyclable remediation material for removing hexavalent chromium from groundwater. Chemosphere 303:135187. https://doi.org/10.1016/j.chemosphere.2022.135187

Lofty J, Muhawenimana V, Wilson CAME, Ouro P (2022) Microplastics removal from a primary settler tank in a wastewater treatment plant and estimations of contamination onto European agricultural land via sewage sludge recycling. Environ Pollut 304:119198. https://doi.org/10.1016/jenvpol2022119198

Loganathan K, Chelme-Ayala P, El-Din MG (2016) Pilot-scale study on the treatment of basal aquifer water using ultrafiltration, reverse osmosis and evaporation/crystallization to achieve zero-liquid discharge. J Environ Manag 165:213–223. https://doi.org/10.1016/j.jenvman.2015.09.019

Mengesha A, Sahu O (2022) Sustainability of membrane separation technology on groundwater reverse osmosisprocess. Clean Eng Technol 12:100457. https://doi.org/10.1016/jclet2022100457

Mohammadi A, Malakootian M, Dobaradaran S, Hashemi M, Jaafarzadeh N (2022) Occurrence, seasonal distribution, and ecological risk assessment of microplastics and phthalate esters in leachates of a landfill site located near the marine environment: Bushehr port, Iran as a case. Sci Total Environ 842:156838

Monira S, Bhuiyan MA, Haque N, Shah K, Roychand R, Hai FI, Pramanik BK (2021) Understanding the fate and control of road dust-associated microplastics in stormwater. Process Saf Environ Protect 152:47–57

Myszka R, Enfrin M, Giustozzi F (2023) Microplastics in road dust: a practical guide for identification and characterisation. Chemosphere 315:137757

Narevski AC, Novakovic MI, Petrovic MZ, Mihajlovic IJ, Maodus NB, Vujic GV (2021) Occurrence of bisphenol A and microplastics in landfill leachate: lessons from South East Europe. Environ Sci Pollut Res 28:42196–42203

Natesan U, Vaikunth R, Kumar P, Ruthra R, Srinivasalu S (2021) Spatial distribution of microplastic concentration around landfill sites and its potential risk on groundwater. Chemosphere 277:130263

Nayahi NT, Ou B, Liu Y, Janjaroen D (2022) Municipal solid waste sanitary and open landfills: contrasting sources of microplastics and its fate in their respective treatment systems. J Clean Prod 380:135095

Orona-Navar C, García-Morales R, Loge FJ, Mahlknecht J, AguilarHernandez I, Ornelas-Soto N (2022) Microplastics in Latin America and the Caribbean: a review on current status and perspectives. J Environ Manag 309:114698

Patchaiyappan A, Dowarah K, Ahmed SZ, Prabakaran M, Jayakumar S, Thirunavukkarasu C, Devipriya SP (2021) Prevalence and characteristics of microplastics present in the street dust collected from Chennai metropolitan city. India. Chemosphere 269:128757

Praagh MV, Hartman C, Brandmyr E (2018) Microplastics in landfill leachates in the nordic countries

Prata JC, da Costa JP, Lopes I, Duarte AC, Rocha-Santos T (2020) Environmental exposure to microplastics: an overview on possible human health effects. Sci Total Environ 702:134455. https://doi.org/10.1016/j.scitotenv.2019.134455

Puthcharoen A, Leungprasert S (2019) Determination of microplastics in soil and leachate from the landfills. Thai Environ Eng J 33(3):39–46

Rahat MHH, Massarra DC, Wang DG (2022) Using plastic wastes in construction: opportunities and challenges. EPiC Series in Built Environ 3:785–794

Rahman MM, Sultan MB, Alam M (2022) Microplastics and adsorbed micropollutants as emerging contaminants in landfill: a mini-review. Curr Opin Environ Sci Health 31:100420

Rajendran S, Priya AK, Kumar PS, Hoang TK, Sekar K, Chong KY, Khoo KS, Ng HS, Show PL (2022) A critical and recent developments on adsorption technique for removal of heavy metals from wastewater: a review. Chemosphere 139:135146. https://doi.org/10.1016/j.chemosphere.2022.135146

Rubio L, Marcos R, Hernández A (2020) Potential adverse health effects of ingested micro-and nanoplastics on humans. Lessons learned from in vivo and in vitro mammalian models. J Toxicol Environ Health B 23(2):51–68

Shaik F, Mohammed N, Rao LN (2022) Treatment technologies for the removal of micro plastics from aqueous medium. In: AIP conference proceedings. AIP Publishing LLC, p 030009

Sharma S, Sharma V, Chatterjee S (2023) Contribution of plastic and microplastic to global climate change and their conjoining impacts on the environment-A review. Sci Total Environ 875:162627

Shen M, Song B, Zhu Y, Zeng G, Zhang Y, Yang Y, Wen X, Chen M, Yi H (2020) Removal of microplastics via drinking water treatment: current knowledge and future directions. Chemosphere 251:126612

Shen M, Xiong W, Song B, Zhou C, Almatrafi E, Zeng G, Zhang Y (2022) Microplastics in landfill and leachate: occurrence, environmental behavior and removal strategies. Chemosphere 305:135325

Shi X, Chen Z, Wu L, Wei W, Ni BJ (2022) Microplastics in municipal solid waste landfills: detection, formation and potential environmental risks. Curr Opin Environ Sci Health 12:100433

Shirazi MMGF, Shekoohiyan S, Moussavi G, Heidari M (2023) Microplastics and mesoplastics as emerging contaminants in Tehran landfill soils: the distribution and induced-ecological risk. Environ Pollut 324:121368

Sholokhova A, Denafas G, Ceponkus J, Omelianenko T (2023) Microplastics in landfill bodies: abundance, spatial distribution and effect of landfill age. Sustainability 15(6):5017

Silva AL, Prata JC, Duarte AC, Soares AM, Barceló D, Rocha-Santos T (2021) Microplastics in landfill leachates: the need for reconnaissance studies and remediation technologies. Case Stud Chem Environ Eng 3:100072

Singh N, Khandelwal N, Ganie ZA, Tiwari E, Darbha GK (2021) Eco-friendly magnetic biochar: an effective trap for nanoplastics of varying surface functionality and size in the aqueous environment. Chem Eng J 418:129405. https://doi.org/10.1016/jcej2021129405

Singh S, Malyan SK, Maithani C, Kashyap S, Tyagi VK, Singh R, Pandey RP (2023) Microplastics in landfill leachate: occurrence, health concerns, and removal strategies. J Environ Manag 342:118220

Sun J, Zhu ZR, Li WH, Yan X, Wang LK, Zhang L, Jin J, Dai X, Ni BJ (2021) Revisiting microplastics in landfill leachate: unnoticed tiny microplastics and their fate in treatment works. Water Res 190:116784

Thornton Hampton LM, Lowman H, Coffin S, Darin E, De Frond H, Hermabessiere L, Miller E, de Ruijter VN, Faltynkova A, Kotar S, Monclús L (2022) A living tool for the continued exploration of microplastic toxicity. Microplast Nanoplast 2(1):1–11. https://doi.org/10.1186/s43591-022-00032-4

Trihadiningrum Y, Wilujeng SA, Tafaqury R, Radita DR, Radityaningrum AD (2023) Evidence of microplastics in leachate of Randegan landfill, Mojokerto City, Indonesia, and its potential to pollute surface water. Sci Total Environ 874:162207

UN report (2022) United nations report on World populations prospects. https://population.un.org/wpp/

Wan Y, Chen X, Liu Q, Hu H, Wu C, Xue Q (2022) Informal landfill contributes to the pollution of microplastics in the surrounding environment. Environ Pollut 293:118586

Wang Z, Sedighi M, Lea-Langton A (2020) Filtration of microplastic spheres by biochar: removal efficiency and immobilisation mechanisms, Water Res 184:116165. https://doi.org/10.1016/jwatres2020116165

Wu P, Lin S, Cao G, Wu J, Jin H, Wang C, Wong MH, Yang Z, Cai Z (2022) Absorption, distribution, metabolism, excretion and toxicity of microplastics in the human body and health implications. J Hazard Mater 38:129361

Xiang Y, Jiang L, Zhou Y, Luo Z, Zhi D, Yang J, Lam SS (2022) Microplastics and environmental pollutants: key interaction and toxicology in aquatic and soil environments. J Hazard Mater 422:126843. https://doi.org/10.1016/j.jhazmat.2021.126843

Xu Z, Sui Q, Li A, Sun M, Zhang L, Lyu S, Zhao W (2020) How to detect small microplastics (20–100 μm) in freshwater, municipal wastewaters and landfill leachates? A trial from sampling to identification. Sci Total Environ 733:139218

Xu JL, Lin X, Wang JJ, Gowen AA (2022) A review of potential human health impacts of micro- and nanoplastics exposure. Sci Total Environ 14:158111. https://doi.org/10.1016/j.scitotenv.2022.158111

Yang M, Huang DY, Tian YB, Zhu QH, Zhang Q, Zhu HH, Xu C (2021) Influences of different source microplastics with different particle sizes and application rates on soil properties and growth of Chinese cabbage (*Brassica chinensis* L.). Ecotoxicol Environ Saf 222:112480

Yu F, Wu Z, Wang J, Li Y, Chu R, Pei Y, Ma J (2022) Effect of landfill age on the physical and chemical characteristics of waste plastics/microplastics in a waste landfill sites. Environ Pollut 306:119366

Zhao Y, Jia G, Shang Y, Zhao P, Cui X, Guo Q (2022) Chlorine migration during hydrothermal carbonization of recycled paper wastes and fuel performance of hydrochar. Process Saf Environ Prot 158:495–502

Zhou Y, Kumar M, Sarsaiya S, Sirohi R, Awasthi SK, Sindhu R, Awasthi MK (2022) Challenges and opportunities in bioremediation of micro-nano plastics: a review. Sci Total Environ 802:149823

Correction to: Landfills in Developing Economies: Drivers, Challenges, and Sustainable Solutions

Pablo Emilio Escamilla-García

Correction to:
Chapter 6 in: A. Anouzla and S. Souabi (eds.), *Technical Landfills and Waste Management*, Springer Water, https://doi.org/10.1007/978-3-031-52633-6_6

In the original version of the book, the name Pablo Emilio & Escamilla-García has been replaced by Pablo Emilio Escamilla-García in Chapter 6. The book and the chapter have been updated with the changes.

The updated version of this chapter can be found at
https://doi.org/10.1007/978-3-031-52633-6_6

A. Anouzla and S. Souabi (eds.), *Technical Landfills and Waste Management*,
Springer Water, https://doi.org/10.1007/978-3-031-52633-6_11

MIX
Papier aus verantwortungsvollen Quellen
Paper from responsible sources
FSC® C105338

If you have any concerns about our products,
you can contact us on
ProductSafety@springernature.com

In case Publisher is established outside the EU,
the EU authorized representative is:
Springer Nature Customer Service Center GmbH
Europaplatz 3, 69115 Heidelberg, Germany

Printed by Libri Plureos GmbH
in Hamburg, Germany